Vue.js 3 高级编程
UI组件库开发实战

杨海民◎著

电子工业出版社
Publishing House of Electronics Industry
北京·BEIJING

内 容 简 介

本书系统全面地介绍了 UI 组件库的开发过程，包括 UI 组件库的开发规范、底层逻辑、开发思维，以及 UI 组件库的运作原理、版本号管理、Markdown 文件、npm 发布等。本书配有 UI 组件库设计稿源文件和 UI 组件库源码，并附带 300 个左右的 Git 分支以及与本书代码清单一一对应的实例代码。本书可以帮助读者进一步提升开发能力、业务理解能力，让开发人员更深入地了解 CSS 的应用，掌握 Vue.js 3 的新特性，如 Composition API、provide、inject、teleport、slot 等关于 Vue.js 3 的开发技巧。

本书适合 Web 前端开发人员，需要提升自身开发能力或想开发属于自己的 UI 组件库的读者，以及有兴趣从事 Web 前端工作、想深入了解 UI 组件库底层逻辑的读者。本书也可作为高等院校计算机相关专业的师生用书和培训学校的教材。

未经许可，不得以任何方式复制或抄袭本书之部分或全部内容。
版权所有，侵权必究。

图书在版编目（CIP）数据

Vue.js 3 高级编程：UI 组件库开发实战 / 杨海民著.
北京：电子工业出版社，2024. 11. -- ISBN 978-7-121-49095-8

Ⅰ．TP393.092.2

中国国家版本馆 CIP 数据核字第 20249721R8 号

责任编辑：张　爽
印　　刷：三河市君旺印务有限公司
装　　订：三河市君旺印务有限公司
出版发行：电子工业出版社
　　　　　北京市海淀区万寿路 173 信箱　邮编：100036
开　　本：787×980　1/16　印张：20.5　字数：492 千字
版　　次：2024 年 11 月第 1 版
印　　次：2025 年 5 月第 2 次印刷
定　　价：89.00 元

凡所购买电子工业出版社图书有缺损问题，请向购买书店调换。若书店售缺，请与本社发行部联系，联系及邮购电话：(010) 88254888，88258888。
质量投诉请发邮件至 zlts@phei.com.cn，盗版侵权举报请发邮件至 dbqq@phei.com.cn。
本书咨询联系方式：faq@phei.com.cn。

前言

为什么要学习开发 UI 组件库

在前端开发领域,技术发展非常迅速——由最初的 HTML 到现在的前端框架、模块化、组件化、UI 组件库等。UI 组件库已成为现代前端开发人员的首选工具之一,如 AntDesign、Element Plus。第三方的 UI 组件库设计美观、交互功能强大,甚至有些神秘。也正因如此,掌握 UI 组件库的开发可以帮助前端开发人员大幅提高自身能力及生产力,是从初级开发人员向高级开发人员过渡的过程。开发 UI 组件库不但可以揭秘 UI 组件库的底层逻辑,而且可以更深入地学习 UI 组件库的开发思维、功能交互、CSS 技巧等,进而开发一套属于自己的 UI 组件库。

为什么要写作本书

笔者从前端领域的"小白"一路走到今天,建立了多个自己的前端学习交流群。在学习群中交流时,笔者发现大多数前端开发人员仍然停留在 UI 组件库的应用层面,并没有深入了解 UI 组件库的底层逻辑。当然,也有很多群友想开发 UI 组件库,但无奈于对 UI 组件库不了解,导致无法深入学习。

因此,为了让更多的前端开发人员和爱好者学习 UI 组件库的开发技巧,并体验 UI 组件库的开发过程,笔者决定撰写本书。如果你跟随本书的内容安排认真地开发 UI 组件库,那么笔者相信,随着学习过程的逐步深入,你会发现 UI 组件库原来没有那么神秘,你会有豁然开朗之感,并且越来越有成就感,学习动力也会越来越强。

读者对象

◎ 计算机专业在校学生、初入职场等对 UI 组件库不熟悉的 Web 开发人员。
◎ 具有一定的 Vue.js 3 技术基础,希望进一步学习 Vue.js 3 技术、提升自身能力的 Web 开发人员。

本书内容

本书共 18 章,各章简介如下。

第 1 章 "初识 UI 组件库" 介绍 UI 组件库是什么,以及其对项目和团队起到的作用。

第 2 章 "UI 组件库的开发规范" 介绍开发 UI 组件库过程中需要注意的规范,如命名规范、目录结构、样式规范等，有助于读者理解并管理组件。

第 3 章 "使用 Monorepo 构建组件库"介绍如何初始化 UI 组件库，以及组件的目录结构、演示库的初始化及测试。

第 4~16 章是本书的重点内容，介绍 UI 组件库的具体开发过程，包括各种组件的开发思路和思想、CSS 的模式、组件的方法封装，通过不同的属性配置实现不同模式的组件渲染效果，以及组件底层逻辑和 Vue.js 3 技术的新方法、新应用。

第 17 章 "构建 UI 组件库文档"介绍如何使用 VitePress 快速构建组件库说明文档，如何使用 Markdown 编写组件库文档内容，包括使用 Markdown 解析组件、渲染组件源码、渲染组件示例，以及描述组件的属性、方法等。

第 18 章 "UI 组件库的打包和发布"介绍使用 Rollup 实现 UMD、CJS、ESM 等不同模式的打包，以及全量和按需打包 CSS，并采用 Gulp 多任务模式执行指令打包 UI 组件库，最终将打包后的组件库发布至 npm 平台。

使用示例

本书中的示例代码基于 Windows 10 及以上版本操作系统运行，也可以在 macOS 操作系统中运行，需要安装下列软件。

- ◎ Chrome 23 或更高版本、Safari 6 或更高版本、Firefox 21 或更高版本、Opear 15 或更高版本。
- ◎ Visual Studio Code 最新版本。
- ◎ Node.js 16 或更高级别的稳定版本。
- ◎ Git 最新版本。

项目源码及课件

在学习本书中的示例代码时，既可以手动输入代码，也可以使用 Git 工具拉取 Gitee 仓库源码文件。仓库源码拉取成功后，可一次性拉取项目的所有远程分支，指令如下。

- ◎ 拉取所有分支：git branch -r | grep -v '\->' | while read remote; do git branch --track "${remote#origin/}" "$remote"; done。
- ◎ 创建并更新所有本地远程分支：git fetch --all。
- ◎ 同步所有远程分支内容：git pull --all。

源码拉取完成后，可使用 git branch -a 指令查看项目的本地和远程分支，项目的分支和本书的"代码清单"一一对应。此外，读者可在本书读者群中下载与本书配套的所有资料，包括 UI 组件库稿件及项目源码等。

在跟随本书学习 UI 组件库开发的过程中，为了确保读者自行开发的 UI 组件库与本书的逻辑一致并正常运行，读者安装的依赖包应尽可能与本书示例代码中安装的依赖包版本号保持一致。

勘误信息

笔者已尽最大努力确保正文和项目源代码正确，但因水平有限，在 UI 组件库开发中难免出现纰漏。若读者在阅读本书的过程中发现任何错误，都请及时与笔者联系，同时可以在本书项目仓库中提交问题（见链接 0-1）。

读者服务

读者在阅读本书时若遇到无法理解的知识点或示例代码无法有效运行，可以通过读者群（详见本书封底处的"读者服务"）与笔者联系，也可以在读者群中讨论关于 Vue.js 3 的知识和问题，与其他读者交流和分享经验，共同进步。

注意事项

本书代码清单中所声明定义的"变量"及"方法"均可自定义，并非业界标准。

本书中的所有代码，无论是业务逻辑，还是函数方法的封装，均遵循常规语法逻辑，并没有使用复杂的逻辑。

本书项目仓库所提供的代码仅供学习参考，请勿用于商业用途，因商业用途而导致的任何问题，笔者概不负责。

目录

第 1 章　初识 UI 组件库 1
　　1.1　UI 组件库是什么 1
　　1.2　UI 组件库的作用 1
　　1.3　UI 组件库的应用 2
　　本章小结 2

第 2 章　UI 组件库的开发规范 3
　　2.1　命名规范 3
　　2.2　目录结构 3
　　2.3　组件结构 4
　　2.4　样式规范 5
　　2.5　组件文档 6
　　2.6　版本管理 6
　　本章小结 7

第 3 章　使用 Monorepo 构建组件库 8
　　3.1　安装 pnpm 8
　　3.2　初始化 UI 组件库 9
　　　　3.2.1　建立工作区 9
　　　　3.2.2　建立 UI 组件库包 9
　　　　3.2.3　调用 UI 组件库包 10
　　3.3　初始化演示库 11
　　3.4　构建组件的目录结构 13
　　　　3.4.1　构建按钮组件 13
　　　　3.4.2　按需加载并导出组件 13
　　　　3.4.3　全局注册导出组件 14
　　3.5　演示库测试组件 15

		3.5.1 全局注册	15
		3.5.2 按需加载	16
	本章小结		16

第 4 章 CSS 架构模式 ... 17

- 4.1 UI 组件库元素设计规范 ... 17
- 4.2 BEM 命名规则 ... 18
- 4.3 命名空间 ... 19
- 4.4 封装 BEM 命名规则 ... 20
  - 4.4.1 生成 Block ... 21
  - 4.4.2 生成 Element 和 Modifier ... 22
- 本章小结 ... 23

第 5 章 按钮组件 ... 24

- 5.1 theme：主题包 ... 24
  - 5.1.1 全量引入 ... 25
  - 5.1.2 按需引入 ... 25
- 5.2 渲染 button 组件 ... 26
- 5.3 theme：主题色渲染 ... 27
- 5.4 round：圆角按钮 ... 30
- 5.5 disabled：禁用按钮 ... 31
- 5.6 text：文字按钮 ... 32
- 5.7 link：链接按钮 ... 33
- 5.8 border：边框按钮 ... 35
- 5.9 block：块级按钮 ... 36
- 5.10 size：按钮尺寸 ... 37
- 5.11 circle：圆形按钮 ... 38
- 5.12 icon：图标按钮 ... 40
- 5.13 loading：加载按钮 ... 41
  - 5.13.1 手动触发加载 ... 41
  - 5.13.2 自动触发加载 ... 43
- 5.14 group：按钮组 ... 44
  - 5.14.1 插槽实现按钮组 ... 45
  - 5.14.2 父级组件的属性 ... 46

本章小结 47

第 6 章　Sass 制定组件库全局变量 48
6.1　deep-merge：定义主题色 48
6.2　mix：生成主题色层次 49
6.2.1　定义@mixin 方法 50
6.2.2　each、for：循环生成层次色调 51
6.3　定义中性色及其他元素 52
6.4　:root 伪类选择器 54
6.5　生成:root 变量 54
6.5.1　定义前缀、块、修改器变量 55
6.5.2　:root 变量名称的生成规则 56
6.5.3　生成层次色 57
6.5.4　获取:root 变量名称 58
6.6　UI 组件库全局规范 59
6.7　UI 组件库应用规范 61
6.7.1　button 组件可变化的属性 61
6.7.2　私有变量 62
6.7.3　绑定变量 64
6.7.4　主题 64
6.7.5　尺寸 67
6.7.6　文字尺寸 68
6.7.7　链接按钮 69
6.7.8　文字按钮 71
6.8　遵循 BEM 命名规则生成组件类名 72
6.8.1　生成块的类名 72
6.8.2　生成元素的类名 73
6.8.3　生成修改器的类名 73
6.8.4　生成状态的类名 74
本章小结 75

第 7 章　图标组件 76
7.1　构建 icon 组件 76
7.2　渲染 icon 组件 77

7.3 尺寸与颜色	78
7.4 SVG 图标	79
7.5 button 组件使用 SVG 图标	80
本章小结	81

第 8 章 输入框组件 ... 82

8.1 渲染 input 组件	82
8.1.1 构建组件	82
8.1.2 渲染组件	83
8.1.3 样式变量	84
8.2 disabled：禁用	86
8.3 placeholder：占位符	87
8.4 maxlength：长度限制	88
8.5 size：尺寸	89
8.6 round：圆角	90
8.7 icon：图标	91
8.7.1 渲染 UI 组件库内置的 SVG 图标	92
8.7.2 渲染 iconfont 字体图标	94
8.8 slot：前置、后置	96
8.8.1 渲染前置、后置组件	97
8.8.2 渲染前置、后置标识	99
8.8.3 渲染前缀、后缀标识	101
8.9 password：密码	102
8.10 value：数据双向绑定	104
8.11 clear：清除数据	105
8.12 count：统计字符数	107
8.13 width：宽度	109
8.14 event：事件	110
8.14.1 focus 和 blur	110
8.14.2 mouseenter 和 mouselevel	112
8.14.3 compositionstart、compositionupdate 和 compositionend	113
8.14.4 change、keydown 和 keyup	115
8.15 expose：暴露对象	116
8.16 textarea：文本域	117

本章小结 118

第9章 布局组件 119

9.1 grid：栅格分栏 119
9.1.1 渲染 grid 组件 119
9.1.2 CSS 弹性布局 121
9.1.3 gutter：间距分隔 122
9.1.4 offset：偏移 124
9.1.5 justif：对齐 125
9.1.6 gap：行间距 126

9.2 container：容器组件 127
9.2.1 容器组件的结构 127
9.2.2 容器外层 128

本章小结 130

第10章 多选框组件 131

10.1 渲染 checkbox 组件 131
10.1.1 构建组件 131
10.1.2 渲染组件 133
10.1.3 样式变量 134

10.2 theme：主题 135
10.2.1 生成主题变量 136
10.2.2 渲染主题 136

10.3 size：尺寸 138
10.3.1 定义 size 变量 138
10.3.2 配置 size 变量 140

10.4 composables：组合式函数 140
10.4.1 定义状态模块 141
10.4.2 应用状态模块 141

10.5 disabled：禁用 142

10.6 group：多选框组 144
10.6.1 provide、inject：通信 145
10.6.2 size：尺寸 146

10.7 model：数据绑定 147

　　　　10.7.1　数据模块定义 ... 147
　　　　10.7.2　数据测试 ... 148
　　10.8　checked：勾选 ... 149
　　　　10.8.1　checkbox 组件 ... 150
　　　　10.8.2　checkboxGroup 组件 ... 151
　　10.9　event：事件 ... 152
　　10.10　async：异步 ... 154
　　　　10.10.1　渲染加载效果 ... 154
　　　　10.10.2　事件交互 ... 155
　　　　10.10.3　数据交互 ... 157
　　10.11　all：全选 ... 158
　　　　10.11.1　渲染全选组件 ... 158
　　　　10.11.2　渲染部分选中状态 ... 159
　　　　10.11.3　存储选项数据 ... 161
　　　　10.11.4　全选交互 ... 162
　　本章小结 ... 163

第 11 章　开关组件 ... 164

　　11.1　渲染 switch 组件 ... 164
　　　　11.1.1　构建组件 ... 164
　　　　11.1.2　渲染组件 ... 165
　　　　11.1.3　私有全局变量 ... 167
　　　　11.1.4　私有样式变量 ... 168
　　11.2　theme：主题 ... 169
　　11.3　size：尺寸 ... 171
　　11.4　text：文字 ... 172
　　11.5　icon：图标 ... 173
　　11.6　centerIcon：中心圆图标 ... 173
　　11.7　disabled：禁用 ... 174
　　11.8　model：数据绑定 ... 176
　　11.9　value：值 ... 177
　　11.10　async：异步 ... 177
　　11.11　transition：过渡动画 ... 179
　　本章小结 ... 180

第 12 章　表单组件 181

12.1　渲染 form 组件 181
12.1.1　构建组件 181
12.1.2　渲染组件 182
12.1.3　文本区域 184
12.2　AsyncValidator：校验库 185
12.3　rules：数据规则 186
12.4　validate：校验函数 187
12.5　trigger：校验规则类型 189
12.6　merge：合并校验规则 191
12.7　validate：数据校验 192
12.8　submit：提交校验 193
12.8.1　存储 formItem 组件数据 194
12.8.2　调用 form 组件校验 195
12.8.3　指定字段校验 196
12.9　reset：重置 197
12.10　required：必填标识 198
12.11　size：尺寸 198
本章小结 199

第 13 章　消息提示组件 200

13.1　createVNode 函数 200
13.1.1　基本语法 200
13.1.2　属性及事件 201
13.2　渲染 message 组件 202
13.2.1　构建组件 202
13.2.2　渲染组件 204
13.3　transition：过渡动画 206
13.3.1　transition 组件 206
13.3.2　动画实现 206
13.3.3　动画过程 207
13.3.4　钩子函数 209
13.4　attribute：初始化属性 210

13.5　z-index：层级顺序 211
13.6　top：顶部偏移 212
　　13.6.1　存储 message 组件 213
　　13.6.2　计算 top 偏移 214
13.7　autoClose：自动关闭 216
13.8　handleClose：手动关闭 218
13.9　allClose：全部关闭 219
13.10　theme：主题 219
13.11　background：背景颜色 220
13.12　主题方法 222
本章小结 223

第 14 章　模态框组件 224
14.1　mask：遮罩层 224
　　14.1.1　构建组件 225
　　14.1.2　渲染组件 226
14.2　modal：对话框 228
　　14.2.1　构建结构 228
　　14.2.2　渲染组件 229
14.3　teleport：传送 231
14.4　transition：过渡动画 231
14.5　footer：脚部 233
　　14.5.1　按钮属性 233
　　14.5.2　脚部插槽 234
　　14.5.3　按钮事件 235
14.6　loading：加载 236
　　14.6.1　confirmLoading 属性 236
　　14.6.2　beforeChange 属性 238
14.7　event：事件回调 239
14.8　maskClose：遮罩关闭 240
14.9　unmount：销毁 241
14.10　width：宽度 242
14.11　fixedScreen：固定屏 243
本章小结 245

第 15 章 　对话框组件 ... 246
15.1 　构建组件 ... 246
15.2 　title：标题 ... 248
15.3 　content：内容描述 ... 250
本章小结 ... 252

第 16 章 　抽屉组件 ... 253
16.1 　构建组件 ... 253
16.2 　placement：方向 ... 254
16.2.1 　absolute：绝对定位 ... 255
16.2.2 　position：位置 ... 256
16.3 　size：尺寸 ... 257
16.4 　transition：过渡动画 ... 258
本章小结 ... 259

第 17 章 　构建 UI 组件库文档 ... 260
17.1 　VitePress ... 260
17.1.1 　初始化文档 ... 260
17.1.2 　配置导航栏 ... 263
17.1.3 　配置侧边栏 ... 264
17.2 　解析 Markdown 文件 ... 265
17.2.1 　主题入口 ... 266
17.2.2 　注册全局组件 ... 267
17.2.3 　markdown-it-container ... 267
17.2.4 　tokens 容器 ... 269
17.3 　UI 组件库解析 ... 271
17.3.1 　定义文档组件 ... 271
17.3.2 　读取容器信息 ... 272
17.3.3 　读取文档组件 ... 273
17.3.4 　渲染组件 ... 274
17.3.5 　代码高亮 ... 277
17.3.6 　展开/收起源码 ... 279
17.4 　撰写组件库文档 ... 281
17.4.1 　Markdown 语法 ... 281

  17.4.2　Markdown 扩展功能 281
  17.4.3　Markdown 表格 282
本章小结 284

第 18 章　UI 组件库的打包和发布 285

18.1　了解 Rollup 285
  18.1.1　初始化 Build 打包目录 286
  18.1.2　Rollup 的基础配置 287
  18.1.3　配置打包路径 288
18.2　UMD 打包 289
  18.2.1　输出 UMD 组件包 290
  18.2.2　测试 UMD 组件包 291
18.3　ESM、CJS 模块化打包 292
  18.3.1　ESM、CJS 打包输出 292
  18.3.2　测试模块化组件包 296
18.4　Gulp 打包 scss 文件 298
  18.4.1　全量打包 CSS 298
  18.4.2　按需加载打包 CSS 299
18.5　Gulp 多任务 301
  18.5.1　series()和 parallel() 301
  18.5.2　删除组件包 302
  18.5.3　生成 package.json 文件 303
18.6　npm 发布 305
  18.6.1　package.json 文件 305
  18.6.2　version 306
  18.6.3　peerDependencies 307
  18.6.4　发布组件库 309
  18.6.5　打包组件库文档 310
  18.6.6　按需引入组件样式 311
本章小结 312

第1章
初识UI组件库

现代前端开发项目中有非常多"开箱即用"的框架，典型的是管理后台，开源项目 vue-element-admin 便是其中之一。vue-element-admin 其实并不是一个管理后台项目，而是结合 Element UI 实现的一个管理后台界面模板，通过集成 Element UI 组件库实现了各种不同的模块效果，如表格列表数据、表格数据的筛选、表单数据的提交、卡片列表、弹窗等模块。用户在使用时只需填充数据，即可轻松实现界面 UI 渲染，无须关心 UI 组件库底层的逻辑，这便是 UI 组件库的魅力。

1.1 UI 组件库是什么

UI 组件库是一种预先设计和编码的用户界面元素集合，用于快速构建具有一致性和易于使用的界面。UI 组件库通常包含各种常见的界面元素，如按钮、表单、导航栏、模态框等，可以帮助开发人员节省时间，并确保整体设计风格一致。通过使用组件库，不但可以快速构建用户界面，而且无须从头开始编写代码。

UI 组件库通常具有以下特点。

（1）**可重用性**：UI 组件库的组件可以在不同的项目中重复使用，提高开发效率。

（2）**一致性**：使用 UI 组件库构建界面，可以确保整体设计风格、交互方式和视觉效果的一致性。

（3）**易用性**：UI 组件库的组件经过设计和测试，具有良好的用户体验，可以帮助开发人员快速构建用户友好的界面。

（4）**定制性**：大多数 UI 组件库允许开发人员根据自己的需求和设计风格进行定制，以满足特定项目的要求。

常见的 UI 组件库有 Bootstrap、Material-UI、Ant Design、Semantic UI、Element Plus 等。在选择 UI 组件库时，开发人员通常会考虑组件的质量、文档的完整性、社区支持和适应性等因素。通过合理选择和使用 UI 组件库，开发人员可以加速项目开发过程，并提高用户界面的质量和一致性。

1.2 UI 组件库的作用

UI 组件库通常包含样式、布局和交互行为的预定义规范，使开发人员可以轻松地创建符合设计准则和用户期望的界面。UI 组件库的作用如下。

（1）**提高开发效率**：UI 组件库提供了预定义的可重用组件，开发人员可以直接使用这些组件，而不需要从头开始编写和设计每个界面元素。这样可以大大节省开发时间，加快项目的开发进度。

（2）**统一界面风格**：UI 组件库定义了一套共享的设计交互和样式规范，确保整个应用程序的用户界面具有一致的外观和交互方式。这样有助于提高用户的使用体验，降低用户的学习成本，提高应用的专业度和可信度。

（3）**提高代码一致性**：通过使用 UI 组件库，开发人员可以遵循一致的代码结构和开发模式，提高代码的可读性、可维护性和可重用性。组件库还提供了一致的 API 和接口，使不同团队之间的协作更加流畅。

（4）**管理和更新方便**：UI 组件库将所有的 UI 组件集中在一处，使组件的管理和更新更方便。如果需要对界面元素进行修改或更新，只需要在组件库中更改一次，所有使用该组件的界面元素都会自动更新，减少了手动修改代码的工作量。

（5）**促进团队协作**：UI 组件库使得团队成员之间的协作更加紧密和高效。通过共享组件库，团队成员可以更好地理解和使用彼此编写的代码。同时，团队成员可以对组件库进行扩展和改进，以满足特定项目的需求，从而推动技术的共享和创新。

1.3 UI 组件库的应用

UI 组件库具有广泛的应用，以下是一些常见的应用场景。

（1）**前端开发**：在构建网页和 Web 应用程序时，开发人员可以使用 UI 组件库来快速创建交互式用户界面。组件库中的预定义组件可以用于构建按钮、表单、菜单、轮播图等常见的 UI 元素，从而加快开发速度并提高一致性。

（2）**移动应用开发**：在开发移动应用程序时，UI 组件库可以提供可重用的组件，以构建用户友好的界面。开发人员可以使用组件库中的预定义组件来构建视图、导航、列表和各种交互元素，从而简化开发过程。

（3）**用户体验设计**：UI 组件库是用户体验设计师的有力工具，他们可以使用组件库的组件来创建原型和模型，以便与相关者共享和验证设计概念。这有助于提高设计的一致性和可重用性，并促进团队合作和反馈。

（4）**后台管理系统**：UI 组件库可用于开发后台管理系统，如数据仪表板、内容管理系统等。这些系统通常需要大量的交互组件和功能模块，使用组件库可以加速开发过程，提高用户界面的一致性和易用性。

本章小结

UI 组件库是一种预先设计和编码的用户界面元素集合，用于快速构建易于使用的界面。使用 UI 组件库可以提高开发效率，保持一致性，提高可维护性，促进团队协作，并为不同的设备提供良好的用户体验。

第 2 章 UI组件库的开发规范

定制开发规范可以确保组件库中的所有组件都遵循相同的规则和约定，从而保证代码的一致性，使组件库的使用和维护更加简单和高效。

UI组件库开发规范中定义了组件库的开发和维护标准，包括代码结构、命名方式、文档等方面。开发规范可以使代码具备良好的可读性和可维护性，降低后续修改和维护的难度。开发团队成员之间共同遵守规范也有助于更好地协作和交流。

2.1 命名规范

UI 组件库的命名规范是在为组件库中的各个组件和相关的样式类命名时应遵循的一套规范。具体来说，命名规范有助于确保代码的可读性、可维护性和一致性。常用的 UI 组件库命名规范的概念如下。

（1）**驼峰命名法**：使用驼峰命名法为组件和相关的样式类命名。驼峰命名法指将多个单词连接起来，每个单词的首字母大写，其他字母小写，如 Button、FormInput 等。

（2）**短横线命名法**：使用短横线命名法为组件和相关的样式类命名。短横线命名法指将多个单词连接起来，使用短横线分隔，所有字母小写，如 form-input 等。

（3）**组件前缀**：为了避免命名冲突和提高代码可读性，可以为每个组件的名称添加特定的前缀。例如，使用"ui-"作为 UI 组件库中组件的统一前缀，如 ui-button、ui-form-input 等。

（4）**语义化命名**：组件的名称应该具有一定的含义，能够准确描述组件的内容和功能，这样有助于开发人员理解和使用组件，提高代码的可读性。例如，使用 button 表示按钮组件，使用 checkbox 表示多选框组件。

（5）**组件变种**：当组件存在不同的变种或样式时，可以在组件名称中使用后缀或修饰词来区分。例如，使用 primary、secondary 来表示按钮的不同样式，如 button-primary、button-secondary。

2.2 目录结构

UI 组件库的目录结构是指在组件库的代码仓库中，组织和管理不同组件和相关文件的文件夹层次结构，如图 2-1 所示。一个良好的目录结构能够提高代码的可维护性和可读性，使开发人员轻松地找到所需的组件和相关文件。

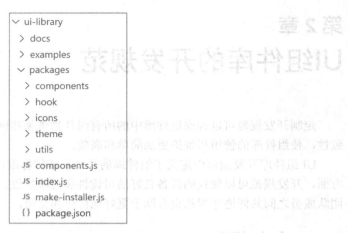

图 2-1 UI 组件库的目录结构

- docs：组件库的说明文件，包括入门指南、组件文档、参考 API 等。
- examples：演示库，用于在开发过程中演示和调试组件。
- packages：组件包，存放所有与组件相关的文件。
- packages/components：存放每个组件的文件夹。每个组件有一个独立的文件夹，包含该组件的 JavaScript 代码、CSS 样式、单元测试相关文件。
- packages/hook：存放与每个组件相关的公共钩子函数。
- packages/icons：存放所有的图标、图形。
- packages/theme：存放每个组件的 CSS 样式表。
- packages/utils：存放可复用的工具函数或模块，供组件使用。

除了上述示例中的文件和文件夹，也可以根据组件库和团队约定调整和扩展目录结构，但始终保持层次清晰、易于理解且遵循一定的命名约定是非常重要的。这样能够提高开发人员的工作效率，减少混乱和错误的发生。

2.3 组件结构

UI 组件库的组件结构是指一个组件包含的各个部分和层级关系。每个组件都由若干个部分组成，这些组成部分共同实现组件的功能和外观。

图 2-2 是 button 组件的目录结构，包含以下部分。

- button/index.js：组件的主入口，通过该文件可访问组件。
- button/src/index.vue：组件的主体逻辑，包含组件的渲染、逻辑等。
- button/src/style/index.js：组件的样式，引入外部的样式、CSS 变量等。

图 2-2　button 组件的目录结构

除了上述基本部分，具体的组件结构还可能包含其他文件和文件夹，视具体的组件需求而定。组件结构的设计应遵循以下 3 个原则。

- **单一职责**：每个文件或文件夹应该只包含一个组件的相关代码。这样可以使代码模块化且易于维护。
- **分离关联点**：将组件的逻辑、样式、测试文档分离，使各个方面的代码相对独立，降低代码耦合度。
- **可扩展性**：组件的结构应该具备良好的可扩展性，可以灵活地修改组件的样式，增强组件的功能，同时保持代码的简洁性和可读性。

2.4　样式规范

UI 组件库的样式规范是指在设计和开发 UI 组件库时定义和遵循的规范，用于统一组件库中各个组件的样式和外观。[①]

下面是 UI 组件库样式规范中的一些常见概念。

- **尺寸和间距**：定义组件的尺寸和内外间距的规范，使组件在不同的场景下保持一致的外观和排列方式。
- **颜色**：定义组件库中使用的颜色规范，包括文本颜色、背景颜色、边框颜色等，以确保一致的视觉效果和配色方案。
- **字体样式**：定义组件中文本内容的字号、字体家族、行高等样式规范，以确保一致的字体显示效果。
- **样式类命名规范**：定义使用样式类的命名规范，以使开发人员能够快速理解和使用样式类，避免冲突，提高可读性。
- **主题定制**：提供一套可定制的主题规范，使开发人员能够根据项目需求自定义组件的颜色、字体等样式，以提供个性化的外观。

上述样式规范的概念可以根据 UI 组件库和团队约定进行调整和组合，但始终保持一致性和

[①] 有关 UI 组件库的规范详情，可查看本书附带资源中的 UI 组件库设计稿源文件，获取方式见本书封底处的读者服务。

可读性是非常重要的。遵循样式规范可以提高组件库的可维护性和可扩展性，使开发人员更容易地使用和定制组件库。

2.5 组件文档

组件文档是 UI 组件库中的重要组成部分，用于为开发人员提供有关组件的详细信息和使用方法。下面是一些常见的组件文档内容。

- 组件概述：对组件进行简要介绍，说明组件的用途和主要功能，描述组件的作用、场景和优势。
- 使用方法：提供使用组件的示例代码和详细说明。说明如何引入组件、设置属性、处理事件和使用其他相关功能。
- 属性列表：列出组件支持的属性及其说明。对于每个属性，提供说明和默认值，以及可能的取值范围和用法示例。
- 事件列表：列出组件触发的事件及其说明。对于每个事件，提供说明、参数和用法示例，以及如何使用事件处理程序。
- 插槽列表：如果组件支持插槽，则需列出组件的插槽及其说明。说明如何在插槽中插入内容，并提供示例代码。
- 样式类列表：列出组件的样式类及其说明，以便开发人员自定义组件的样式。提供示例代码，并说明如何使用这些样式类。
- 注意事项：提供一些使用组件时需注意的事项，例如组件的兼容性、依赖关系和已知问题。如果有一些常见的错误或陷阱，也可以在这里提醒开发人员。
- 示例和演示：提供一些示例代码和演示效果，用于展示组件的不同用法和状态。可以使用代码片段、屏幕截图或在线演示链接来展示示例。

通过提供详细和清晰的组件文档，开发人员可以更好地理解和使用组件。文档应该易于理解和搜索，以便开发人员快速找到所需的信息并解决问题。

2.6 版本管理

版本管理是一种用于跟踪和控制不同版本的软件或组件库的过程，旨在确保开发团队有效地管理变更、解决问题并发布更新版本。以下是版本管理的一些常见做法。

- 版本控制系统：使用版本控制系统，如 Git、Subversion 或 Mercurial，来跟踪代码的变化。通过提交代码并使用版本控制系统的分支和合并功能，开发团队可以安全地并行开发和修复不同的功能。
- 分支管理：在版本控制系统中，使用分支来开展不同的工作。主分支（如"master"或"main"）用于发布稳定版本，而开发分支用于开发新功能和实验。此外，还可以为修复 bug 和处理紧急问题创建临时分支。

- ◎ 标签和版本号：为每个发布的版本打上标签或版本号，以方便识别和跟踪。
- ◎ 版本发布：在团队确定一个版本稳定并准备发布时，执行版本发布流程。这可能涉及编译、打包文件、生成文档、运行测试，以及发布到指定的位置。
- ◎ 更新说明：在发布每个版本时，提供清晰、详细的更新说明。说明版本的变更内容、修复的问题、添加的功能和可能的兼容性问题。
- ◎ 回归测试：在发布每个版本前进行回归测试，以验证修改和新功能是否引入了新问题。这有助于确保发布的版本是稳定和可靠的。
- ◎ 版本更新：及时更新使用组件库的应用程序或项目，以获取最新的稳定版本、修复安全问题和错误的相关信息。

版本管理有助于确保组件库的稳定性、可靠性和可维护性。它提供了一种可追溯和可控的方式来管理变更，并为开发人员和用户提供了清晰的信息和指南。

本章小结

本章介绍了 UI 组件库的开发规范，包括代码结构、命名方式、文档等方面形成的开发规范，规范可以使代码具备良好的可读性和可维护性，降低后续修改和维护的难度，并且使开发团队之间共同遵守规范及更好地协作和交流。

第 3 章
使用Monorepo构建组件库

Monorepo 是将多个软件项目或代码库存储在单个版本控制存储库中的软件开发模型。按照传统方法，每个项目或代码库都有自己独立的存储库，然而使用 Monorepo，可以使所有的项目或代码库共享同一个存储库。这种模型的优势包括更易于实现代码共享、跨项目的原子性更改，以及更好地进行协作和一致性管理。它还提供更简单的构建、测试和部署过程，并能够更好地支持跨项目的重构和代码重用。

Monorepo 具有以下特点。

◎ 单一代码库：所有相关的项目或代码库都存储在同一个版本控制存储库中，使得代码管理集中且一致。

◎ 代码共享：不同的项目或代码库可以轻松共享代码，避免重复编写和维护功能相似的代码。

◎ 原子性更改：对于涉及多个项目的更改，Monorepo 可以确保更改的原子性，可以一次提交和审查整个更改，避免项目之间的不一致问题。

◎ 简化构建和测试：由于所有代码都在同一个存储库中，构建和测试过程得以简化，可以通过一次构建和测试来确保整个代码库的一致性。

◎ 更好地协作：开发人员可以更轻松地共享代码和协作开发，减少项目之间的隔阂和重复工作。

◎ 可重构性和代码重用：Monorepo 模型有助于促进代码的重构和重用，使得跨项目的代码调整和重构更加容易。

总的来说，Monorepo 提供了一种更集中、更一致的代码管理和协作方式，尤其适用于大型项目和具有多个相互依赖部分的应用程序。

3.1 安装 pnpm

pnpm 的全称是 performant npm，指高性能的 npm。与 npm 和 yarn 一样，pnpm 是一款包管理工具。pnpm 根据自身独特的包管理方法解决了 npm、yarn 内部潜在的安全及性能问题，在多数情况下拥有更快的安装速度，占用更小的存储空间。与 npm 和 yarn 不同，pnpm 采用一种名为"虚拟存储"的新颖方法来管理依赖项。它通过在一个统一的位置安装依赖项，并在项目之间共享它们，以减少使用的磁盘空间和安装时间。pnpm 还支持并发安装和锁定使用版本，以确保项目的依赖关系始终保持一致。同时，pnpm 还提供了一些其他功能，如本地缓存、快速恢复等，以提高依赖项的管理效率。

使用 pnpm 前需要全局安装，执行指令：npm install -g pnpm@8.15.0。然后使用 pnpm -v 检查安装版本，如出现版本号，则表示安装成功，如图 3-1 所示。

```
终端    问题    输出    调试控制台
PS D:\ui-library> pnpm -v
8.15.0
```

图 3-1　安装 pnpm 并检查版本

3.2　初始化 UI 组件库

初始化 UI 组件库是为了更好地管理文件模块的目录结构，区分不同文件模块的功能。UI 组件库将使用 pnpm + workspace 的方式来搭建，为了统一管理 UI 组件库的文件，可以任意选择一个磁盘新建文件夹，并命名为 ui-library，作为 UI 组件库的"根"目录，然后在 ui-library 目录下新建以下 3 个文件夹。

（1）packages：UI 组件库。该目录用于存放所有组件，在 packages 目录中还可以建立其他与组件相关的目录，如组件包 components、主题包 theme、工具包 utils 等。

（2）examples：演示库。用于调试开发环境。

（3）docs：组件文档。即使用 VitePress 构建的 UI 组件库使用说明文档。

3.2.1　建立工作区

采用 pnpm+Monorepo 管理多个项目，需要用到 pnpm-workspace.yaml 配置文件，在 workspace 中指定需要包含的目录，也就是 packages、examples 和 docs 三个目录。在 ui-library 根目录下创建 pnpm-workspace.yaml 配置文件，并写入需要包含的目录，如代码清单 library-03-1 所示。

代码清单 library-03-1
```
1.  > ui-library/pnpm-workspace.yaml:
2.  packages:
3.    - examples # 存放组件测试的代码
4.    - docs # 存放组件文档
5.    - packages/* # packages 目录下都是组件包
```

3.2.2　建立 UI 组件库包

UI 组件库的所有组件以及与组件相关的业务都放在 packages 目录中，如 components 组件包、utils 工具包、hook 钩子函数包等，因此需要单独建立文件夹，并在每个目录下初始化 package.json 文件，将其声明为 npm 包，视为独立的"包"来应用。

以 components 组件包为例，进入 components 目录中执行 pnpm init 生成 package.json 文件，并将文件中的 name 属性自定义为@ui-library/components，如代码清单 library-03-2 所示。

代码清单 library-03-2

```
1.  > packages/components/package.json:
2.  {
3.    "name": "@ui-library/components",
4.    "version": "1.0.0",
5.    "description": "",
6.    "main": "index.js",
7.    "scripts": {
8.      "test": "echo \"Error: no test specified\" && exit 1"
9.    },
10.   "keywords": [],
11.   "author": "",
12.   "license": "ISC"
13. }
```

对于 utils 工具包、hook 钩子函数包，同样是在初始化 package.json 文件后，将包的名称自定义为 @ui-library/utils、@ui-library/hook，如图 3-2 和图 3-3 所示，对应分支为本书配套代码中的 library-03-2。

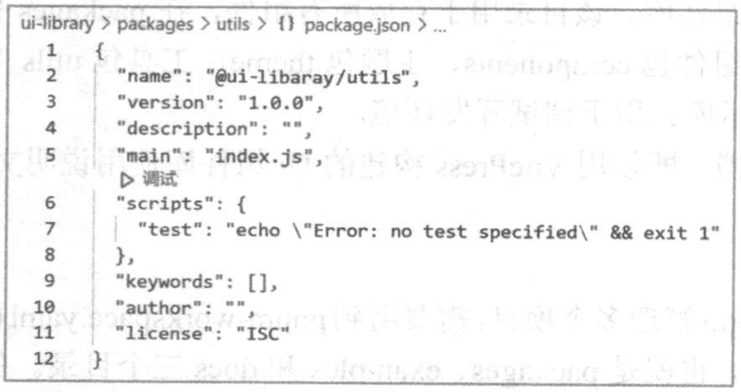

```
ui-library > packages > utils > {} package.json > ...
1   {
2     "name": "@ui-libaray/utils",
3     "version": "1.0.0",
4     "description": "",
5     "main": "index.js",
        ▷ 调试
6     "scripts": {
7       "test": "echo \"Error: no test specified\" && exit 1"
8     },
9     "keywords": [],
10    "author": "",
11    "license": "ISC"
12  }
```

图 3-2　utils 工具包的 package.json 文件

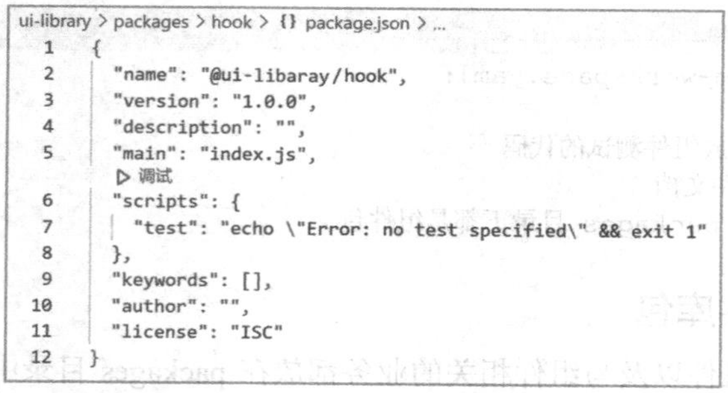

```
ui-library > packages > hook > {} package.json > ...
1   {
2     "name": "@ui-libaray/hook",
3     "version": "1.0.0",
4     "description": "",
5     "main": "index.js",
        ▷ 调试
6     "scripts": {
7       "test": "echo \"Error: no test specified\" && exit 1"
8     },
9     "keywords": [],
10    "author": "",
11    "license": "ISC"
12  }
```

图 3-3　hook 钩子函数包的 package.json 文件

3.2.3　调用 UI 组件库包

使用 pnpm init 指令为 packages 目录下的每个文件夹创建 package.json 文件后，它们已经变成了独立的包。要使各个包之间可以相互调用，只需要在 ui-library 根目录下安装依赖即可。进入 ui-library 根目录，在终端执行以下依赖指令，便可安装包。

◎　安装 components 组件包：pnpm install @ui-library/components -w

◎　安装 utils 工具包：pnpm install @ui-library/utils -w

◎　安装 hook 钩子函数包：pnpm install @ui-library/hook -w

安装完成后，可以在根目录的 packags.json 文件的 dependencies 属性中查看已安装的包，根目录中生成了 node_modeuls 依赖包文件夹，如图 3-4 所示。

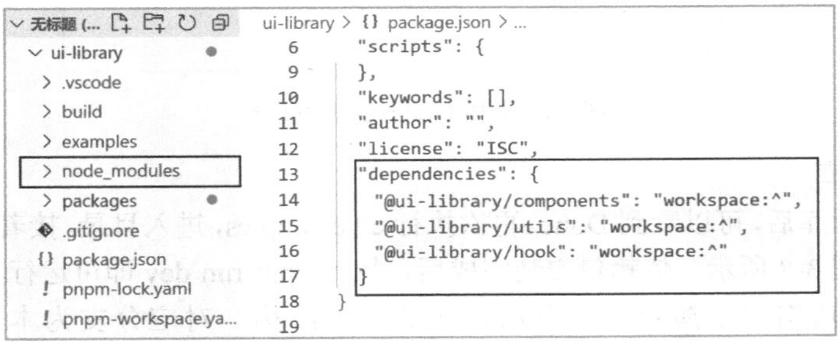

图 3-4　依赖 UI 组件库安装包

提示：-w 表示安装到公共模块的 packages.json 中，也就是根目录下的 packages.json 中。

3.3　初始化演示库

演示库用在组件开发的调试过程中，开发人员可以直观查看组件库的开发效果。3.2 节已经定义 examples 目录为演示库的包，所以在使用 Vite 脚手架构建演示库项目时，可以直接执行指令 npm create vite@latest examples。其中，examples 是自定义的演示库目录，如图 3-5 所示。

```
PS D:\ui-library> npm create vite@latest examples
Need to install the following packages:
  create-vite@4.4.1
Ok to proceed? (y)
```

图 3-5　安装 Vite 脚手架

如果你的计算机中没有安装过 Vite，就会出现“Need to install the following packages: create-vite@4.4.1”，也就是提示你需要先安装 Vite，直接按回车键即可。

1. 选择开发框架

选择 Vue.js 3 框架，采用 Vue.js 3+setup 语法糖开发，并使用 Vue.js 3.4 版本，如图 3-6 所示。

```
PS D:\ui-library> npm create vite@latest examples
? Select a framework: » - Use arrow-keys. Return to submit.
    Vanilla
>   Vue
    React
    Preact
```

图 3-6　选择开发框架

2. 选择开发技术

如图 3-7 所示，这里选择 JavaScript，原因在于使用 JavaScript 开发组件库比使用 TypeScript 的门槛更低。先使用 JavaScript 开发组件库，再使用 TypeScript 二次开发，不但可以扩展组件库开发的思路，还可以更熟悉组件库。

```
? Select a variant: » - Use arrow-keys. Return to submit.
    TypeScript
>   JavaScript
    Customize with create-vue ↗
    Nuxt ↗
```

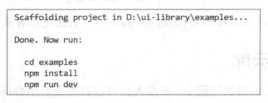

图 3-7　选择 JavaScript 开发技术

3. 构建演示库

成功构建演示库后，可以看到 Done。再次执行 cd examples，进入目录，接着执行 npm install，安装依赖包，如图 3-8 所示。依赖包安装完成后，执行 npm run dev 即可运行演示库，如图 3-9 所示，到此便成功构建了演示库以及演示包的目录结构。对应分支为本书配套代码中的 library-03-3。

```
Scaffolding project in D:\ui-library\examples...

Done. Now run:

  cd examples
  npm install
  npm run dev
```

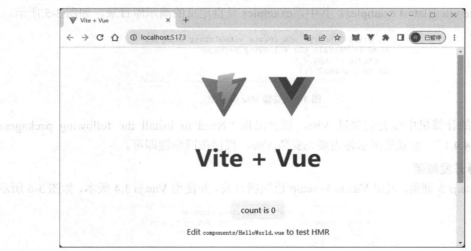

图 3-8　examples 演示库构建完成

图 3-9　运行 examples 演示库

3.4 构建组件的目录结构

在 2.4 节中,已详细介绍 UI 组件库中的组件结构及其设计原则。因此,本节将以按钮(button)组件为例构建组件的目录结构,后续的组件开发过程将遵循 button 组件的目录结构。

3.4.1 构建按钮组件

packages 是 UI 组件库的整个包,该目录下已经建立了 components 目录,也就是放置所有组件的目录。构建 button 组件,就是在 components 目录下建立 button 文件夹,button 组件的目录结构如图 3-10 所示。

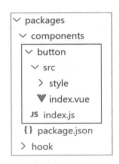

图 3-10 button 组件的目录结构

图 3-10 中用方框标记的部分是 button 组件的基础结构。button 目录是组件的根目录,index.js 是访问 button 组件的入口文件,src 目录下的 index.vue 文件是 button 组件的 UI 渲染文件,style 目录存放 button 组件的样式文件,button 组件的 index.vue 文件如代码清单 library-03-4 所示。

代码清单 library-03-4
```
> packages/components/button/src/index.vue:
<template>
    <button>这是一个测试的按钮</button>
</template>
<script>export default { name: 'a-button' };</script>
```

其中,定义 button 组件的 name 为"a-button","a-"是组件前缀。使用前缀的目的是避免命名冲突,提高代码的可读性,例如 Element Plus 的"el-"、Ant Design 的"a-"。

提示:UI 组件库的前缀可根据需要自定义。

3.4.2 按需加载并导出组件

在项目中,导出组件的操作可以多次使用,减少重复编写代码的工作量,有助于模块化开发,提高项目的可维护性和可扩展性。此外,也可以将特定功能或界面的实现细节封装起来,使其他开发人员可以直接使用组件,而不必关心内部的实现细节。因此,我们只需要将

src/index.vue 文件引入 index.js,使用 export default 导出组件,并提供按需加载的方式,如代码清单 library-03-5 所示。

代码清单 library-03-5

```
1.  > packages/utils/install.js
2.  export const componentInstall = (com) => {
3.    com.install = (app) => {
4.      app.component(com.name, com)
5.    }
6.    return com
7.  }
8.
9.  > packages/components/button/index.js:
10. import { componentInstall } from "@ui-library/utils"
11. import Button from "./src/index.vue"
12. // 提供按需加载的方式
13. export const AButton = componentInstall(Button)
14. // 导出组件
15. export default AButton
16.
17. > packages/components/index.js
18. export * from "./button"
```

在 utils 工具类中建立 install.js 文件,定义 componentInstall 方法用于注册组件(第 2~7 行)。该方法中的 app.component 是 Vue.js 3 的实例对象(第 4 行),第一个参数是组件名称,会自动获取组件内的 name 属性值,第二个参数是要注册的组件,然后返回组件即可(第 6 行)。

在 button 组件中引入工具类的 componentInstall 方法(第 10 行),在调用 componentInstall 方法时传入 button 组件,然后将返回的组件赋给变量 AButton,并使用 export 关键字导出组件,提供按需加载的功能(第 13 行),同时使用 export default 导出组件(第 15 行)。

最后从 components/index.js 文件中导出 button 组件。components/index.js 汇总了所有"按需加载"的组件,后续增加新的组件,也都需要从该文件中导出。

3.4.3　全局注册导出组件

全局注册导出组件是将所有组件汇聚到一个文件中,使用循环的方式批量注册组件,如代码清单 library-03-6 所示。

代码清单 library-03-6

```
1.  > packages/components.js
2.  import { AButton } from "./components/button/index.js"
3.  export default [ AButton ]
4.
5.  > packages/index.js:
6.  // 按需加载
7.  export * from "./components/index"
8.  // 全局注册
```

```
9.  import components from "./components.js";
10. // 全局安装
11. export const install = function (app) {
12.   // 判断是否安装
13.   if (install.installed) return;
14.   // 安装组件
15.   components.forEach((c) => app.use(c))
16. };
17. export default install
```

在 packages 目录下新建 components.js 文件，引入 button 组件（第 2 行），然后使用 Array 数组导出 button 组件（第 3 行）。将 components.js 文件引入 packages/index.js 文件（第 9 行），然后定义函数 install，并定义 app 作为参数接收 Vue.js 3 实例对象注册的所有组件（第 11~16 行）。首先判断 install.installed 是否已安装，如果是，则返回（第 13 行），否则使用 forEach 方法遍历数据（第 15 行），并调用 app.use 对象注册组件（第 15 行），最后使用 export default 导出函数 install（第 17 行）。

需要注意的是，packages/index.js 文件同时导出了按需加载的组件（第 7 行）。这是因为组件库无论是"按需加载"还是"全局注册"，都通过 packages/index.js 作为入口打包所有组件，因此需要在该文件中导出按需加载的组件。

按需加载目前仅仅可以加载组件，而对于组件样式，需要在打包组件库的过程中重新处理样式的引用路径，并达到自动按需加载组件样式的目的。打包组件库采用的是 Rollup，详见第 18 章。

3.5 演示库测试组件

演示库用于 UI 组件库开发过程的调试、UI 效果渲染等，在此之前需要测试导出的组件库是否可以在演示库中使用。

3.5.1 全局注册

全局注册是最常见的一种开发方法，好处在于一次注册、任意位置可用，缺点在于增加了项目的整体体积，如代码清单 library-03-7 所示。

代码清单 library-03-7
```
1.  > examples/src/main.js:
2.  import { createApp } from 'vue'
3.  import App from './App.vue'
4.  // 相对路径引入组件库
5.  import Azong from "../../packages"
6.  const app = createApp(App)
7.  app.use(Azong).mount('#app')
8.
9.  > examples/src/App.vue
10. <template>
```

```
11.     <a-button></a-button>
12. </template>
```

在 examples 演示库的 main.js 文件中使用相对路径引入 packages 组件库（第 5 行），并自定义名称为 Azong，然后使用 Vue.js 3 实例对象的.use 方法全局注册组件库（第 7 行）。这样，App.vue 文件中的 template 模板层便可使用<a-button>组件名称渲染 button 组件（第 10~12 行），渲染效果如图 3-11 所示。

这是一个测试的按钮

图 3-11　全局注册组件库渲染按钮

提示：由于 UI 组件库的开发和演示库均在本地环境下预览，因此采用相对路径的方式引用。在第 18 章中完成组件库打包，发布至 npm 后便可使用 npm install 方法安装我们自己开发的 UI 组件库。

3.5.2　按需加载

按需加载是指要用到的组件应采用 ES Modules 方法引入，好处在于可以减小项目的体积，提升性能，如代码清单 library-03-8 所示。

代码清单 library-03-8
```
1. > examples/src/App.vue
2. <template>
3.     <a-button></a-button>
4. </template>
5. <script setup>
6. import { AButton } from '../../packages';
7. </script>
```

按需加载同样使用相对路径的方式引入，通过 import 关键字导入 AButton 组件，无须在 main.js 中全局注册组件，渲染效果与图 3-11 一致。

本章小结

使用 Monorepo 构建并管理组件库，可以有效地管理 UI 组件库的不同模块、不同功能，并初步了解如何建立、导出组件目录结构等。在 UI 组件的开发过程中，通过演示库调试组件的渲染效果、功能，并实现全局注册、按需加载等功能。

在后续的组件开发过程中，新增组件的目录结构与 button 组件一致，并将新增的组件导入 packages/components/index.js 和 packages/components.js 两个文件夹，便可自动完成组件的按需加载和全局注册。

第4章 CSS架构模式

在现代前端项目应用中，UI 组件库是必不可少的工具。作为一名高级前端开发人员，是否有必要拥有一套属于自己的 UI 组件库呢？答案是否定的。但笔者认为，学习 UI 组件库的开发很有必要，目前已有很多非常优秀的 UI 组件库，如 Element Plus、AntDesign 等，在工作中直接安装即可使用，同样可以提高开发效率，达到事半功倍的效果。那我们为什么还要开发属于自己的 UI 组件库呢？

笔者认为有以下几点原因。

（1）**学习优秀的开发思维和逻辑**：阅读优秀的框架源码是前端开发人员的学习内容之一，从优秀的事物中可以学到更好的思路。

（2）**加深对组件的理解**：深入理解如何更规范地开发组件库及开发模式，如解耦、组合式函数、设计规范和 CSS 样式等。

（3）**提高 CSS 水平**：设计 UI 组件库对 CSS 水平有比较高的要求，需要考虑全局的、私有的、可变化的、组合性及其他各方面的知识点。

（4）**为工作加分**：有一套自己开发的 UI 组件库，自然会有自己的思路、想法，这是工作中的一个亮点。

4.1 UI 组件库元素设计规范

UI 组件库元素的设计规范是指提供一套标准化的界面元素和交互模式。UI 组件库元素设计通常包括各种常见的界面元素，如按钮、输入框、单选框、多选框、模态框等，以及定义它们的外观、行为和样式。具体包括以下内容。

（1）**尺寸和间距**：定义组件的标准尺寸和内部间距，确保在不同的设备上展现一致的外观。

（2）**颜色和样式**：确定组件的标准颜色、辅助色、背景、边框和阴影效果，以及悬停或激活状态下的样式变化。

（3）**字体和文字**：规定组件中文字的标准字体、大小、颜色及对齐方式，包括常见的标题、正文和链接样式。

（4）**图标和图像**：定义组件中应用的图标和图像的样式，包括大小、填充色等。

（5）**交互行为**：描述组件的交互行为，例如按钮的点击效果、输入框的聚焦状态等，确保统一的用户体验。

（6）**可访问性规范**：确保组件在视力障碍用户或使用辅助技术的用户群体中具有良好的可访问性，例如合适的焦点状态和文字说明。

经过规范化设计的 UI 组件库能够为开发人员提供清晰的指导,确保在开发过程中准确地实现设计师所期望的视觉和交互效果。

为了让读者更好地了解 UI 组件库的开发过程,笔者借鉴了国内外的 UI 组件库,设计了一套相对简单、美观的 UI 组件库设计稿,供读者学习使用,其中部分内容如图 4-1~图 4-3 所示。UI 组件库设计稿的获取方式详见本书封底的读者服务。

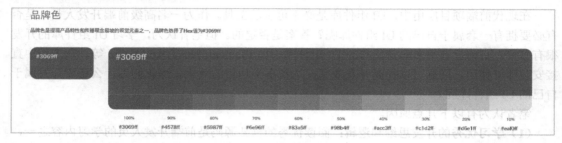

图 4-1　UI 组件库主色调

图 4-2　button 组件

图 4-3　form 组件

4.2　BEM 命名规则

BEM 命名规则是由 Yandex 团队提出的一种 CSS 命名方法论。BEM 是 Block(块)、Element(元素)、Modifier(修改器)的简称,它是 OOCSS 方法论的一种实现模式,本质上是面向对象的思想。下面以 Tabs 组件为例,介绍 BEM 命名规则的使用方法和核心思想,如图 4-4 所示。

图 4-4　Tabs 组件

Tabs 是一个完整的组件，可以将其理解为一个 Block，代表一个逻辑或功能独立的组件，是结构、表现和行为的整体，可定义类名（classname）为 a-tabs。其中，每个切换的标签是一个 Element，可以理解为 Block 中的元素，可定义类名为 a-tabs__item。Modifier 用于描述一个 Block 或 Element 的表现、行为，例如微调图 4-4 中文字的颜色样式。

BEM 命名规则下类名的命名格式为：

<block-name>__<element-name>--<modifier-name>_<modifier_value>: a-tabs__item--size_mini。

其中：

- 所有组件的命名均使用小写字母，复合词使用短横线"-"连接。
- Block 与 Element 之间使用双下画线"__"连接。
- Modifier 与 Block/Element 使用双短横线"--"连接。
- modifier-name 和 modifier_value 之间使用下画线"_"连接。

提示：BEM 命名规则只是一种 CSS 命名方法论，不必严格遵守，可根据你的团队风格进行修改。

4.3 命名空间

UI 组件库的命名通常基于项目或者公司的命名规则。命名规则可以根据具体的应用场景和组织内部的代码规范进行定义。通常情况下，命名应该简洁明了、符合规范，并且准确反映该组件库的功能和用途。例如，一个名为"Acme UI"的组件库可能使用"acme"作为其命名前缀，即命名空间，但命名通常需要根据项目的具体要求和规范来确定。

UI 组件库命名空间的作用是确保在大型应用程序中，不同部分的 UI 组件可以被正确识别和引用。使用命名空间可以隔离不同部分的 UI 组件，避免命名冲突和重复，从而提高代码的可维护性和可读性。命名空间还可以帮助开发人员更清晰地组织和管理 UI 组件，使其易于定位和使用。

根据上述介绍，笔者将以"a"作为 UI 组件库的命名空间，也就是所有组件的"前缀"，如 button 组件为"a-button"、checkbox 组件为"a-checkbox"。确定了命名空间的前缀为"a"后，需要将此前缀提取为公共逻辑，作为所有组件自动生成的规则，如代码清单 library-04-1 所示。

代码清单 library-04-1

```
1.  > packages/hook/use-namespace/index.js
2.  // 默认命名前缀
3.  export const defaultNamespace = "a";
4.  // 命名空间
5.  export const useNamespace = (block) => {
6.    // 命名前缀就是命名空间
7.    const namespace = defaultNamespace;
8.    return {
9.      namespace,
10.   };
```

```
11.   };
12.
13.   > packages/hook/ndex.js
14.   export * from './use-namespace'
15.
16.   > packages/components/button/src/index.vue
17.   <script setup>
18.   import { useNamespace } from '@ui-library/hook';
19.   const ns = useNamespace()
20.   console.log('命名空间>>>', ns.namespace)
21.   </script>
```

在 hook 目录下新建 use-namespace 目录和 index.js 文件。在 index.js 文件中定义变量 defaultNamespace，并定义值为 "a"（第 3 行），接着定义 useNamespace 函数（第 5 行），在该函数中将命名空间 defaultNamespace 赋给新的变量 namespace（第 7 行），并将其返回（第 8~10 行）。

最后从 hook/ndex.js 中导出 use-namespace/index.js 文件，供所有组件使用（第 14 行）。为了测试命名空间是否可用，可以将其引入 button 组件并打印到控制台中查看，如图 4-5 所示。

图 4-5 命名空间

4.4 封装 BEM 命名规则

在编写组件时，如果手写类名，会经常写到 BEM 命名规则所提及的 "-" "--" "__" 和"前缀"，写法烦琐。这时可以把 BEM 命名规则封装成函数，动态生成 UI 组件库的类名，如代码清单 library-04-2 所示。

代码清单 library-04-2
```
1.   > packages/hook/use-namespace/index.js
2.   const _bem = (namespace, block, blockSuffix, element, modifier, modifierValue) => {
3.     // 默认 Block
4.     let className = `${namespace}-${block}`;
5.     // 如果存在子级块
6.     blockSuffix && (className += `-${blockSuffix}`)
7.     // 如果存在元素
8.     element && (className += `__${element}`)
9.     // 如果存在修改器
10.    modifier && (className += `--${modifier}`)
11.    // 如果存在修改器的值
12.    modifierValue && (className += `_${modifierValue}`);
13.    // 返回
```

```
14.     return className;
15.   };
```

在 use-namespace/index.js 文件中定义_bem 函数，分别接收 6 个参数：namespace（命名空间）、block（块）、blockSuffix（子级块）、element（元素）、modifier（修改器）、modifierValue（修改器的值）。

定义变量 className，获取块的类名，也就是将参数 namespace 和 block 用"-"拼接生成的类名，如 a-tabs（第 4 行）；**子级块**由变量 className 和参数 blockSuffix 用"-"拼接生成，如 a-tabs-wrap（第 6 行）；**元素**由变量 className 和参数 element 用"__"拼接生成，如 a-tabs-wrap__item（第 8 行）；**修改器**由变量 className 和参数 modifier 用"--"拼接生成（第 10 行）；**修改器的值**则使用"_"，如 a-tags-wrap__item--size_mini（第 12 行）。最后，返回处理完成的变量 className（第 14 行）。

4.4.1 生成 Block

生成 Block 前需指定当前组件使用的类名，根据定义的_bem 函数生成 Block 类名，如代码清单 library-04-3 所示。

代码清单 library-04-3
```
1.  > packages/hook/use-namespace/index.js
2.  export const useNamespace = (block) => {
3.    const namespace = defaultNamespace;
4.    // 生成 Block
5.    const block = (blockSuffix = '') => _bem(namespace, block, blockSuffix)
6.    return {
7.      namespace,
8.      block
9.    };
10. };
11.
12. > packages/components/button/src/index.vue: template
13. <button :class="ns.b()">这是一个测试的按钮</button>
14.
15. > packages/components/button/src/index.vue: script
16. import { useNamespace } from '@ui-library/hook';
17. const ns = useNamespace('button');
18. console.log('命名空间>>>', ns.b());
```

在 useNamespace 函数中定义 b 方法，调用_bem 函数，并传入 namespace、block、blockSuffix 参数，便可获取到 Block 类名（第 5 行），然后将其返回（第 8 行）。

button 组件调用 useNamespace 方法并传入"button"（第 17 行），表明当前组件是一个按钮的 Block，接着调用"b"方法生成类名（第 18 行），可以直接为按钮绑定类名（第 13 行），如图 4-6 所示。

```
<div id="app" data-v-app>
  <button class="a-button">这是一个测试的按钮</button>
</div>
```

图 4-6 为按钮生成 Block 类名

4.4.2 生成 Element 和 Modifier

生成 Element 和 Modifier 与生成 Block 的方式一致，定义元素和修改器两个方法生成类名并返回，在调用时就会生成对应的类名，如代码清单 library-04-4 所示。

代码清单 library-04-4

```
1.  > packages/hook/use-namespace/index.js
2.  export const useNamespace = (block) => {
3.    const namespace = defaultNamespace;
4.    // 生成 Block（块）
5.    const b = (blockSuffix = "") => _bem(namespace, block, blockSuffix);
6.    // 生成 Element（元素）
7.    const e = (element) => element ? _bem(namespace, block, "", element, "") : "";
8.    // 生成 Modifier（修改器）
9.    const m = (modifier, value) => modifier ? _bem(namespace, block, "", "", modifier, value) : "";
10.   return {
11.     namespace,
12.     b,
13.     e,
14.     m,
15.   };
16. };
17.
18. > packages/components/button/src/index.vue: template
19. <button :class="[ns.b(), ns.m('primary')]">这是一个测试的按钮</button>
```

在 useNamespace 函数中定义 e 和 m 两个方法，分别表示"元素"和"修改器"。两个方法均使用三元运算判断参数 element 或 modifier 是否存在，如果存在，则执行 _bem 函数获取类名（第 7、9 行）。在 button 组件中调用 useNamespace 函数的 e 或 m 方法生成指定的类名（第 19 行），如图 4-7 所示。

```
<div id="app" data-v-app>
  <button class="a-button a-button--primary">这是一个测试的按钮</button>
</div>
```

图 4-7 调用 useNamespace 函数的 m 方法生成类名 a-button--primary

提示：useNamespace 函数还可以定义其他的方法，生成对应功能的类名，如 loading（加载）、disabled（禁用）、active（激活）等，我们将在开发组件的过程中进一步完善 useNamespace 函数。

本章小结

在介绍了 UI 组件库的作用、开发过程,并采用 Monorepo 建立工作区后,我们完成了 UI 组件库的基础搭建工作,也实现了 button 组件的初始化示例开发。上述开发过程只是 UI 组件库的准备工作,我们还需要根据 UI 组件库的"主题色"和"组件设计规范"不断地完善组件库,包括主题色、组件类型、useNamespace 函数、组件样式等功能,形成相对完整的 UI 组件库。

CSS 的 BEM 命名规则与封装是为了更好地维护 UI 组件库的全局样式,通过封装函数,可以在开发过程中使用各个组件实现统一的命名规则,为后期维护组件打下基础。

第 5 章
按钮组件

按钮（button）组件是 UI 界面中常见的元素，用于触发特定的操作或提交表单。它通常包括一个可点击的区域，用于产生点击后的反馈效果，并支持适当的样式和状态变化（如普通状态、悬停状态和禁用状态）。在设计 button 组件时需要考虑到可访问性、视觉吸引力及一致性，以确保它在不同的情境下能够提供良好的使用体验。

从 UI 组件库设计稿中可以看到 button 组件的多样性，如色调、悬停效果、图标、线框、圆角、加载状态等，我们将根据 UI 组件库设计稿逐步完成 button 组件的开发工作。

本章开发 button 组件的 CSS 样式，将以原生的 CSS 语法书写样式，第 6 章将介绍如何使用 Sass 语法通过变量的形式为组件设定主题效果。

5.1　theme：主题包

根据 3.2.2 节中建立 UI 组件库包的方法建立主题包。在 packages 目录下新建 theme 目录，作为 UI 组件库的主题样式包。在 theme 目录下执行 pnpm init 指令，生成 package.json 文件，再将 name 属性的值改为@ui-library/theme。最后，在 ui-library 根目录下执行 pnpm install @ui-library/theme -w，安装主题包，如代码清单 library-05-1 所示。

代码清单 library-05-1
```
1.  > packages/theme/package.json
2.  {
3.    "name": "@ui-library/theme",
4.    "version": "1.0.0",
5.    "description": "",
6.    "main": "index.js",
7.    "scripts": {
8.      "test": "echo \"Error: no test specified\" && exit 1"
9.    },
10.   "keywords": [],
11.   "author": "",
12.   "license": "ISC"
13. }
14.
15. > ui/library/package.json
16. {
17.   ...省略代码
18.   "dependencies": {
19.     "@ui-library/components": "workspace:^",
20.     "@ui-library/hook": "workspace:^",
21.     "@ui-library/theme": "workspace:^",
```

```
22.    · "@ui-library/utils": "workspace:^"
23.    }
24.  }
```

5.1.1 全量引入

主题包安装完成后，要为每个组件建立独立的样式文件。在 theme 目录下新建 src 目录，并在 src 目录下新建 index.scss 和 button.scss 文件，然后将 button.scss 引入 index.scss 文件，最后在 examples 演示包中导入 theme 目录下的 index.scss 文件，如代码清单 library-05-2-1 所示。

代码清单 library-05-2-1
```
1.  > packages/theme/src/index.scss
2.  @use "./button.scss"
3.
4.  > examples/src/main.js
5.  // UI 组件库
6.  import UILibrary from "../../packages"
7.  import "@ui-library/theme/src/index.scss"
```

当前的主题包是基础型的，我们将在开发组件的过程中一步步完善，并介绍如何处理主题包的变量，以及有关 Sass 预解析的应用。

如果在 examples 演示库中出现 Sass 预解析错误，如图 5-1 所示，那么需要在 examples 演示库终端中执行以下指令，将 sass 和 sass-loader 解析器安装到开发环境中。

◎ 安装 sass：npm install sass@1.55.0 -D。
◎ 安装 sass-loader：sass-loader@13.2.0 -D。

```
[plugin:vite:css] Preprocessor dependency "sass" not found. Did you install it? Try `pnpm
add -D sass`.
```

图 5-1 Sass 预解析错误

5.1.2 按需引入

按需引入是指在处理数据时，根据需要逐步引入。例如，在按需引入 button 组件时，会自动按需引入 button 组件所需的 CSS 样式，而不会引入其他组件的 CSS 样式，如代码清单 library-05-2-2 所示。按需引入能够根据实际需求灵活地导入数据，节约资源，提高效率。

代码清单 library-05-2-2
```
1.  > packages\components\button\src\style\index.js:
2.  import "@ui-library/theme/src/button.scss"
3.
4.  > examples/src/main.js
5.  // UI 组件库
6.  import UILibrary from "../../packages"
7.  import "@ui-library/components/button/src/style"
```

在 button 组件的 src/style/index.js 文件中引入主题包的 button.scss 样式文件（第 2 行），然

后在 examples 演示包的 main.js 文件中引入 button 组件 src/style 下的 index.js 文件。该做法相当于从 button 组件中引入样式，也就是使用什么组件，就引入该组件下的 src/style。

问题在于，现在是手动按需引入，也就是在引入组件时需要手动引入对应的组件样式，这对用户来说是很不友好的。为了避免用户手动引入样式，需要在按需引入组件时自动引入对应组件的样式。自动引入样式的动作需要在打包组件库时完成，因此将在打包组件库时解决这个问题，详见第 18 章。

5.2 渲染 button 组件

在开发任何一个组件时，都要用 CSS 样式将组件渲染出来，以便于在 examples 演示包中呈现渲染效果。渲染组件时的一个至关重要的过程就是"由简至繁"——将最基础的效果先呈现出来，再通过"CSS 组合"的方式使组件变得越来越复杂。

根据 UI 组件库设计稿中的"基础 basis"类型，按钮组件分为 5 种类型的主题：默认、主要、成功、警告和错误。其中，"默认"类型的按钮是最基础的，另外 4 种类型则是在"默认"类型的基础上增加了不同的主题色，这就是"由简至繁"的演变。因此，根据 UI 组件库 button 组件的"尺寸标注"示意图，先渲染"默认"类型的按钮，如代码清单 library-05-3 所示。

代码清单 library-05-3
```scss
1.  > packages/theme/src/button.scss
2.  .a-button {
3.    display: inline-flex;              // 盒模型，弹性布局
4.    align-items: center;               // 子元素垂直居中
5.    justify-content: center;           // 子元素水平居中
6.    height: 32px;                      // 高度（默认）
7.    min-width: 80px;                   // 最小宽度
8.    padding: 0 16px;                   // 上下内边距为 0
9.    background-color: #fff;            // 默认白色背景
10.   border-radius: 12px;               // 圆角
11.   border: 2px solid #e3e5f1;         // 2 像素，实线边框
12.   box-sizing: border-box;            // 忽略 padding 和 border 的计算
13.   line-height: 1;                    // 根据该元素本身的字体大小设置行高
14.   color: #4e5158;                    // 文字颜色
15.   text-align: center;                // 文字居中
16.   font-size: 14px;                   // 字体大小为 14
17.   white-space:nowrap;                // 不换行
18.   transition: .3s;                   // 300 毫秒过渡
19.   outline: none;                     // 去除外轮廓
20.   cursor: pointer;                   // 鼠标指针变为"手"
21.   user-select: none;                 // 禁止选中文本内容
22.   vertical-align: middle;            // 当前元素的垂直中心点和父元素的基线上
23.   span {
24.     line-height: 1;                  // 根据该元素自身的字体大小设置行高
25.     display: inline-flex;            // 盒模型，弹性布局
```

```
26.     align-items: center;            // 子元素垂直居中
27.   }
28. }
```

在上述代码清单中，类名.a-button 是通过 useNamespace 函数中的 b 方法生成的，在定义样式时使用.a-button 即可。要注意的是，UI 组件库设计稿中的组件设计多数是有"边框"和"背景颜色"的，如 input 组件、checkbox 组件、radio 组件等。因此，在书写 CSS 样式时，将这两个属性分开，以便于处理"边框类型"的按钮，渲染效果如图 5-2 所示。

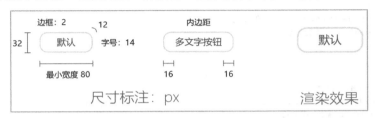

图 5-2 渲染默认类型的按钮

提示：button 组件的边框通过 border 属性实现，当然也可以采用 box-shadow 内阴影的方式处理。

5.3 theme：主题色渲染

主题色是对按钮的一种修饰，需要用"修改器"生成不同主题的类名。UI 组件库设计稿中有 5 种主题类型，分别定义为：default、primary、success、warning、error。因此，可使用命名空间 useNamespace 函数中的 m 方法生成对应的类名：a-button--default、a-button--primary、a-button--success、a-button--warning、a-button--error。通过 CSS 组合的方式覆盖主题色即可，如代码清单 library-05-4 所示。

代码清单 library-05-4
```
1.  > packages/theme/src/button.scss
2.  // 默认
3.  .a-button--default {
4.    background-color: #fff;
5.    border-color: #e3e5f1;
6.    color: #4d5059;
7.    &:hover {
8.      background-color: #f5f6fa;
9.    }
10. }
11. // 主要
12. .a-button--primary {
13.   background-color: #3069ff;
14.   border-color: #3069ff;
15.   color: #fff;
16.   &:hover {
```

```
17.      background-color: #6e96ff;
18.      border-color: #6e96ff;
19.    }
20.  }
21.  // 成功
22.  .a-button--success {
23.    background-color: #14cd70;
24.    border-color: #14cd70;
25.    color: #fff;
26.    &:hover {
27.      background-color: #5adc9b;
28.      border-color: #5adc9b;
29.    }
30.  }
31.  // 警告
32.  .a-button--warning {
33.    background-color: #ffa81a;
34.    border-color: #ffa81a;
35.    color: #fff;
36.    &:hover {
37.      background-color: #ffc25e;
38.      border-color: #ffc25e;
39.    }
40.  }
41.  // 错误
42.  .a-button--error {
43.    background-color: #ff4a5b;
44.    border-color: #ff4a5b;
45.    color: #fff;
46.    &:hover {
47.      background-color: #ff808c;
48.      border-color: #ff808c;
49.    }
50.  }
```

在 4.4.2 节中，我们调用 useNamespace 函数的 m 方法传入参数"primary"，生成了类名 a-button--primary。button.scss 文件定义了与类名 a-button--primary 相对应的主题色，此时渲染出来的按钮效果是蓝色的，如图 5-3 所示。

主要

图 5-3　a-button--primary "主要" 按钮（蓝色）

此时回到 button 组件，将 ns.m 修改器的值改为 "success"，便会看到绿色的按钮，如代码清单 library-05-5 所示，渲染如图 5-4 所示。

代码清单 library-05-5
```
1.  > packages/components/button/src/index.vue
2.  <template>
```

```
3.    <button :class="[ns.b(), ns.m('success')]">
4.      <span>默认</span>
5.    </button>
6. </template>
```

图 5-4　a-button--success "成功" 按钮（绿色）

通过配置 ns.m 修改器，可以很方便地生成不同主题色的按钮。因此，在 button 组件提供的指定属性中更新 ns.m 修改器的值，并让用户根据指定的属性从外部传入指定的值，便能得到不同主题的按钮，如代码清单 library-05-6 所示。

代码清单 library-05-6
```
1.  > packages/components/button/src/index.vue: template
2.  <button :class="[ns.b(), ns.m(type)]">
3.      <span><slot v-if="$slots.default" /></span>
4.  </button>
5.
6.  > packages/components/button/src/index.vue: script
7.  import { useNamespace } from '@ui-library/hook';
8.  const ns = useNamespace("button");
9.  /** props */
10. const props = defineProps({
11.   type: {
12.     type: String,
13.     default: "default", // 默认值
14.   }
15. });
16.
17. > examples\src\App.vue
18. <a-button>默认</a-button>
19. <a-button type="primary">主要</a-button>
20. <a-button type="success">成功</a-button>
21. <a-button type="warning">警告</a-button>
22. <a-button type="error">错误</a-button>
```

在 button 组件的 defineProps 对象中定义 type 属性（第 10~15 行），将 type 属性绑定到 ns.m 方法上（第 2 行），按钮的文本依然是从外部传入的，可采用 "默认插槽" slot 的方式定义（第 3 行）。最后，在 examples 演示包中调用<a-button>按钮，使用 type 属性传入不同的类型及文本（第 18~22 行），渲染效果如图 5-5 所示。

图 5-5　不同主题色的按钮

5.4　round：圆角按钮

圆角按钮是指按钮的左右两侧为半圆形，可查看 UI 组件库设计稿中的"圆角按钮/round"效果。button 组件的圆角效果只有"是"和"否"两种情况，选择"是"则设置圆角，反之则不设置。在设置圆角效果时，只需为按钮添加 CSS 属性 border-radius。因此，可根据是否设置圆角的情况，为 button 组件添加类名 is-round，如代码清单 library-05-7 所示。

代码清单 library-05-7
```
1.  > packages/hook/use-namespace/index.js
2.  export const useNamespace = (block) => {
3.    ...省略代码
4.    // 状态设置
5.    const is = (activeName, state) => activeName && state ? `is-${activeName}` : '';
6.    return { namespace, b, e, m, is };
7.  };
8.
9.  > packages/components/button/src/index.vue: template
10.   <button :class="[ns.b(), ns.m(type), ns.is('round', round)]"></button>
11.
12. > packages/components/button/src/index.vue: script
13. /** props */
14. const props = defineProps({
15.   type: { ...省略代码 },
16.   round: Boolean, // 布尔类型，默认为 false
17. });
18.
19. > packages/theme/src/button.scss
20. .a-button {
21.   ...省略代码
22.   &.is-round { border-radius: 100px; }  // 圆角按钮
23. }
24.
25. > examples/src/App.vue
26. <a-button type="primary" round>主要</a-button>
```

业务中常见的交互按钮，如 disabled 禁用按钮、loading 加载按钮、text 文字按钮、link 链接按钮等，也都以"是"与"否"两种状态决定是否需要添加类名 is-round 实现"圆角"。因此，可通过命名空间将其封装成统一方法，传入相应的参数即可。

在 useNamespace 函数中添加 is 方法并接收两个参数：activeName 动作类型名称、state 状态（第 5 行）。用三元运算判断 activeName 和 state 是否都存在，如果是，则返回"is-"拼接动作类型名称，否则返回"空"字符串（第 5 行）。最后，在 button 组件中调用 ns.is 方法时，传入第一个固定值的参数'round'，第二个参数表示 defindProps 对象的 round 状态（第 10、16 行）。

在 examples 演示库中为"主要"按钮添加 round 属性（第 26 行），便会为"主要"按钮组合类名 is-round。在 button.scss 文件中也添加类名 is-round，并设置 border-radius 属性值为 100px（第 22 行），渲染效果如图 5-6 所示。

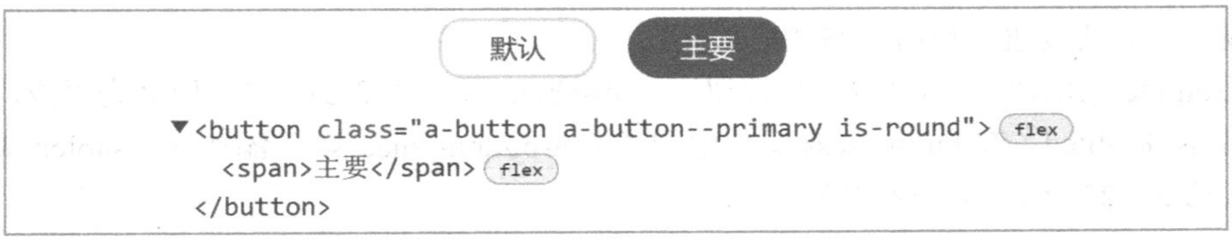

图 5-6　将 button 组件设置为圆角按钮

5.5　disabled：禁用按钮

禁用按钮和圆角按钮一致，也只有"是"和"否"两种情况，选择"是"则设置禁用，反之则不设置。同样采用 ns.is 方法处理，仅需要改变 CSS 样式，如代码清单 library-05-8 所示。

代码清单 library-05-8
```
1.  > packages/components/button/src/index.vue: template
2.  <button
3.      :class="[...省略代码, ns.is('disabled', disabled)]"
4.      :disabled="disabled">
5.  </button>
6.
7.  > packages/components/button/src/index.vue: script
8.  /** props */
9.  const props = defineProps({
10.     type: { ...省略代码 },
11.     round: Boolean, // 布尔类型，默认为 false
12.     disabled: Boolean, // 布尔类型，默认为 false
13. });
14.
15. > packages/theme/src/button.scss
16. ...省略代码
17. // 成功按钮
18. .a-button--success {
19.     ...省略代码
20.     &:hover { ...省略代码 }
21.     &.is-disabled {
22.         cursor: not-allowed;
23.         &, &:hover, &:focus {
24.             background-color: #89e6b7;
25.             border-color: #89e6b7;
26.         }
27.     }
28. }
29.
30. > examples/src/App.vue
31. <a-button type="success" disabled>成功</a-button>
```

调用 ns.is 方法，传入第一个固定值参数 'disabled'，第二个参数表示 defindProps 对象的 disabled 状态（第 3、12 行），为按钮添加自身的 disabled 属性并绑定 defindProps 对象的 disabled，这样才能真正激活按钮的禁用状态（第 4 行）。

在 examples 演示库中为"成功"按钮添加 disabled 属性（第 31 行），便会为"成功"按钮组合类名 is-disabled。在 button.scss 文件中的类名 a-button--success 下添加 is-disabled 类名的属性样式（第 21~27 行），渲染效果如图 5-7 所示。

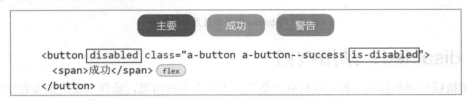

图 5-7　将 button 组件设置为禁用按钮

提示：这里仅展示了"成功"按钮的禁用效果，如需了解其他按钮的禁用效果，可查看本书配套代码中的分支 library-05-8。

5.6　text：文字按钮

文字按钮和圆角按钮一致，也只有"是"和"否"两种情况，同样采用 ns.is 方法处理。在设置 CSS 样式时，去除"背景颜色"和"边框颜色"，再将文字颜色改为对应的主题色即可，如代码清单 library-05-9 所示。

代码清单 library-05-9
```
1.  > packages/components/button/src/index.vue: template
2.    <button :class="[...省略代码, ns.is('text', text)]"></button>
3.
4.  > packages/components/button/src/index.vue: script
5.  /** props */
6.  const props = defineProps({
7.    ...省略代码
8.    text: Boolean, // 布尔类型，默认为 false
9.  });
10.
11. > packages/theme/src/button.scss
12. .a-button {
13.   ...省略代码
14.   &.is-text {
15.     background-color: transparent; // 背景颜色透明
16.     border-color: transparent;     // 边框颜色透明
17.   }
18. }
19. // 警告按钮
20. .a-button--warning {
21.   ...省略代码
22.   &.is-disabled { ...省略代码 }
23.   &.is-text {
```

```
24.      color: #ffa81a;
25.      &:hover {
26.        background-color: #fff6e8;
27.        border-color: #fff6e8;
28.      }
29.    }
30.  }
31.
32.  > examples/src/App.vue
33.  <a-button type="warning" text>警告</a-button>
```

调用 ns.is 方法，传入第一个固定值参数'text'，第二个参数表示 defindProps 对象的 text 状态（第 2、8 行）。在 examples 演示库中为"警告"按钮添加 text 属性（第 33 行），便会为"警告"按钮组合类名 is-text，在 button.scss 文件中添加类名 is-text 的属性样式（第 14~17、23~29 行），渲染效果如图 5-8 所示。

图 5-8　将 button 组件设置为文字按钮

提示：这里仅展示了"警告"按钮的文字效果，如需了解其他按钮的文字效果，可查看本书配套代码中的分支 library-05-9。

5.7　link：链接按钮

链接按钮和文字按钮基本一致，同样是将"背景颜色"和"边框颜色"去除，将文字颜色改为对应的主题色，将按钮的高度设置为 auto、内边距设置为 0，再为文字加上底部实线，如代码清单 library-05-10 所示。

代码清单 library-05-10
```
1.  > packages/components/button/src/index.vue: template
2.  <button :class="[...省略代码, ns.is('link', link)]"></button>
3.
4.  > packages/components/button/src/index.vue: script
5.  /** props */
6.  const props = defineProps({
7.    ...省略代码
8.    link: Boolean, // 布尔类型，默认为 false
9.  });
```

```
10.
11.  > packages/theme/src/button.scss
12.  .a-button {
13.    ...省略代码
14.    &.is-link {
15.      &, &:hover {
16.        padding: 0;                        // 清除内边距
17.        height: auto;                      // 高度自动
18.        min-width: auto;                   // 最小宽度自动
19.        background-color: transparent;     // 背景颜色透明
20.        border-color: transparent;         // 边框颜色透明
21.        text-decoration: underline;        // 文字底部实线
22.      }
23.    }
24.  }
25.  // 警告按钮
26.  .a-button--primary {
27.    &.is-text { ...省略代码 }
28.    &.is-link {
29.      color: #4d5059;
30.      &:hover {
31.        color: #767a87;
32.      }
33.    }
34.  }
35.
36.  > examples/src/App.vue
37.  <a-button type="primary" link>主要</a-button>
```

调用 ns.is 方法，传入第一个固定值参数为'link'，第二个参数是 defindProps 对象的 link 状态（第 2、8 行）。在 examples 演示库中为所有按钮添加 link 属性（第 37 行），便会为按钮组合类名 is-link，在 button.scss 文件中添加类名 is-link 的属性样式（第 14~23、28~33 行），渲染效果如图 5-9 所示。

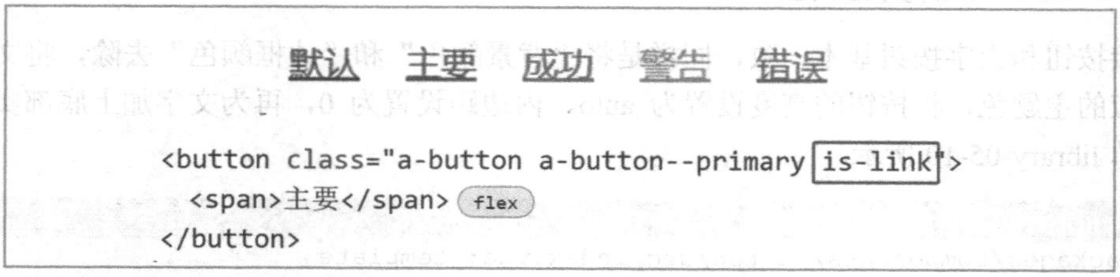

图 5-9 将 button 组件设置为链接按钮

提示：这里仅展示了"主要"按钮的链接效果，如需了解其他按钮的链接效果，可查看本书配套代码中的分支 library-05-10。

5.8　border：边框按钮

在 5.2 节中渲染 button 组件时，已经拆分了"背景颜色"和"边框"两个属性，所以在实现边框按钮时，只需将"背景颜色"设置为白色，并将文字颜色改为对应的主题色即可，如代码清单 library-05-11 所示。

代码清单 library-05-11
```
1.  > packages/components/button/src/index.vue: template
2.    <button :class="[...省略代码, ns.is('border', border), ns.is('dashed',
   dashed)]"></button>
3.
4.  > packages/components/button/src/index.vue: script
5.  /** props */
6.  const props = defineProps({
7.    ...省略代码
8.    border: Boolean, // 实线边框。布尔类型，默认为 false
9.    dashed: Boolean, // 虚线边框。布尔类型，默认为 false
10. });
11.
12. > packages/theme/src/button.scss
13. .a-button {
14.   ...省略代码
15.   &.is-border {
16.     &, &:hover {
17.       background-color: transparent;    // 背景颜色透明
18.     }
19.   }
20.   &.is-dashed {
21.     border-style: dashed; // 虚线边框
22.   }
23. }
24. // 主要按钮
25. .a-button--primary {
26.   &.is-link { ...省略代码 }
27.   &.is-border {
28.     color: #3069ff;
29.   }
30. }
31.
32. > examples/src/App.vue
33. <a-button type="primary" border>主要</a-button>
34. <a-button type="ssuccexx" border dashed>成功</a-button>
```

调用 ns.is 方法，传入第一个固定值参数'border'，第二个参数表示 defindProps 对象的 border 状态（第 2、8 行）。在 examples 演示库中为所有按钮组件添加 border 属性，便会为按钮组合类名 is-border，在 button.scss 文件中添加类名 is-border 的属性样式（第 15~19、27~29 行）。

虚线边框的实现方法是，在调用 ns.is 方法时传入'dashed'（第 2 行），便会生成类名 is-dashed，在 button.scss 文件中添加类名 is-dashed 的属性样式（第 20~22 行），渲染效果如图 5-10 所示。

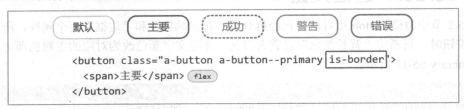

图 5-10　将 button 组件设置为边框按钮

提示：这里仅展示了"主要"按钮的边框效果，如需了解其他按钮的边框效果，可查看本书配套代码中的分支 library-05-11。

5.9　block：块级按钮

块级按钮的处理方式是将元素自身的盒模型改为"block"，block 类型的盒模型元素宽度会自动适应父元素。同样采用 ns.is 方法为元素添加类名 is-block，如代码清单 library-05-12 所示。

代码清单 library-05-12
```
1.  > packages/components/button/src/index.vue: template
2.  <button :class="[...省略代码, ns.is('block', block)]"></button>
3.
4.  > packages/components/button/src/index.vue: script
5.  /** props */
6.  const props = defineProps({
7.    ...省略代码
8.    block: Boolean, // 布尔类型，默认为 false
9.  });
10.
11. > packages/theme/src/button.scss
12. .a-button {
13.   ...省略代码
14.   &.is-block {
15.     display: block;    // 块级元素
16.     width: 100%;       // 宽度100%，适应父元素宽度
17.     margin-left: 0;    // 左边边距为 0
18.   }
19. }
20.
21. > examples/src/App.vue
22. <a-button type="primary" block>主要</a-button>
```

调用 ns.is 方法，传入第一个固定值参数'block'，第二个参数表示 defindProps 对象的 block 状态（第 2、8 行）。在 examples 演示库中为所有按钮添加 block 属性（第 22 行），便会为按钮组合类名 is-block，在 button.scss 文件中添加类名 is-block 的属性样式（第 14~18 行），渲染效果如图 5-11 所示。

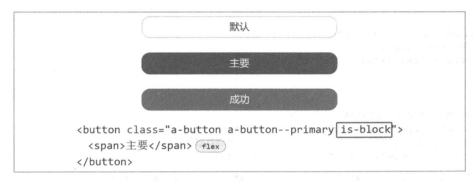

图 5-11 将 button 组件设置为边框按钮

提示：这里仅展示了"主要"按钮的块级效果，如需了解其他按钮的块级效果，可查看本书配套代码中的分支 library-05-12。

5.10 size：按钮尺寸

按钮尺寸的开发方式与 5.3 节主题色的方式一致。根据 UI 组件库设计稿，分别定义 3 种尺寸类型：small（小的）、default（默认的）、large（大的）。使用命名空间 useNamespace 函数中的 m 方法生成对应的类名：a-button--size_small、a-button--size_default、a-button--size_large。通过 CSS 组合的方式修改按钮的高度，如代码清单 library-05-13 所示。

代码清单 library-05-13
```
1.  > packages/components/button/src/index.vue: template
2.  <button :class="[...省略代码, ns.m(type), ns.m('size', size)]"></button>
3.
4.  > packages/components/button/src/index.vue: script
5.  /** props */
6.  const props = defineProps({
7.    type: { ...省略代码 },
8.    size: {
9.      type: String,
10.     default: "",
11.   }
12. });
13.
14. > packages/theme/src/button.scss
15. .a-button {
16.   ...省略代码
17.   &.a-button--size_small {    // 小的
18.     height: 24px;
19.     padding: 0 10px;
20.     border-radius: 10px;
21.     font-size: 12px;
22.     min-width: auto;
```

```
23.   }
24.   &.a-button--size_large {    // 大的
25.     height: 40px;
26.     font-size: 16px;
27.   }
28. }
29.
30. > examples/src/App.vue
31. <a-button type="primary" size="small">小的</a-button>
32. <a-button type="success" size="default">默认的</a-button>
33. <a-button type="warning" size="large">大的</a-button>
```

调用 ns.m 方法，传入的第一个参数是固定值'size'，表示要修改"尺寸"；第二个参数表示 defindProps 对象的 size 属性（第 2、8~11 行）。在 examples 演示库中为按钮添加 size 属性并传入 small、default、large（第 31~33 行），便会为 3 个按钮分别组合类名 a-button--size_small、a-button--size_default、a-button--size_large，在 button.scss 文件中添加对应类名的属性样式（第 17~27 行）。

要注意的是，类名 a-button--size_small 不需要最小宽度，因此将 min-width 属性的值改为 auto，内边距左右两侧为 10px，圆角 border-radius 为 10px，并设置字号为 12px，渲染效果如图 5-12 所示。

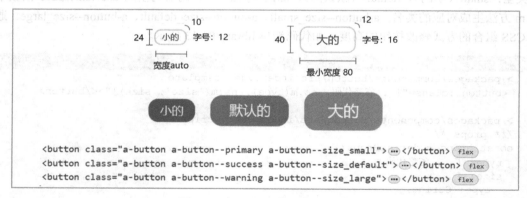

图 5-12　为 button 组件设置按钮尺寸

提示：button 组件的类名 a-button 中已经定义了高度属性 height 为 32px，也就是 button 组件的默认尺寸。而 defindProps 对象属性 size 的默认值是"空"，因此，对于默认尺寸的按钮，无须配置属性 size 的值为 default，也一样可以达到默认尺寸的效果。

5.11　circle：圆形按钮

圆形按钮和圆角按钮一致，都采用 ns.is 方法处理，隐藏按钮的文字并且保持按钮的宽度和高度一致，设置 border-radius 属性的值大于或等于按钮自身的宽高，如代码清单 library-05-14 所示。

代码清单 library-05-14

```
1.  > packages/components/button/src/index.vue: template
2.  <button :class="[...省略代码, ns.is('circle', circle)]">
3.    <span v-if="$slots.default"><slot /></span>
4.  </button>
5.
6.  > packages/components/button/src/index.vue: script
7.  /** props */
8.  const props = defineProps({
9.    size: { ...省略代码 },
10.   circle: Boolean,  // 圆形按钮
11. });
12.
13. > packages/theme/src/button.scss
14. .a-button {
15.   ...省略代码
16.   &.a-button--small {
17.     ...省略代码
18.     &.is-circle {
19.       width: 24px;       // 宽度
20.     }
21.   }
22.   &.a-button--default {
23.     &.is-circle {
24.       width: 32px;       // 宽度
25.     }
26.   }
27.   &.a-button--large {
28.     ...省略代码
29.     &.is-circle {
30.       width: 40px;       // 宽度
31.     }
32.   }
33.   &.is-circle {
34.     border-radius: 100px;
35.     padding: 0;
36.     min-width: auto;
37.   }
38. }
39.
40. > examples/src/App.vue
41. <a-button type="primary" circle size="small">小的</a-button>
42. <a-button type="success" circle>默认的</a-button>
43. <a-button type="warning" circle size="large">大的</a-button>
```

调用 ns.is 方法，传入第一个固定值的参数'circle'，第二个参数表示 defindProps 对象的 circle 状态（第 2、10 行）。在 examples 演示库中为按钮添加 circle 属性（第 41~43 行），使按钮组合类名 is-circle，在 button.scss 文件中添加类名 is-circle 的属性样式（第 14~38 行）。

在<slot />插槽的位置增加对$slots.default 的判断（第 3 行），判断是否使用了"默认"插槽，

如果是，则渲染插槽，否则不渲染，渲染效果如图 5-13 所示。

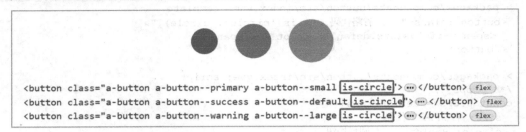

图 5-13　将 button 组件设置为圆形按钮

5.12　icon：图标按钮

图标按钮的开发方式有多种，比较简单的是使用阿里巴巴矢量图标库（后文简称"阿里图库"）的字体图标，或者开发图标组件加载 SVG 图标，再默认内置一些图标供用户使用。本节采用阿里图库的方式开发图标按钮，如代码清单 library-05-15 所示。第 7 章将采用开发图标组件的方式。

代码清单 library-05-15
```
1.  > packages/components/button/src/index.vue: template
2.  <button :class="[...省略代码, ns.is('circle', circle)]">
3.    <i v-if="icon" class="a-icon iconfont" :class="icon"></i>
4.    <span v-if="$slots.default"><slot /></span>
5.  </button>
6.
7.  > packages/components/button/src/index.vue: script
8.  /** props */
9.  const props = defineProps({
10.   size: { ...省略代码 },
11.   icon: {
12.     type: String,
13.     default: "",
14.   }
15. });
16.
17. > packages/theme/src/button.scss
18. &.a-button--size_small {
19.   .a-icon { font-size: 12px; }
20. }
21. &.a-button--size_default {
22.   .a-icon { font-size: 14px; }
23. }
24. &.a-button--size_large {
25.   .a-icon { font-size: 16px; }
26. }
27. .a-icon + span { margin-left: 4px; }
```

```
28.
29.  > examples/src/App.vue
30.  <a-button type="primary" circle size="small" icon="icon-xiazai">小的</a-button>
31.  <a-button type="success" circle icon="icon-xiazai">默认的</a-button>
32.  <a-button type="warning" circle size="large" icon="icon-xiazai">大的</a-button>
```

阿里图库的图标通过 CSS 的伪类生成，因此只需要接收外部传入的类名，再写入指定标签即可。在 defineProps 对象中定义 icon 属性为字符串类型（第 11~14 行），为 button 组件添加<i>标签作为图标的渲染区域，使用:class 绑定 defineProps 对象中的 icon 属性（第 3 行），并且使用 v-if 判断是否存在 icon，存在则显示，反之则不显示。v-if 判断可以确保只有传入了图标才会显示。

阿里图库中的图标是一种字体类型，可以设置 font-size 属性，但自身默认字体大小为 16px，因此需要根据按钮的不同尺寸重写图标的大小（第 18~26 行），最后使用外边距属性 margin 拉开图标和文字之间的距离（第 27 行），如图 5-14 所示。

图 5-14 阿里图库中的按钮图标

提示：阿里图库的图标使用方法相对简单，本书不作讲解，读者可自行搜索相关用法。阿里图库的官方网址见链接 5-1①。

5.13 loading：加载按钮

加载按钮通常用于在网页或应用程序中显示正在加载或处理中的状态。当用户执行某些操作时，系统需要时间来处理请求或加载数据，这时加载按钮通常会显示为动画或指示器，告知用户"系统正在处理中"，防止用户多次点击或操作同一个按钮。加载按钮有助于提升用户体验，让用户知道系统是否正在处理他们的请求，同时有助于减少用户的不确定感。

5.13.1 手动触发加载

手动触发类型的加载按钮与禁用按钮的作用一致，通过外部传入属性的方式控制 button 组件是否启用加载状态，同样采用 ns.is 方法处理，如代码清单 library-05-16 所示。

```
代码清单 library-05-16
1.  > packages/components/button/src/index.vue: template
2.  <button
3.      :class="[...省略代码, ns.is('loading', loading)]"
4.      :disabled="disabled || loading"
5.  >
```

① 参考链接的获取方法详见本书封底的读者服务。

```
6.      <template v-if="loading">
7.        <i class="a-icon iconfont icon-loading"
8.          :class="[`${ns.is('loading-transition', loading)}`]"> // is-loading-transition
9.        </i>
10.     </template>
11.     <i v-else-if="icon" class="a-icon iconfont" :class="icon"></i>
12.     <span v-if="$slots.default && !circle"><slot /></span>
13.   </button>
14.
15. > packages/components/button/src/index.vue: script
16. /** props */
17. const props = defineProps({
18.   size: { ...省略代码 },
19.   loading: Boolean, // 加载按钮
20. });
21.
22. > packages/theme/src/button.scss
23. // 默认、主要、成功、警告、错误
24. .a-button--default {
25.   &.is-disabled, &.is-loading { ...省略代码 }
26. }
27.
28. > packages/theme/src/loading.scss
29. .is-loading-transition {
30.   animation: rotate 1s linear 0s infinite;
31. }
32. @keyframes rotate {
33.   0% { transform: rotate(0deg); }
34.   100% { transform: rotate(360deg); }
35. }
36.
37. > examples/src/App.vue: template
38. <a-button type="primary" :loading="loading" size="small">小的</a-button>
39. <a-button type="success" :loading="loading">默认的</a-button>
40. <a-button type="warning" :loading="loading" size="large">大的</a-button>
```

调用 ns.is 方法，传入第一个固定值的参数'loading'，第二个参数表示 defindProps 对象的加载状态（第 3、19 行），同时为按钮自身的属性 disabled 使用"或"运算绑定 loading（第 4 行）。由于 loading 按钮和 icon 按钮都存在图标，因此需要有单独渲染 loading 的区域使两者互斥，新增<template>单独渲染 loading 按钮（第 6~10 行），同样使用<i>标签渲染一个固定的"加载"图标（第 8 行），并使用 v-if 条件判断，确保其为 loading 时才开始渲染<template>（第 6 行），将其他 icon 图标的条件改为 v-else-if（第 11 行）。

由于加载按钮的渲染效果和禁用按钮一致，因此使类名 is-loading 与 is-disable 并行即可（第 25 行）。为了实现加载按钮的旋转效果，可定义类名 is-loading-transition，采用 CSS3 的动画效果实现连续性的 360 度旋转（第 29~35 行），然后使用 ns.is 方法为 loading 按钮绑定 is-loading-transition 类名（第 8 行），渲染效果如图 5-15 所示。

图 5-15 将 button 组件设置为加载按钮（手动触发）

5.13.2 自动触发加载

自动触发的加载按钮也是一种业务型操作，自动触发根据 API 接口请求的过程实现 loading 的自动变更，而不需要手动改变状态，如代码清单 library-05-17 所示。

代码清单 library-05-17
```
1.  > packages/components/button/src/index.vue: template
2.  <button
3.    :class="[...省略代码, ns.is('loading', loading || _loading)]"
4.    :disabled="disabled || loading || _loading"
5.    @click="handlerChange"
6.  >
7.    <template v-if="loading || _loading">
8.      <i
9.        class="a-icon iconfont icon-loading"
10.       :class="[`${ns.is('loading-transition', loading || _loading)}`]"> //
is-loading-transition
11.     </i>
12.   </template>
13. </button>
14.
15. > packages/components/button/src/index.vue: script
16. import { ref } from "vue"
17. /** props */
18. const props = defineProps({
19.   ...省略代码
20.   beforeChange: Function
21. });
22. /** emits */
23. const emit = defineEmits(['click']);
24. /** loading */
25. const _loading = ref(false)
26. /**
27.  * 点击事件
28.  */
29. const handlerChange = (ev) => {
30.   const isFunction = Object.prototype.toString.call(props.beforeChange) === '[object Function]'
31.   if(!isFunction) {
32.     emit('click', ev)
33.     return false
34.   }
35.   // 启用加载
36.   _loading.value = true
```

```
37.    props.beforeChange().finally(() => _loading.value = false)
38.  }
39.
40. > examples/src/App.vue: template
41. <a-button type="warning" :before-change="handlerChange">提交按钮</a-button>
42.
43. > examples/src/App.vue: script
44. const handlerChange = () => {
45.   return new Promise((resolve, reject) => {
46.     setTimeout(() => {
47.       resolve()
48.     }, 2000)
49.   })
50. }
```

API 接口的请求一般有 4 个过程，分别是：客户端发送请求、服务器接收请求、服务器处理请求、服务器返回响应。从 API 业务请求交互的角度来说，以上 4 个过程可以理解为 3 个阶段，分别是：请求前、请求中（服务器）、请求完成。而我们在开发加载效果时，只需要知道 2 个过程即可，分别是：请求前、请求完成。

请求前是发起 API 请求之前要做的事情，也就是开启加载效果；请求完成是服务器返回响应，无论是成功响应还是失败响应，都属于请求完成，只是结果不同。因此，可以使用 Promise 对象完成自动触发加载的整个过程。

在 defineProps 对象中定义 beforeChange 属性为 Function 类型，也就是一个函数（第 20 行）。为 button 组件添加 @click 事件，绑定 handlerChange 方法（第 5 行）。在 handlerChange 方法中优先判断 props.beforeChange 是否是一个函数（第 30 行），如果不是，则调用 emit 对象回调父级的 @click 事件并阻止（第 32、33 行），反之则继续执行。此时便是"请求前"的状态，把变量 _loading 的值设置为 true，表示启用加载状态（第 25、36 行）。接着调用 props.beforeChange 方法，并在回调 finally 状态时，重置 _loading 的值为 false（第 37 行）。最后，把变量 _loading 绑定到 button 组件中，实现加载状态（第 3、4、7、10 行）。

在 examples 演示库中添加属性 :before-change 并绑定 handlerChange 方法（第 41 行），在该方法中必须返回 Promise 对象 resolve 或 reject 的回调，否则无法自动触发 button 组件的加载效果，渲染效果如图 5-16 所示。

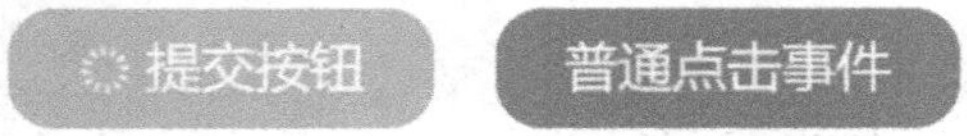

图 5-16　将 button 组件设置为加载按钮（自动触发）

5.14　group：按钮组

按钮组是一种用于将多个按钮组合在一起的组件，通常用于有效地组织和控制用户界面中的按钮。将相关的按钮放在同一组中，用户可以更容易地识别它们。

5.14.1 插槽实现按钮组

按钮组的开发相对比较简单，仅需要在 button 组件外包裹一层元素，这会用到"默认插槽"功能，如代码清单 library-05-18 所示。

代码清单 library-05-18
```
1.  > packages/components/buttonGroup/src/index.vue:
2.  <template>
3.    <div :class="[ns.b()]"><slot /></div>
4.  </template>
5.  <script>export default { name: "a-button-group" };</script>
6.  <script setup>
7.  import { useNamespace } from '@ui-library/hook';
8.  const ns = useNamespace("button-group");
9.  </script>
10.
11. > packages/components/button/src/index.vue: template
12. <button :class="[...省略代码, ns.is('button-group', isGroup)]>
13.
14. > packages/components/button/src/index.vue: script
15. import { ref, computed, getCurrentInstance } from "vue"
16. ...省略代码
17. /** computed */
18. const isGroup = computed(() => {
19.   return getCurrentInstance().parent.proxy.$options.name === 'a-button-group'
20. })
21.
22. > packages/theme/src/button.scss
23. .a-button {
24.   ...省略代码
25.   &.is-button-group {
26.     border-radius: 0;
27.     + .is-button-group { margin-left: -2px; }
28.     &:first-child { border-radius: 12px 0 0 12px; }
29.     &:last-child { border-radius: 0 12px 12px 0; }
30.   }
31. }
```

按钮组相当于一个新的组件，根据 3.4 节中构建组件目录的方式，在 packages/components 目录下创建 buttonGroup 组件，并在 index.vue 文件中添加代码清单（第 2~9 行），其中第 3 行是"包裹"层，并将该组件定义名称为 a-button-group（第 5 行）。

在 button 组件中引入 Vue.js 3 实例的 getCurrentInstance 对象（第 15 行），并获取父级组件的 name，判断其是否为 a-button-group（第 19 行），将结果返回给变量 isGroup（第 18~20 行），将变量 isGroup 绑定到 :class 中（第 12 行）。如果 isGroup 的值为 true，则为按钮追加类名 is-button-group，否则不追加，如图 5-17 所示。

```
▼<div class="a-button-group">
  ▶<button class="a-button a-button--default is-button-group">…</button>
  ▶<button class="a-button a-button--default is-button-group">…</button>
  ▶<button class="a-button a-button--primary is-button-group">…</button>
</div>
```

图 5-17　为 button 组件追加类名 is-button-group

最后重写 CSS 样式，如果存在类名 is-button-group，则设置所有按钮的圆角属性 border-radius 为 0（第 26 行），并从第 2 个按钮开始，将左侧边距往左移动-2px，与边框重叠（第 27 行）；设置第 1 个按钮的左侧为两个圆角，右侧为 0；设置最后一个按钮的右侧为两个圆角，左侧为 0（第 28、29 行），渲染效果如图 5-18 所示。

图 5-18　将 button 组件设置为按钮组

5.14.2　父级组件的属性

父级组件的属性是指可以被子级组件继承的属性。比如，当子级组件和父级组件同时存在相同的属性时，子级组件应该用自身配置的属性；如果只在父级组件中配置了属性，那么所有子级组件应该用父级组件配置的属性，如代码清单 library-05-19 所示。

代码清单 library-05-19

```
1.  > packages/components/buttonGroup/src/index.vue: template
2.  <template>
3.    <div :class="[
4.      ns.m(isGroupSize),
5.      ns.is('button-group', isGroup)
6.    ]"><slot /></div>
7.  </template>
8.
9.  > packages/components/buttonGroup/src/index.vue: script
10. import { useNamespace, useParent } from '@ui-library/hook';
11. const ns = useNamespace("button");
12. const parent = useParent("button-group");
13. const isGroup = parent.group()
14. const isGroupSize = props.size || parent.props('size')
15.
16. > packages/hook/use-parent/index.js
17. import { getCurrentInstance } from "vue"
18. // 默认命名前缀
19. export const defaultNamespace = "a";
20.
21. export const useParent = (parentName) => { // parentName 接收父级组件定义的名称
22.   // 获取父级
23.   const parent = getCurrentInstance().parent
```

```
24.     // 获取父级组件
25.     const group = () => {
26.       return parent.proxy.$options.name === `${defaultNamespace}-${parentName}`
27.     }
28.     // 获取父级属性
29.     const props = (attr) => {
30.       return parent.props?.[attr] || false
31.     }
32.     return {
33.       group, props
34.     }
35.   }
36.
37.   > examples/src/App.vue
38.   <a-button-group size="small">
39.     <a-button>默认</a-button>
40.     <a-button icon="icon-baocun">默认</a-button>
41.     <a-button type="primary" icon="icon-bofangfmbofangliebiao">主要</a-button>
42.   </a-button-group>
```

子级组件在继承父级组件的数据前，需要先获取父级组件的属性配置信息，可通过 Vue.js 3 实例的 getCurrentInstance 对象的 parent 属性获取。由于父级组件应用的场景比较多，如 checkbox 多选框、radio 单选框、form 表单等，因此可将获取父级组件的方法抽离成函数，抽离方式与 useNamespace 完全一致。

在 hook 目录下新建 use-parent 目录和 index.js 文件，在 index.js 文件中写入第 17~35 行代码。其中，group 方法的 parent.proxy.$options.name 获取父级组件定义的名称，再与传入的 parentName 对比是否一致，如果一致，就表示存在父级组件，反之则不存在（第 26 行）。使用 props 方法获取父级组件 props 对象的属性，如果能获取到指定的属性，则返回属性，反之则返回 false（第 30 行）。调用 group 和 props 方法（第 13、14 行），变量 isGroup 保持不变，变量 isGroupSize 使用"或"运算判断 props.size 是否存在，如果不存在，则获取父级组件的 size 属性（第 14 行）。最后为 ns.m 方法绑定变量 isGroupSize（第 4 行），渲染效果如图 5-19 所示。

图 5-19　按钮组中父级组件的 size 属性

提示：按钮组通过递归的逻辑获取父级组件的属性，会消耗比较多的性能。在 10.6.1 节中将使用 provide、inject 通信的方式来优化性能消耗的问题。

本章小结

本章初步实现了 button 组件的开发配置，通过属性配置的方式，结合 CSS 的命名规则，为 button 组件渲染各种不同的效果，从而体现一个组件"由简至繁"的过程。

第 6 章
Sass制定组件库全局变量

UI 组件库的主题色是指在设计中用于呈现整体视觉风格的核心颜色。在 UI 组件库中，通常使用主题色突出重要信息、动作或交互元素。主题色在 UI 设计中非常重要，它有助于确立品牌或应用程序的整体外观和视觉，并为用户提供一致的界面体验。

除主题色外，UI 组件库中还有组件尺寸、文字大小、边框颜色、圆角、边距等一系列的样式属性规范，即 UI 组件库的全局样式属性规范。因此，可使用 Sass 以变量名称的形式生成这些属性，将其统一管理起来。在后续的组件开发过程中，只需读取生成的全局样式属性变量名称，即可使所有的组件达到统一的 UI 效果。

Sass 封装 CSS 样式是指使用 Sass（一种 CSS 预处理器）来组织和管理 CSS 样式代码。通过 Sass，开发人员可以使用变量、嵌套规则、Mixin（混合）、逻辑运算等功能来书写更具模块化和可维护性的 CSS 代码。Sass 允许开发人员定义可重用的样式，并通过引用它们来避免代码重复。此外，Sass 还提供了对嵌套规则的支持，这使得结构清晰的样式表更容易阅读和维护。Sass 通过其强大的功能和结构化的方法，帮助开发人员更高效地管理和封装 CSS 样式。

6.1 deep-merge：定义主题色

在第 5 章中，button 组件的主题色虽然是根据 UI 组件库设计稿的颜色定制的，但是我们将主题色的色值固定写在了 button.scss 文件中，这样会导致 button 组件的主题色无法变化，并且其他用户在使用我们开发的组件库时也无法修改主题色。因此，在开发 UI 组件库时，需要将主题色、尺寸等公共属性抽离到单独的文件中，作为变量的形式引用，使 UI 组件库根据变量的变化而变化。

sass:map 是 Sass 提供的一种数据结构地图，用于存储键值对。Sass 的 map 常常被称为数据地图，因为它总是以 key:value 的形式成对出现，与 JSON 相似。它类似于其他编程语言中的字典或哈希表，如代码清单 library-06-1 所示。

代码清单 library-06-1
```
1.  > packages/theme/src/index.scss
2.  @use "./common/var.scss";
3.
4.  > packages/theme/src/common/var.scss
5.  @use "sass:map";
6.
7.  /** colors */
8.  $colors: () !default;
9.  // 合并
```

```scss
10.    $colors: map.deep-merge(
11.        (
12.            'white': #ffffff,            // 白色
13.            'black': #000000,            // 黑色
14.            'primary': ('base': #3069ff),    // 主要
15.            'success': ('base': #14cd70),    // 成功
16.            'warning': ('base': #ffa81a),    // 警告
17.            'error': ('base': #ff4a5b),      // 错误、危险
18.        ),
19.        $colors
20.    );
21.
22.    // 色调
23.    $color-white: map.get($colors, 'white') !default;
24.    $color-black: map.get($colors, 'black') !default;
25.    $color-primary: map.get($colors, 'primary', 'base') !default;
26.    $color-success: map.get($colors, 'success', 'base') !default;
27.    $color-warning: map.get($colors, 'warning', 'base') !default;
28.    $color-error: map.get($colors, 'error', 'base') !default;
```

map 是 Sass 内置的方法，使用 map 前需要先引入它（第 5 行）。定义变量$colors 存储主题色，初始化默认值为"空"，并设置属性!default，表示属性$colors 存在默认值，但可以被重写（第 8 行）。接着使用 Sass 的合并对象方法 map.deep-merge 将两个值合并，再重新赋值给变量$colors（第 10~20 行）。然后定义$color-white、$color-black、$color-primary、$color-success、$color-warning、$color-error 等变量，使用 Sass 的 map.get 对象设置主题色（第 23~28 行）。map.get 的第一个参数是变量$colors，从第二个参数开始则是一层层往下走的参数，例如 map.get($colors, 'primary', 'base')，表示找到变量$colors 下的 primary 下的 base，最终获取到颜色值"#3069ff"，其他变量以此类推。最后将 var.scss 变量文件引入 index.scss 即可（第 2 行）。

提示：本节定义的主题色为 primary、success、warning、error，这 4 种颜色也是 UI 组件库的全局规范颜色，后续的组件开发过程涉及的主题色均围绕这 4 个值变化。

6.2　mix：生成主题色层次

主题色层次是指一种视觉层次结构，用于确定设计中各个元素的重要性和关联性。主题色除了用来突出重要信息、动作或交互元素，还能在整个设计中起到引导和强调的作用。

UI 组件库设计稿中的品牌色（主题色）以及辅助色都存在"色调层次"，也就是由 10%到 100%变化。据此，UI 组件的设计上也应用了不同层次的颜色，因此需要对这些层次颜色进行初始化定义，如下代码所示。

代码

```scss
1.    // 层次色调
2.    $colors: map.deep-merge(
3.        (
```

```
4.          'primary': (
5.              "light-1": mix(#ffffff, $color-primary, 90), // #eaf0ff
6.          )
7.      ),
8.      $colors
9.  );
```

由于每种主题色都有不同层次的颜色，手动逐个添加颜色比较耗时，且容易出错。因此，可借助 Sass 的 mix 对象混合模式自动生成层次颜色。在上述代码中，"light-1" 使用 mix 方法将 #ffffff（白色）和变量$color-primary 两个颜色混合，并以 90% 的比例生成颜色#eaf0ff，对应 UI 组件库设计稿"品牌色"的 10%，如图 6-1 所示。

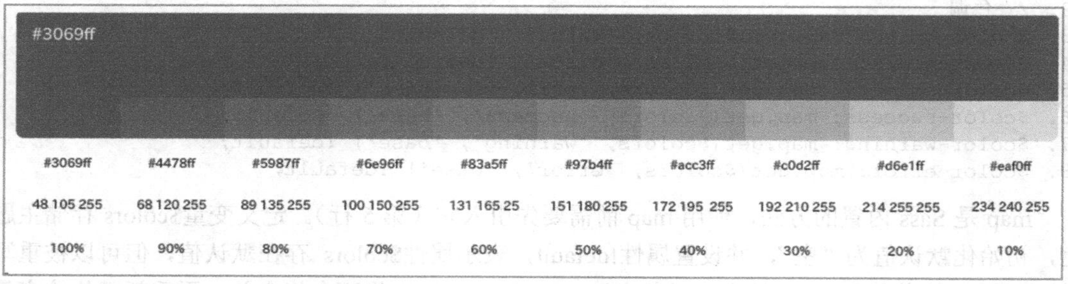

图 6-1 品牌色百分比

层次色调的处理过程也是将 map.deep-merge 方法与变量$colors 合并，最终与主色调合并后的结果如下代码所示。

```
代码
1.  $colors: (
2.      ...省略代码
3.          'primary': (
4.              'base': #3069ff,          // 主要
5.              "light-1": #eaf0ff,       // 10%
6.          ),
7.  );
```

6.2.1 定义@mixin 方法

了解了 Sass 的 mix 对象逻辑后，我们便可通过"循环"的方式自动生成所有主题色的层次颜色，但因主题色有多种，因此需要结合"函数"的方式进行处理。Sass 提供了@mixin（混入）的指令用于创建可重用的样式集合，然后使用@include 引用这些样式，如代码清单 library-06-2 所示。

代码清单 library-06-2

```
1.  > packages/theme/src/common/var.scss
2.  $types: primary, success, warning, error;
3.
```

```
4.  /** colors */
5.  $colors: () !default;
6.  ...省略代码
7.
8.  /** mixin 设置颜色层次 */
9.  @mixin set-color-level($type, $number, $mode: 'light', $mix-color) {
10.     $colors: map.deep-merge(
11.         (
12.             $type: (
13.                 '#{$mode}-#{$number}': mix(
14.                     $mix-color,
15.                     map.get($colors, $type, 'base'),
16.                     $number * 10
17.                 ),
18.             ),
19.         ),
20.         $colors
21.     ) !global;
22. }
```

在上述代码清单中，定义属性$types 并设置 primary、success、warning、error（第 2 行），这 4 个值与 6.1 节主题色的全局规范关键字的值一一对应。使用@mixin 指令自定义 set-color-level 的混入对象，并传入 4 个参数$type、$number、$mode 和$mix-color（第 9 行），在@mixin 中调用 map.deep-merge 方法进行合并。

提示：本节定义的$types 属性的值与 6.1 节中的主题色 primary、success、warning、error 一一对应，后续的组件主题色均使用$types 循环生成不同的主题色。

6.2.2 each、for：循环生成层次色调

生成主题色层次时需要使用 Sass 提供的 each、for 等循环模式，使用 each 循环一连串的值，使用 for 循环指定范围的逻辑，如代码清单 library-06-3 所示。

代码清单 library-06-3
```
1.  > packages/theme/src/common/var.scss
2.  $types: primary, success, warning, error;
3.  @mixin set-color-level($type, $number, $mode: 'light', $mix-color) { ...省略代码 }
4.
5.  // 生成层次颜色
6.  @each $type in $types {
7.      @for $i from 1 through 9 {
8.          @debug '打印 $type';    // 观察数据
9.          @debug $type;           // 观察数据
10.         @include set-color-level($type, $i, 'light', $color-white);
11.     }
12. }
13. @debug $colors;
```

使用@each 方法循环$types 的值（第 6 行），在@each 的每一次循环中又使用@for 从 1 至 9

循环 9 次（第 7 行），对应 10%、20%、30%、……、90%。在@for 循环的过程中使用@include 指令调用 set-color-level 并传入参数，即可自动生成不同主题色调的层次颜色（第 10 行）。

在@for 循环的过程中，如果无法确定数据的准确性，可使用 Sass 的@debug 指令在终端输出数据（第 8、9 行），最后把属性$colors 打印至终端，查看数据是否正确，如图 6-2 所示。

```
..\packages\theme\src\common\var.scss:60 Debug: (error: ("light-9": #ffedef, "light-8": #ffbdbe, "light-7": #ffc9ce, "light-6": #ffb7bd, "light-5": #ffa5ad, "light-4": #ff929d, "light-3": #ff808c, "light-2": #ff6e7c, "light-1": #ff5c6b, "base": #ff4a5b), warning: ("light-9": #fff6e8, "light-8": #ffeed1, "light-7": #ffe5ba, "light-6": #ffdca3, "light-5": #ffd48d, "light-4": #ffcb76, "light-3": #ffc25f, "light-2": #ffb948, "light-1": #ffb131, "base": #ffa81a), success: ("light-9": #e8faf1, "light-8": #d0f5e2, "light-7": #b9f0d4, "light-6": #a1ebc6, "light-5": #8ae6b8, "light-4": #72e1a9, "light-3": #5bdc9b, "light-2": #43d78d, "light-1": #2cd27e, "base": #14cd70), primary: ("light-9": #eaf0ff, "light-8": #d6e1ff, "light-7": #c1d2ff, "light-6": #acc3ff, "light-5": #98b4ff, "light-4": #83a5ff, "light-3": #6e96ff, "light-2": #5987ff, "light-1": #4578ff, "base": #3069ff), "white": #ffffff, "black": #000000)
```

图 6-2　输出自动生成的层次颜色并打印

6.3　定义中性色及其他元素

除了主题色、辅助色，UI 组件库中还有文字、边框、尺寸等不同元素的设计规范，如 input 输入框、button 按钮尺寸大小等，因此也要对这些设计规范的定义进行完善，如图 6-3 所示。

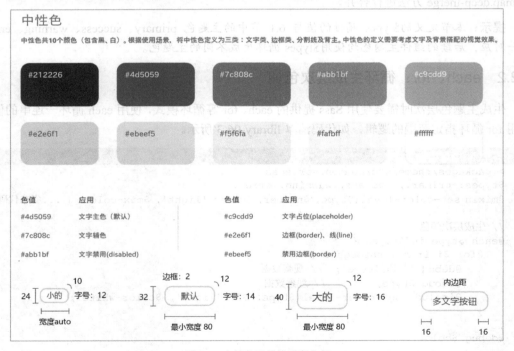

图 6-3　中性色及其他元素定义

图 6-3 中共有 10 种"中性色",并且标注了"色值"和"应用",以及按钮的尺寸、字体和多文字的情况。根据 UI 组件库设计稿中的描述来初始化定义规范,如代码清单 library-06-4 所示。

代码清单 library-06-4

```scss
> packages/theme/src/common/var.scss
/** 文字颜色 */
$text-color: () !default;
$text-color: map.deep-merge(
    (
        'primary': #4d5059,      // 文字主色
        'secondary': #7c808c,    // 文字辅色
        'disabled': #abb1bf,     // 文字禁用
        'placeholder': #c9cdd9,  // 文字占位
    ),
    $text-color
);

/** 文字尺寸 */
$font-size: () !default;
$font-size: map.deep-merge(
    (
        'small': 12px,      // 小的
        'default': 14px,    // 默认的
        'large': 16px,      // 大的
    ),
    $font-size
);

/** 边框 */
$border-color: () !default;
$border-color: map.deep-merge(
    (
        'default': #e2e6f1,     // 边框默认
        'disabled': #ebeef5,    // 边框禁用
    ),
    $border-color
);

/** 组件尺寸 */
$component-size: () !default;
$component-size: map.deep-merge(
    (
        'small': 24px,      // 小的
        'default': 32px,    // 默认的
        'large': 40px,      // 大的
    ),
    $component-size
)
```

6.4 :root 伪类选择器

在 CSS 中，:root 是一个伪类，用于选择文档的根元素，即 HTML 元素。这意味着:root 选择器会匹配 HTML 文档中的根元素，无论这个元素是什么标签。

通常情况下，:root 伪类可用于自定义全局的 CSS 属性（也称 CSS 变量）。通过在:root 伪类中自定义全局的 CSS 变量，可以确保它们在整个文档中全局可用，如代码清单 library-06-5 所示。

```
代码清单 library-06-5
1.  > packages/theme/src/index.scss
2.  @use "./initRoot.scss";
3.  @use "./button.scss";
4.  @use "./loading.scss";
5.
6.  > packages/theme/src/initRoot.scss
7.  @use "./common/var.scss" as *;
8.  :root {
9.      --a-color-primary: #{$color-primary};   // 主要
10.     --a-color-success: #{$color-success};   // 成功
11. }
12.
13. > packages/theme/src/button.scss
14. // 主要
15. .a-button--primary {
16.   background-color: var(--a-color-primary);
17.   border-color: var(--a-color-primary);
18.   ...省略代码
19. }
20. // 成功
21. .a-button--success {
22.   background-color: var(--a-color-success);
23.   border-color: var(--a-color-success);
24.   ...省略代码
25. }
```

:root 全局伪类以双短横线 "--" 开头定义全局属性，如--a-color-primary、--a-color-success（第 8、9 行）。通过 var 读取:root 定义的全局属性，如 background-color: var(--a-color-primary)、background-color: var(--a-color-success)（第 16、17、22、23 行），如图 6-4 所示。

图 6-4　使用 var 读取:root 全局属性

6.5 生成:root 变量

在 6.4 节中，我们手动定义了--a-primary-color 和--a-success-color 两个变量，并使按钮组件

读取到了变量。但手动定义变量的方式效率比较低,尤其是当"层次"颜色有几十种时。如果使用手动定义的方式,那么完全没必要使用 Sass 的合并、混入、mix 等方法,只要一个个写就可以了。因此,为了避免手动定义的弊端,下面采用自动生成变量的方式。

假设要生成"主色""成功""警告""错误"和"层次"这些 UI 组件库的主色和辅助色,我们需要优先规范好变量的规则并加上"前缀",如下代码所示。

```
1.  // 主色
2.  :root {
3.      --a-color-primary: ...;    // 主要
4.      --a-color-successr: ...;   // 成功
5.      --a-color-warning: ...;    // 警告
6.      --a-color-danger: ...;     // 错误
7.  }
8.
9.  // 层次色
10. :root {
11.     // 主色层次
12.     --a-color-primary-light-1: ...;  // 10%
13.     --a-color-primary-light-2: ...;  // 20%
14.     ...
15.     --a-color-primary-light-9: ...;  // 90%
16.
17.     // 成功色层次
18.     --a-color-success-light-1: ...;  // 10%
19.     --a-color-success-light-2: ...;  // 20%
20.     ...
21.     --a-color-success-light-9: ...;  // 90%
22. }
```

6.5.1 定义前缀、块、修改器变量

以"--a"开头定义"主色"和"成功色"的变量名称的规则,其中"a"是整个 UI 组件库的前缀,是在为组件绑定类名时已经定义好的。现在也要为生成变量定义相同的规则,包括"块"和"修改器"等,如代码清单 library-06-6 所示。

代码清单 library-06-6

```
1.  > packages/theme/src/common/config.scss
2.  $namespace: 'a' !default;                    // 前缀
3.  $connect: '-' !default;                      // 块、子级
4.  $element-connect: '__' !default;             // 元素
5.  $modifier-connect: '--' !default;            // 修改器
6.  $modifier-value-connect: '_' !default;       // 修改器的值
7.  $state-prefix: 'is-' !default;               // 状态前缀,如禁用、圆角、加载等
```

上述代码清单中定义的变量的值与 hook/use-namespace 的 _bem 函数拼接类名的字符基本一

致,这是为了确保 UI 组件库定义类名的规则统一。

6.5.2 :root 变量名称的生成规则

:root 变量名称的生成规则是,使用 Sass 中的@each 方法把一连串的字符拼接起来,组合上"前缀"产生新的变量名称,再为新的变量名称赋上颜色值,如代码清单 library-06-7 所示。

代码清单 library-06-7

```scss
1.  > packages/theme/src/common/function.scss
2.  @use "config.scss" as *;
3.  /** 组合变量名称 */
4.  @function createVarName($list) {
5.    $name: '--' + $namespace;           // --拼接前缀
6.    @each $item in $list {              // each 循环
7.      @if $item != '' {                 // 判断不为空时执行
8.        $name: $name + '-' + $item;     // 开始拼接
9.      }
10.   }
11.   @debug $name;
12.   @return $name;                      // 返回结果
13. }
14.
15. > packages/theme/src/common/mixins.scss
16. @use "sass:map";
17. @forward "function";
18. @use "function" as *;
19. @use "var.scss" as *;
20. /** 主色、辅助色 */
21. @mixin set-color() {
22.   @each $type in $types {  // $types: primary、success、wraning、error
23.     $color: map.get($colors, $type, "base");        // 获取颜色值
24.     #{createVarName(('color', $type))}: #{$color};  // 产生新的变量名称
25.   }
26. }
27.
28. packages/theme/src/initRoot.scss
29. @use "./common/var.scss" as *;
30. @use "./common/mixins.scss" as *;
31. :root {
32.   // 主色、辅助色
33.   @include set-color();
34. }
```

生成变量名称就是要得到一个新的变量名称,因此可使用 Sass 的@function 指令返回变量名称。在 common 目录下新建 function.scss 文件,用于存放所有@function 指令函数的规则。首先引用 config.scss(第 2 行),再定义 createVarName 函数名称并接收参数$list。在函数内定义变量$name,默认赋值"--",拼接变量$namespace 形成"--a",也就是我们所需要的格式(第 5 行)。

使用@each 循环参数$list，在每次循环时使用@if 判断$item 是否为空（第 7 行），如果不为空，则与变量$name 再次结合，用"-"拼接（第 8 行）。最后返回最终的结果$name（第 12 行）。如果不确定最终$name 的值是否正确，可使用@debug 输出变量$name（第 11 行）。

在 common 目录下新建 mixins.scss 文件，使用@mixin 混合指令定义 set-color。同样使用@each 循环来自 var.scss 的变量$types（第 22 行）。在每次@each 循环过程中，使用 map.get 方法一层层获取变量$colors 的值（第 23 行），最后调用 function.scss 的 createVarName 函数传入$type 和固定值'color'，获取新的变量名称，并赋值为$colors（第 24 行）。

在 initRoot.scss 文件中调用@include set-color()生成变量名称，在 function.scss 中使用@debug 输出变量名称，如图 6-5 所示。如果无法确定生成变量的颜色值是否准确，可在浏览器控制台中查看:root 变量的内容，如图 6-6 所示。

```
..\packages\theme\src\common\function.scss:11 Debug: --a-color-primary
..\packages\theme\src\common\function.scss:11 Debug: --a-color-success
..\packages\theme\src\common\function.scss:11 Debug: --a-color-warning
..\packages\theme\src\common\function.scss:11 Debug: --a-color-error
```

图 6-5　输出变量名称

```
:root {
    --a-color-primary: #3069ff;
    --a-color-success: #14cd70;
    --a-color-warning: #ffa81a;
    --a-color-danger:  #ff4a5b;
}
```

图 6-6　:root 变量的内容

6.5.3　生成层次色

生成层次色的方法是使用@mixin 混合指令定义名称，并增加一层@for 循环，如代码清单 library-06-8 所示。

代码清单 library-06-8

```scss
1.  > packages/theme/src/common/mixins.scss
2.  ...省略代码
3.  @mixin set-color-light() {
4.    @each $type in $types {
5.      @for $i from 1 through 9 {
6.        $color: map.get($colors, $type, "light-" + $i);
7.        #{createVarName(('color', $type, 'light', $i))}: #{$color};
8.      }
9.    }
10. }
11.
12. packages/theme/src/initRoot.scss
13. @use "./common/mixins.scss" as *;
14. :root {
```

```
15.    // 主色、辅助色
16.    @include set-color();
17.    // 层次色
18.    @include set-color-light();
19.  }
```

@mixin 混合指令的 set-color-light 方法使用 @each 循环变量 $types（第 4 行），在每次 @each 循环过程中再执行 @for 循环，同样使用 map.get 方法一层层获取变量 $colors 的值（第 6 行），最后调用 function.scss 的 createVarName 函数传入参数，获取新的变量名称（第 7 行）。使用 @debug 获取输出变量名称，如图 6-7 所示。浏览器控制台 :root 变量的内容，如图 6-8 所示。

```
..\packages\theme\src\common\function.scss:11 Debug: --a-color-primary-light-1
..\packages\theme\src\common\function.scss:11 Debug: --a-color-primary-light-2
..\packages\theme\src\common\function.scss:11 Debug: --a-color-primary-light-3
..\packages\theme\src\common\function.scss:11 Debug: --a-color-primary-light-4
..\packages\theme\src\common\function.scss:11 Debug: --a-color-primary-light-5
..\packages\theme\src\common\function.scss:11 Debug: --a-color-primary-light-6
..\packages\theme\src\common\function.scss:11 Debug: --a-color-primary-light-7
..\packages\theme\src\common\function.scss:11 Debug: --a-color-primary-light-8
..\packages\theme\src\common\function.scss:11 Debug: --a-color-primary-light-9
..\packages\theme\src\common\function.scss:11 Debug: --a-color-success-light-1
```

图 6-7 输出变量名称

```
:root {
  --a-color-primary: #3069ff;
  --a-color-success: #14cd70;
  --a-color-warning: #ffa81a;
  --a-color-danger:  #ff4a5b;
  --a-color-primary-light-1: #4578ff;
  --a-color-primary-light-2: #5987ff;
  --a-color-primary-light-3: #6e96ff;
  --a-color-primary-light-4: #83a5ff;
  --a-color-primary-light-5: #98b4ff;
  --a-color-primary-light-6: #acc3ff;
  --a-color-primary-light-7: #c1d2ff;
  --a-color-primary-light-8: #d6e1ff;
  --a-color-primary-light-9: #eaf0ff;
  --a-color-success-light-1: #2cd27e;
  --a-color-success-light-2: #43d78d;
  --a-color-success-light-3: #5bdc9b;
  --a-color-success-light-4: #72e1a9;
  --a-color-success-light-5: #8ae6b8;
  --a-color-success-light-6: #a1ebc6;
```

图 6-8 :root 变量的内容

6.5.4 获取 :root 变量名称

我们已经通过 @mixin 混合指令定义的 set-color 和 @mixin set-color-light 两个方法生成了主色、辅助色和层次色，在浏览器控制台中看到 :root 伪类生成了大量的全局变量名称，此时便可以直接使用:root 的变量名称。但如果直接使用，仍然要写"--a"前缀，另外，在变量名称非常多的情况下，可能无法知道哪些变量名称可以使用。因此，需要定义方法通过传参的方式获取:root 变量名称，使:root 变量名称的应用更简单，如代码清单 library-06-9 所示。

代码清单 library-06-9

```scss
> packages/theme/src/common/function.scss
...省略代码
/** 获取变量名称 */
@function getVarName($args...) {
  @return createVarName(($args));
}

packages/theme/src/button.scss
@use "./common/mixins.scss" as *;
/ 主要
.a-button--primary {
  $var: getVarName('color', 'primary'); // --a-color-primary
  background-color: var(#{$var});
  border-color: var(#{$var});
}
// 成功
.a-button--success {
  $var: getVarName('color', 'success'); // --a-color-success
  background-color: var(#{$var});
  border-color: var(#{$var});
}
```

在 function.scss 文件中定义 getVarName 方法并接收参数（第 4 行），在 getVarName 方法内调用 createVarName 方法返回组合后的变量名称（第 5 行）。在 button.scss 文件中调用 getVarName 方法并传入参数，将获取的变量名称赋给变量$var（第 12、18 行），在需要使用变量名称的位置，使用 Sass 的插值语句"#{}"赋值（第 13、14、19、20 行）。

6.6 UI 组件库全局规范

UI 组件库全局规范包括统一的元素命名规则、一致的尺寸和间距定义、明确的颜色和样式规范定义。这些规范有助于确保整个 UI 组件库的一致性和可维护性。

元素的设计规范有文字尺寸为 12px、14px、16px，组件的尺寸为 24px、32px、40px 等。我们在 6.3 节中定义了部分的 UI 组件库规范变量，现在用这些规范自动生成:root 全局变量，以便统一，如代码清单 library-06-10 所示。

代码清单 library-06-10

```scss
> packages/theme/src/common/var.scss
/** 文字颜色 */ $text-color
/** 文字尺寸 */ $font-size
/** 边框颜色 */ $border-color
```

```scss
5.  /** 组件尺寸 */ $component-size
6.  /** 全局配置 */
7.  $global: (
8.      'text-color': $text-color,           // 文字颜色
9.      'font-size': $font-size,             // 文字尺寸
10.     'border-color': $border-color,       // 边框颜色
11.     'component-size': $component-size,   // 组件尺寸
12. );
13.
14. > packages/theme/src/common/mixins.scss
15. @mixin set-global-var(){
16.   @each $key, $data in $global {
17.     @if $data { // 判断是否存在数据
18.       @each $type, $value in $data {
19.         #{createVarName(($key, $type))}: #{$value};
20.       }
21.     }
22.   }
23. }
24.
25. > packages/theme/src/initRoot.scss
26. :root {
27.     @include set-global-var();
28. }
```

在 var.scss 文件中定义变量$global，将变量$text-color、$font-size、$border-color、$component-size 全部集合到变量$global 中（第 7~11 行）。如果后续有新增的全局规范，也集合于此。在 mixins.scss 文件中定义 set-global-var 指令，使用@each 循环变量$global。其中，$key 和$data 是映射变量$global 中的各个键值对，也就是 key 和 value（第 16 行）。

首先判断$data 是否存在（第 17 行），如果存在，则继续执行@each 循环，此次循环的是 UI 组件库配置的变量数据，依然调用 createVarName 方法，将$key 和$type 作为参数传入后生成变量名称，$value 是变量名称对应的值。最后在 index.scss 文件中执行 set-global-var()，生成全局规范变量，如图 6-9 所示。

```
:root {
    --a-text-color-primary:     #4d5059;
    --a-text-color-secondary:   #7c808c;
    --a-text-color-disabled:    #abb1bf;
    --a-text-color-placeholder: #c9cdd9;
    --a-font-size-small:   12px;
    --a-font-size-default: 14px;
    --a-font-size-large:   16px;
    --a-border-color-default:  #e2e6f1;
    --a-border-color-disabled: #ebeef5;
    --a-component-size-small:   24px;
    --a-component-size-default: 32px;
    --a-component-size-large:   40px;
}
```

图 6-9　生成全局规范变量

6.7 UI 组件库应用规范

在前面几节中，我们通过 Sass 提供的指令、方法和循环实现了 :root 全局变量的自动生成、读取等功能，其中包括 UI 组件库的主色、辅助色、层次色、尺寸、字体大小等变量。接下来，我们要学习如何更好地使用 :root 全局变量，使 UI 组件库开发更轻松、更规范。

回顾第 5 章开发 button 组件的过程，均以最传统的方式书写 CSS 样式。尤其是在书写不同主题色的按钮时，会出现相同的类名，比如每个主题都会有 :hover、.is-disabled、.is-loading、.is-text、.is-link 等。这样会造成多种主题，重复多次书写相同的内容。如果有更多的主题色，后期维护更是难上加难，因此我们要转换思路，从"变量"的方向入手。

从变量的方向入手，首先要分析组件。下面以 button 组件为例，分析如何在 button 组件上使用变量，button 组件的 3 种状态如图 6-10 所示。

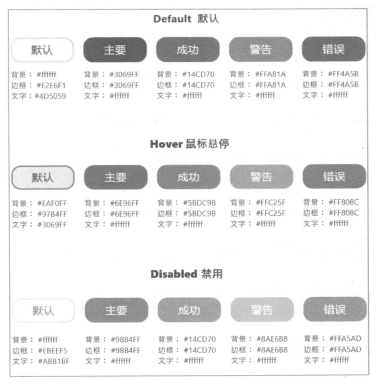

图 6-10　button 组件的 3 种状态

6.7.1　button 组件可变化的属性

UI 组件库设计稿中的 button 组件有很多种类型，如基本按钮（默认的）、圆角按钮、禁用按钮、加载按钮、文字按钮、边框按钮等。在这些不同类型的按钮中，我们要分析会产生"变化"的属性。如图 6-10 所示,从"默认"、"鼠标悬停"和"禁用"3 种按钮类型中,可以分析出"背景颜色"、"边框颜色"和"文字颜色"会产生变化。因此,我们要根据这些可变化的属性,定义属于 button 组件的私有变量,再将 UI 组件库全局规范变量的值赋给 button 组件的私有变量,然后根据不同主题的按钮重写相应主题的变量的值。

6.7.2 私有变量

在为 button 组件定义私有变量时,因 UI 组件库不只有 button 组件这一种组件,因此需要用单独的文件来定义 button 组件的私有变量,如代码清单 library-06-11 所示。

代码清单 library-06-11

```scss
> packages/theme/src/common/componentVar.scss
@use "function" as *;
// 按钮组件配置
@function buttonVar($type: ''){
  $button: (
    // 默认
    'border-color': (
      'default': getVarName('border-color', 'default'),
    ),
    'text-color': (
      'default': getVarName('text-color', 'primary'),
    ),
    'bg-color': (
      'default': getVarName('color', 'white'),
    ),
    // 鼠标悬停
    'hover-text-color': (
      'default': getVarName('color', 'primary'),
    ),
    'hover-bg-color': (
      'default': getVarName('color', 'primary', 'light-9'),
    ),
    'hover-border-color': (
      'default': getVarName('color', 'primary', 'light-7'),
    ),
    // 禁用
    'disabled-text-color': (
      'default': getVarName('text-color', 'disabled'),
    ),
    'disabled-bg-color': (
      'default': getVarName('color', 'white'),
    ),
    'disabled-border-color': (
      'default': getVarName('border-color', 'disabled'),
    ),
  );
  @return $button;
```

```
38.  }
39.
40.  > packages/theme/src/common/mixins.scss
41.  // 设置组件的变量
42.  @mixin set-component-var($name, $var, $varKey: 'default') {
43.    @each $key, $value in $var {
44.      $varName: getVarName($name, $key);
45.      $val: map.get($var, $key, $varKey);
46.      @if ($val) { #{$varName}: var(#{$val}); }
47.    }
48.  }
49.
50.  > packages/theme/src/button.scss
51.  @use "./common/mixins.scss" as *;
52.  @use "./common/componentVar.scss" as *;
53.  .a-button {
54.    @include set-component-var('button', buttonVar());
55.  }
```

在 common 目录下新建 componentVar.scss 文件，用于放置各种组件的私有变量。由于现在处理的是 button 组件的变量，可根据图 6-10 使用@function 指令定义"默认"类型按钮的边框颜色变量 border-color、字体颜色变量 text-color、背景颜色变量 bg-color（第 7~15 行）。需要注意的是，这 3 个变量绑定的是 button 组件中最基础的"默认"类型按钮。

针对"鼠标悬停"和"禁用"两种类型按钮，只需要在每个变量前面添加 hover 和 disabled 标识即可，如鼠标悬停文字按钮的颜色变量为 hover-text-color、禁用按钮的文字颜色变量为 disabled-text-color，其他变量以此类推（第 17~35 行），然后返回变量$button（第 37 行）。

在 mxixns.scss 文件中定义 set-component-var 混合指令，用于自动生成组件私有变量。指令的第一个参数$name 是自定义"块"的名称；第二个参数$var 是"变量"；第三个参数$varKey 是读取 button 组件的私有变量，默认值是 default（第 42 行），因此会读到 button 组件私有变量下的所有 default 属性的变量。@each 循环来自 componentVar.scss 文件定义的变量 buttonVar()，也就是$var 传入的数据，调用 getVarName 方法生成.a-button 所需的变量名称并赋给$varName（第 44 行）。使用 map.get 方法一层层获取 buttonVar()的数据并赋值给变量$var，因为$varKey 的默认值是 default，因此 map.get 获取的数据是$button 下的 default 数据（第 45 行）。最后使用@if 判断变量$var 存在时即生成变量（第 46 行），生成的私有变量如图 6-11 所示。

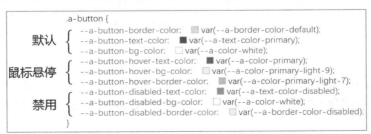

图 6-11 button 组件的私有变量

在图 6-11 的类名.a-button 中,左侧是 button 组件的私有变量,并且标识了"--a-button",右侧是读取的 UI 组件库全局规范变量。

6.7.3 绑定变量

类名.a-button 已经生成了 button 组件的私有变量,在图 6-11 中,"默认"类型的变量是 button 组件最基础的按钮所需要使用的变量,将这些生成的变量与类名.a-button 可变化的属性一一绑定即可,如代码清单 library-06-12 所示。

```scss
> packages/theme/src/button.scss
.a-button {
    ...省略代码
    background-color: var(#{getVarName('button', 'bg-color')});
    border: 1px solid var(#{getVarName('button', 'border-color')});
    color: var(#{getVarName('button', 'text-color')});
    &:hover {
      color: var(#{getVarName('button', 'hover-text-color')});
      background-color: var(#{getVarName('button', 'hover-bg-color')});
      border-color: var(#{getVarName('button', 'hover-border-color')});
    }
    &.is-disabled, &.is-loading {
      color: var(#{getVarName('button', 'disabled-text-color')});
      background-color: var(#{getVarName('button', 'disabled-bg-color')});
      border-color: var(#{getVarName('button', 'disabled-border-color')});
      cursor: not-allowed;
    }
}
```

目前,我们定义的可变化的属性只有背景颜色、边框颜色和文字颜色,在类名.a-button 中只绑定了 background-color、border 和 color 三个属性。在:hover 伪类中,则使用带有"hover"的变量;在类名.is-disabled 和.is-loading 中,则使用带有"disabled"的变量,渲染效果如图 6-12 所示。

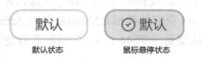

图 6-12 为 button 组件绑定变量

6.7.4 主题

在完成了 button 组件私有变量的定义和绑定后,我们了解了可变化的属性是私有变量的值,通过重写即可改变主题色,因此我们需要找出变量之间的规则,如代码清单 library-06-13 所示。

代码清单 library-06-13

```scss
> packages/theme/src/button.scss
// 主要
.a-button--primary {
  --a-button-border-color: var(--a-color-primary);
  --a-button-bg-color: var(--a-color-primary);
  --a-button-text-color: var(--a-color-white);
  --a-button-hover-text-color: var(--a-color-white);
  --a-button-hover-bg-color: var(--a-color-primary-light-3);
  --a-button-hover-border-color: var(--a-color-primary-light-3);
}
// 成功
.a-button--success {
  --a-button-border-color: var(--a-color-success);
  --a-button-bg-color: var(--a-color-success);
  --a-button-text-color: var(--a-color-white);
  --a-button-hover-text-color: var(--a-color-white);
  --a-button-hover-bg-color: var(--a-color-success-light-3);
  --a-button-hover-border-color: var(--a-color-success-light-3);
};
```

在上述代码中，类名.a-button--primary 和.a-button--success 所使用的私有变量 ":" 的左侧部分完全一致，而 ":" 右侧的 var 所使用的变量来自 UI 全局规范的:root，只有 "primary" 和 "success" 不同，渲染效果也会发生变化，如图 6-13 所示。

图 6-13　primary 和 success 主题渲染

根据上述规则，只需要在 buttonVar()的$button 变量中预先定义其他主题色的规则，再通过@each 循环的方式传入 "primary" 和 "success"，如代码清单 library-06-14 所示。

代码清单 library-06-14

```scss
> packages/theme/src/common/componentVar.scss
@use "function" as *;
// 按钮组件配置
@function buttonVar($type: ''){
  $button: (
    'border-color': (
      'default': ...省略代码, 'type': getVarName('color', $type)
    ),
    'text-color': (
      'default': ...省略代码, 'type': getVarName('color', 'white')
    ),
    'bg-color': (
      'default': ...省略代码, 'type': getVarName('color', $type)
    ),
    'hover-text-color': (
```

```scss
16.        'default': ...省略代码, 'type': getVarName('color', 'white')
17.      ),
18.      'hover-bg-color': (
19.        'default': ...省略代码, 'type': getVarName('color', $type, 'light-3')
20.      ),
21.      'hover-border-color': (
22.        'default': ...省略代码, 'type': getVarName('color', $type, 'light-3')
23.      ),
24.      'disabled-text-color': (
25.        'default': ...省略代码, 'type': getVarName('color', 'white')
26.      ),
27.      'disabled-bg-color': (
28.        'default': ...省略代码, 'type': getVarName('color', $type, 'light-5')
29.      ),
30.      'disabled-border-color': (
31.        'default': ...省略代码, 'type': getVarName('color', $type, 'light-5')
32.      ),
33.    );
34.    @return $button;
35. }
36.
37. > packages/theme/src/button.scss
38. @use "./common/mixins.scss" as *;
39. @use "./common/componentVar.scss" as *;
40. .a-button { ...省略代码 }
41. // 主题按钮
42. @each $type in $types {
43.   $className: '.a-button--' + $type;
44.   #{$className} {
45.     @include set-component-var('button', buttonVar($type), 'type');
46.   };
47. }
```

在 componentVar.scss 文件的 buttonVar()方法中接收参数$type（第 4 行），在变量$button 的每个私有变量下都新增 type（第 6~32 行），所有的 type 都使用 getVarName 方法获取 UI 组件全局规范变量的名称，其中，$type 是要接收的参数。在 button.scss 文件中使用@each 循环 var.scss 文件的变量$types（第 43 行），先拼接类名赋给变量$className（第 44 行），然后调用 set-component-var 传入第三个参数"type"，便会指定获取$button 私有变量下 type 属性的值（第 45 行）。生成不同主题按钮的私有变量如图 6-14 所示，当为按钮添加不同主题的类名时，会读取指定的变量覆盖原有的变量，达到变换主题的效果。

```
.a-button--success {                                          .a-button--primary {
  --a-button-border-color:       var(--a-color-success);        --a-button-border-color:       var(--a-color-primary);
  --a-button-text-color:         □var(--a-color-white);         --a-button-text-color:         □var(--a-color-white);
  --a-button-bg-color:           ■var(--a-color-success);       --a-button-bg-color:           ■var(--a-color-primary);
  --a-button-hover-text-color:   □var(--a-color-white);         --a-button-hover-text-color:   □var(--a-color-white);
  --a-button-hover-bg-color:     ■var(--a-color-success-light-3); --a-button-hover-bg-color:     ■var(--a-color-primary-light-3);
  --a-button-hover-border-color: ■var(--a-color-success-light-3); --a-button-hover-border-color: ■var(--a-color-primary-light-3);
  --a-button-disabled-text-color: □var(--a-color-white);        --a-button-disabled-text-color: □var(--a-color-white);
  --a-button-disabled-bg-color:  ■var(--a-color-success-light-5); --a-button-disabled-bg-color:  ■var(--a-color-primary-light-5);
  --a-button-disabled-border-color: ■var(--a-color-success-light-5); --a-button-disabled-border-color: ■var(--a-color-primary-light-5);
}                                                             }
              success 私有变量                                              primary 私有变量

.a-button--warning {                                          .a-button--error {
  --a-button-border-color:       var(--a-color-warning);        --a-button-border-color:       ■var(--a-color-error);
  --a-button-text-color:         □var(--a-color-white);         --a-button-text-color:         □var(--a-color-white);
  --a-button-bg-color:           ■var(--a-color-warning);       --a-button-bg-color:           ■var(--a-color-error);
  --a-button-hover-text-color:   □var(--a-color-white);         --a-button-hover-text-color:   □var(--a-color-white);
  --a-button-hover-bg-color:     ■var(--a-color-warning-light-3); --a-button-hover-bg-color:     ■var(--a-color-error-light-3);
  --a-button-hover-border-color: ■var(--a-color-warning-light-3); --a-button-hover-border-color: ■var(--a-color-error-light-3);
  --a-button-disabled-text-color: □var(--a-color-white);        --a-button-disabled-text-color: □var(--a-color-white);
  --a-button-disabled-bg-color:  ■var(--a-color-warning-light-5); --a-button-disabled-bg-color:  ■var(--a-color-error-light-5);
  --a-button-disabled-border-color: ■var(--a-color-warning-light-5); --a-button-disabled-border-color: ■var(--a-color-error-light-5);
}                                                             }
              warning 私有变量                                              error 私有变量
```

图 6-14 不同主题按钮的私有变量

6.7.5 尺寸

根据 componentVar.scss 文件中定义的 button 组件私有变量的开发模式，对按钮尺寸的处理就相对简单多了，只需在 componentVar.scss 中添加 "size" 的私有变量，自动生成 --a-button-size 的变量，如代码清单 library-06-15 所示。

代码清单 library-06-15

```
1.  > packages/theme/src/common/componentVar.scss
2.  @use "function" as *;
3.  // 按钮组件配置
4.  @function buttonVar($type: ''){
5.    $button: (
6.      ...省略代码
7.      'size': (
8.        'default': getVarName('component-size', 'default')
9.      )
10.   );
11.   @return $button;
12. }
13.
14. > packages/theme/src/button.scss
15. .a-button {
16.   height: var(#{getVarName('button', 'size')});
17.   &.a-button--size_small {
18.     padding: 0 10px;
19.     font-size: 12px;
20.     min-width: auto;
21.     .a-icon { font-size: 12px; }
```

```
22.   }
23.   &.a-button--size_default {
24.     .a-icon { font-size: 14px; }
25.   }
26.   &.a-button--size_large {
27.     font-size: 16px;
28.     .a-icon { font-size: 16px; }
29.   }
30.   &.is-circle {
31.     border-radius: 100px;
32.     padding: 0;
33.     min-width: auto;
34.     width: var(#{getVarName('button', 'size')})
35.   }
36. }
37. .a-button--size_small {
38.   #{getVarName('button', 'size')}: var(#{getVarName('component-size', 'small')});
39. }
40. .a-button--size_large {
41.   #{getVarName('button', 'size')}: var(#{getVarName('component-size', 'large')});
42. }
```

在 componentVar.scss 文件中新增"size"属性，自动生成 button 组件自身所需要的私有变量--a-button-size，再调用 getVarName 方法获取 UI 组件全局规范变量--a-component-size-default（第 7~9 行）。将变量--a-button-size 赋给类名.a-button 的 height 属性（第 16 行），接着使类名.a-button--size_small 和.a-button--size_large 重写变量--a-button-size 的值，同样调用 UI 组件全局规范变量--a-component-size-small 和--a-component-size-large（第 37~42 行）。由于类名.is-circle 中的属性 width 和 height 的尺寸一致，因此可以直接获取--a-button-size（第 34 行），渲染效果如图 6-15 所示。

图 6-15 button 组件的尺寸效果

6.7.6 文字尺寸

文字尺寸的处理方法和按钮尺寸一致，在 componentVar.scss 中添加"font-size"属性，自动生成--a-button-font-size 的私有变量，如代码清单 library-06-16 所示。

代码清单 library-06-16
```
1. > packages/theme/src/common/componentVar.scss
2. @function buttonVar($type: ''){
3.   $button: (
4.     ...省略代码
```

```
5.       'font-size': (
6.         'default': getVarName('font-size', 'default')
7.       )
8.    );
9.    @return $button;
10. }
11.
12. > packages/theme/src/button.scss
13. .a-button {
14.   font-size: var(#{getVarName('button', 'font-size')});
15.   .a-icon {
16.     font-size: var(#{getVarName('button', 'font-size')});
17.   }
18.   &.a-button--size_small {
19.     padding: 0 10px;
20.     min-width: auto;
21.   }
22.   .a-button--size_small {
23.     #{getVarName('button', 'font-size')}: var(#{getVarName('font-size', 'small')});
24.     #{getVarName('button', 'size')}: var(#{getVarName('component-size', 'small')});
25.   }
26.
27.   .a-button--size_large {
28.     #{getVarName('button', 'font-size')}: var(#{getVarName('font-size', 'large')});
29.     #{getVarName('button', 'size')}: var(#{getVarName('component-size', 'large')});
30.   }
```

在 componentVar.scss 文件中新增"font-size"属性，生成 button 组件自身所需的私有变量--a-button-font-size，再调用 getVarName 方法获取 UI 组件全局规范变量--a-font-size-default（第 5~7 行）。将变量--a-button-font-size 赋给类名.a-button 的 font-size 属性（第 14 行），由于类名.a-icon 中也有属性 font-size，因此可以直接使用变量--a-button-font-size（第 16 行）。接着在类名.a-button--size_small 和.a-button--size_large 新增变量--a-button-size 的重写值，同样调用 UI 组件全局规范变量--a-font-size-small 和--a-font-size-large（第 23、28 行），渲染效果如图 6-16 所示。

图 6-16　button 组件的文字尺寸效果

6.7.7　链接按钮

链接按钮的处理方法和主题按钮一致，存在"默认"、"鼠标悬停"和"禁用"3 种交互类型，可在 componentVar.scss 中添加 is-link-color、hover-is-link-color 和 disabled-is-link-color 这 3 个属性自动生成变量，如代码清单 library-06-17 所示。

代码清单 library-06-17

```scss
1.  > packages/theme/src/common/componentVar.scss
2.  @function buttonVar($type: ''){
3.    $button: (
4.      // 默认
5.      'is-link-color': (
6.        'default': getVarName('text-color', 'primary'),
7.        'type': getVarName('color', $type),
8.      ),
9.      // 鼠标悬停
10.     'hover-is-link-color': (
11.       'default': getVarName('text-color', 'secondary'),
12.       'type': getVarName('color', $type, 'light-4'),
13.     ),
14.     // 禁用
15.     'disabled-is-link-color': (
16.       'default': getVarName('text-color', 'disabled'),
17.       'type': getVarName('color', $type, 'light-5'),
18.     ),
19.   );
20.   @return $button;
21. }
22.
23. > packages/theme/src/button.scss
24. .a-button {
25.   &.is-link {
26.     color: var(#{getVarName('button', 'is-link-color')});
27.     &:hover { color: var(#{getVarName('button', 'hover-is-link-color')}); }
28.     &.is-disabled,
29.     &.is-loading { color: var(#{getVarName('button',
'disabled-is-link-color')}); }
30.     &, &:hover { ...省略代码 }
31.   }
32. }
```

在 componentVar.scss 文件的属性$button 中定义 3 个属性 is-link-color、hover-is-link-color 和 disabled-is-link-color，分别读取 UI 组件库全局规范变量（第 5~18 行）。button.scss 文件的类名&.is-link 的属性 color 默认读取私有变量 is-link-color（第 26 行）,:hover 伪类的属性 color 读取私有变量 hover-is-link-color（第 27 行），is-disabled 和 is-loading 类名的属性 color 则读取私有变量 disabled-is-link-color（第 29 行）。渲染效果如图 6-17 所示。

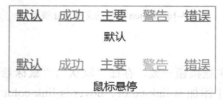

图 6-17 将 button 组件设置为链接按钮

6.7.8 文字按钮

文字按钮的处理方法与主题按钮一致，同样存在"默认"、"鼠标悬停"和"禁用"3 种交互类型，但文字按钮又多了背景颜色的变化，因此在 componentVar.scss 文件中需要定义文字和背景颜色属性，如代码清单 library-06-18 所示。

代码清单 library-06-18

```scss
> packages/theme/src/common/componentVar.scss
@function buttonVar($type: ''){
  $button: (
    // 默认
    'is-text-color': (
      'default': getVarName('text-color', 'primary'),
      'type': getVarName('color', $type),
    ),
    'is-text-bg-color': (
      'default': getVarName('color', 'white'),
      'type': getVarName('color', 'white'),
    ),
    // 鼠标悬停
    'hover-is-text-color': (
      'default': getVarName('text-color', 'primary'),
      'type': getVarName('color', $type),
    ),
    'hover-is-text-bg-color': (
      'default': getVarName('text-color', 'light-8'),
      'type': getVarName('color', $type, 'light-9')
    ),
    // 禁用
    'disabled-is-text-color': (
      'default': getVarName('text-color', 'disabled'),
      'type': getVarName('color', $type, 'light-5'),
    ),
    'disabled-is-text-bg-color': (
      'default': getVarName('color', 'white'),
      'type': getVarName('color', 'white')
    )
  );
  @return $button;
}

> packages/theme/src/button.scss
.a-button {
  &.is-text {
    color: var(#{getVarName('button', 'is-text-color')});
    &:hover {
      background-color: var(#{getVarName('button', 'hover-is-text-bg-color')});
      border-color: var(#{getVarName('button', 'hover-is-text-bg-color')});
    }
```

```
43.       &.is-disabled,
44.       &.is-loading {
45.         color: var(#{getVarName('button', 'disabled-is-text-color')});
46.         background-color: var(#{getVarName('button',
'disabled-is-text-bg-color')});
47.         border-color: var(#{getVarName('button', 'disabled-is-text-bg-color')});
48.       }
49.     }
50. }
```

为 componentVar.scss 文件的属性$button 定义 is-text-color 和 is-text-bg-color 两个属性，分别读取 UI 组件库全局规范变量，为:hover 伪类和禁用添加前缀 hover 和 disabled（第 5~30 行）。button.scss 文件的类名&.is-text 的属性 color 默认读取私有变量 is-text-color（第 38 行）；:hover 伪类的属性 border-color 和 background-color 均读取私有变量 hover-is-text-bg-color（第 40、41 行）；将 is-disabled 和 is-loading 类名改为 color、background-color 和 border-color，3 个属性均读取变量 disabled-is-test-color、disabled-is-text-bg-color（第 45~47 行），渲染效果如图 6-18 所示。

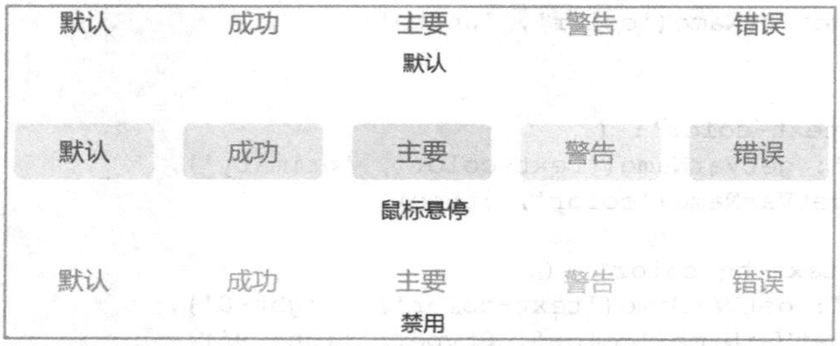

图 6-18　将 button 组件设置为文本按钮

6.8　遵循 BEM 命名规则生成组件类名

在 4.3 节已经详细介绍了 BEM 命名规则，包括在 UI 组件库的应用和命名，并定义了 use-namespace 方法，在组件的标签上自动生成 "-" "--" "__" 等连接符和带有前缀的类名。因此，在书写组件的 CSS 样式的类名时，也应该采用自动生成连接符的方式。

6.8.1　生成块的类名

块（block）的类名使用 Sass 的混合指令@mixin 生成，通过读取 config.scss 配置文件的连接符生成指定的类名，如代码清单 library-06-19 所示。

代码清单 library-06-19
```
1. > packages/theme/src/common/mixins.scss
2. // 生成块
3. @mixin b($block) {
4.     $b: '.' + $namespace + $connect + $block !global; //.a-button !global
5.     #{$b} {
```

```
6.        @content;
7.    }
8.  }
9.
10. > packages/theme/src/button.scss
11. @include b(button) {
12.   @include set-component-var('button', buttonVar());
13.   ...省略代码
14. }
```

在 mixins.scss 文件中使用@mixin 混合指令定义 "b" 名称,并接收参数$block(第 3 行)。指令内定义变量$b,对 "." "$namespace" "$connect" 和 "$block" 进行拼接,生成块的类名,然后为该类名添加!global 关键字,代表变量$b 是全局属性,可在任意位置使用(第 4 行)。接着使用插值语句 "#{}" 动态生成类名,并使用@content 渲染内容(第 5~7 行)。最后使用@include 在调用时传入 button(第 11 行)。

6.8.2 生成元素的类名

元素(element)的类名生成方法与混合指令@mixin b 基本一致,区别在于元素使用了 Sass 的@at-root 指令使生成的 CSS 样式脱离父级,不与父级形成上下级嵌套的模式,如代码清单 library-06-20 所示。

代码清单 library-06-20
```
1.  > packages/theme/src/common/mixins.scss
2.  // 生成 elem
3.  @mixin e($elem) {
4.    $name: $b + $element-connect + $elem;  // $b 是全局变量
5.    @at-root {
6.      #{$name} {
7.        @content;
8.      }
9.    }
10. }
```

6.8.3 生成修改器的类名

修改器(modifier)的类名生成方法与混合指令@mixin b 基本一致,同样使用 Sass 的@at-root 指令生成不与父级形成上下级嵌套的模式,参数是$attr 和$value,如代码清单 library-06-21 所示。

代码清单 library-06-21
```
1.  > packages/theme/src/common/mixins.scss
2.  // 生成修改器
3.  @mixin m($attr, $value: '') {
4.    $modifier: $b + $modifier-connect + $attr + $modifier-value-connect + $value;
5.    @at-root {
6.      #{$modifier} {
7.        @content;
```

```
8.      }
9.    }
10. }
11.
12. > packages/theme/src/button.scss
13. @include b(button) {
14.   @include m(size, small){ // 尺寸 small
15.     padding: 0 10px;
16.     min-width: auto;
17.     #{getVarName('button', 'font-size')}: var(#{getVarName('font-size', 'small')});
18.     #{getVarName('button', 'size')}: var(#{getVarName('component-size', 'small')});
19.   }
20.   @include m(size, large){ // 尺寸 large
21.     #{getVarName('button', 'font-size')}: var(#{getVarName('font-size', 'large')});
22.     #{getVarName('button', 'size')}: var(#{getVarName('component-size', 'large')});
23.   }
24. }
```

6.8.4 生成状态的类名

状态（state）通过拼接"is-"字符生成类名，并且可以同时生成多个类名，因此需要使用 @each 循环组合，如代码清单 library-06-22 所示。

代码清单 library-06-22

```
1.  > packages/theme/src/common/mixins.scss
2.  // 状态
3.  @mixin s($attrs, $and: true) {
4.    $state: '';
5.    @each $attr in $attrs {
6.      $state: if($and, '&', '') + '.' + $state-prefix + $attr + ',' + $state;
7.    }
8.    #{$state} {
9.      @content;
10.   }
11. }
12.
13. > packages/theme/src/button.scss
14. @include b(button) {
15.   @include s((disabled, loading)) {
16.     color: var(#{getVarName('button', 'disabled-text-color')});
17.     background-color: var(#{getVarName('button', 'disabled-bg-color')});
18.     border-color: var(#{getVarName('button', 'disabled-border-color')});
19.     cursor: not-allowed;
20.   }
21. }
```

本章小结

使用 Sass 制定 UI 组件库的全局变量，可以确保整个 UI 组件库的样式一致。通过将相同的颜色、字体、间距、尺寸等设置为全局变量，使所有组件引用全局变量，从而保持一致的风格。

使用全局变量可以更方便地扩展和定制组件库。对于不同的项目，可能需要对组件库进行定制，这时只需覆盖相应的全局变量即可，无须大幅修改组件的代码。开发人员在创建新组件时，可以直接使用已经定义好的全局变量，无须重复定义一些常用样式属性，从而提高开发效率。

第 7 章
图标组件

图标（icon）组件通常用于在应用程序或网站中显示图标，以增强用户界面的视觉吸引力和交互性。它们可以用于各种目的，例如指示操作、导航，以及表示特定的功能或内容。在前端开发中，图标组件库不仅提供了丰富的图标选项，而且支持自定义样式和大小，以满足不同的设计需求。图标组件的开发通常采用"插槽"模式，并提供"尺寸"及"颜色"两种属性定义图标的样式。

7.1 构建 icon 组件

在 3.4 节中已经介绍了如何构建组件的目录结构，并且开发了 button 组件。在构建图标组件的结构时，遵循 3.4 节中的目录结构即可，如代码清单 library-07-1 所示。

```
代码清单 library-07-1
1.  > packages/components/icon/src/index.vue
2.  <template>
3.    <i :class="[ns.b()]">
4.      <slot />
5.    </i>
6.  </template>
7.
8.  <script>export default { name: "a-icon" };</script>
9.  <script setup>
10. import { useNamespace } from '@ui-library/hook';
11. const ns = useNamespace("icon");
12. </script>
13.
14. > examples/src/App.vue
15. <template>
16.   <a-icon></a-icon>
17. </template>
```

在上述代码清单中，采用<i>标签并使用<slot />插槽作为图标的渲染区域（第 3~5 行）。调用命名空间的 useNamespace 方法传入"icon"（第 11 行），并在<i>标签中调用 ns.b 方法自动生成类名（第 3 行），渲染效果如图 7-1 所示。

```
<i class="a-icon"></i>
```

图 7-1 为 icon 组件生成类名

7.2 渲染 icon 组件

图标组件的渲染采用"插槽"模式，只需要在组件的全闭合标签内写上任意字符，就能渲染成功，如代码清单 library-07-2 所示。

代码清单 library-07-2
```
1.  > examples/src/App.vue
2.  <a-icon>
3.    <i class="iconfont icon-eye"></i>
4.  </a-icon>
5.
6.  > packages/theme/src/icon.scss
7.  .a-icon {
8.    height: 1em;
9.    width: 1em;
10.   line-height: 1em;
11.   display: inline-flex;
12.   justify-content: center;
13.   align-items: center;
14.   position: relative;
15.   fill: currentColor;
16.   font-size: inherit;
17.   svg { width: 1em; height: 1em; }
18. }
19.
20. > packages/theme/src/index.scss
21. ...省略代码
22. @use "./icon.scss";
```

`<i class="iconfont icon-eye"></i>` 是阿里图库的图标，在 `<a-icon>` 中写入插槽内容，即可渲染图标，如图 7-2 所示。其中，矩形是 icon 组件的整体范围，图标与组件垂直、水平居中，这是因为.a-icon 的 CSS 样式的属性 width、height、line-height 的值被设置为 "1em"（第 8~10 行），而 1em 会根据父元素的 font-size 变化。例如，如果父元素的 font-size 为 16px，则子级的 1em 会继承父级的 font-size 为 16px，因此可以使图标与组件垂直、水平居中。

要注意的一个属性是 fill，这个属性在常规 CSS 中是没有的，只在 XML-CSS 中存在，用于设置当前元素的填充内容，例如颜色、图片等，常见的有 SVG 图标。

icon 组件的 css 文件处理方式和 button 组件一致，单独抽离一份属于 icon 图标组件的 icon.scss 文件，再将其引入 theme/src/index.scss 文件（第 22 行）。

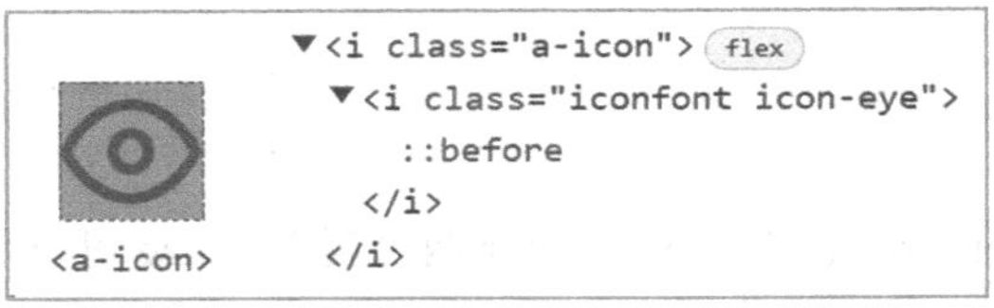

图 7-2 为 icon 组件渲染阿里图库图标

7.3 尺寸与颜色

在处理 icon 组件的尺寸与颜色时，要使用行间样式 style，因为我们不确定使用该组件的用户会定义多大的图标，也不知道用户会使用哪种颜色，所以无法采用类名的方式。尺寸和颜色的处理方法与 use-namespace 一致，如代码清单 library-07-3 所示。

代码清单 library-07-3

```
1.  > examples/src/App.vue
2.  <a-icon size="30" color="red">
3.    <i class="iconfont icon-eye"></i>
4.  </a-icon>
5.
6.  > packages/components/icon/src/index.vue: template
7.  <i :class="[ns.b()]" :style="[styleSize, styleColor]">
8.    <slot />
9.  </i>
10.
11. > packages/components/icon/src/index.vue: script
12. import { computed } from "vue"
13. import { useNamespace, useStyle } from '@ui-library/hook';
14. const ns = useNamespace("icon");
15. const uStyle = useStyle();
16. /** props */
17. const props = defineProps({
18.   size: { type: [String, Number], default: "" },
19.   color: { type: String, default: "" }
20. });
21. const styleSize = computed(() => uStyle.fontSize(props.size))
22. const styleColor = computed(() => uStyle.color(props.color))
23.
24. > packages/hook/use-style/index.js
25. export const useStyle = () => {
26.   // 尺寸
27.   const fontSize = (value) => {
28.     return value ? {'font-size': `${value}px`} : {}
29.   }
30.   // 颜色
31.   const color = (value) => {
32.     return value ? {'color': value} : {}
33.   }
34.   return { size, color }
35. }
36.
37. > packages/hook/index.js
38. export * from './use-style'
```

为 icon 组件的 defineProps 定义 size 和 color 两个属性。size 可以是字符串或数字类型，默认值为空（第 18 行）；color 为字符串类型，默认值也为空（第 19 行）。

在 hook 目录下新建 use-style/index.js 文件并定义 useStyle 方法，其中包括 size 和 color 两个方法，并且都定义参数 value 用于接收数据（第 25~35 行）。在 size 方法中采用三元运算判断参数 value 是否有值，如果有，则返回 font-size 属性，否则返回空的 JSON 对象（第 28 行）。对于 color 方法，同样采用三元运算，但返回的属性只有 color（第 32 行）。

把 useStyle 引入 icon 组件并将其赋给变量 uStyle（第 13、15 行），再声明 styleSize 和 styleColor 两个变量，采用 computed 计算属性调用 uStyle.size 方法传入 props 对象的 size 属性，调用 uStyle.color 方法传入 props 对象的 color 属性（第 21、22 行）。最后将 styleSize、styleColor 与 icon 组件的行间样式 style 绑定（第 7 行）。

使用 examples 渲染库调试 icon 组件时，只需为属性 size 和 color 传入相对应的值，便能实现自定义尺寸和颜色的效果（第 2 行），如图 7-3 所示。

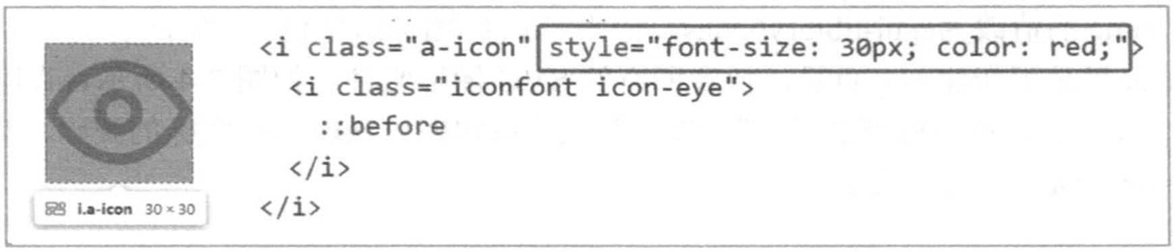

图 7-3　为 icon 组件自定义尺寸和颜色

7.4　SVG 图标

SVG（可缩放矢量图形）是一种基于 XML 的图形格式，可用于在网页上呈现图标、图像和矢量图形。它提供了一种灵活的方式来定义图像的外观，并且可以使图像被缩放，而不会失真。SVG 图标和阿里图库的图标类似，均可实现图标的放大、缩小及颜色变化。在 UI 组件库中，通常由设计师设计好规范，将 SVG 图标导出为.svg 格式的文件，并导入.vue 文件。本书的 SVG 图标来自阿里图库，仅用于演示，如代码清单 library-07-4 所示。

```
代码清单 library-07-4
1.  > packages/icons/index.js
2.  import Eye from "./svg/Eye.vue"
3.  import EyeOff from "./svg/EyeOff.vue"
4.  export {
5.      Eye, EyeOff
6.  }
7.
8.  > packages/icons/svg/Eye.vue
9.  <template>
10.     <svg ...省略代码></svg>
11. </template>
12.
13. > packages/icons/package.json
14. {
15.     "name": "@ui-library/icons",
```

```
16.     ...省略代码
17.   }
18.
19.  > examples/src/App.vue
20.  <template>
21.    <a-icon size="30" color="red">
22.      <Eye />
23.    </a-icon>
24.  </template>
25.
26.  <script setup>
27.  import { Eye } from "@ui-library/icons"
28.  </script>
```

在packages目录下新建icons文件夹，在icons中执行pnpm init指令生成package.json文件，并把属性name的值改为@ui-library/icons，作为"SVG图标组件包"（第15行）。

在icons目录下新建svg目录，放置所有的SVG图标组件。如图7-4所示，在阿里图库中任意选择一个图标，单击底部的"复制SVG代码"按钮即可获取当前图标的源码，然后将图标源码放入<template>模块（第9~11行）。

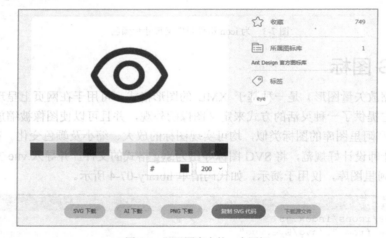

图7-4　阿里图库中的一个图标

在icons目录下新建index.js文件，将svg目录下的所有图标导入index.js文件，然后全部导出，这个操作如同通过packages/index.js导出所有组件。使用SVG图标相当于使用组件，只需导入指定的组件名称（第27行），然后在<a-icon>组件中渲染即可（第21~23行）。

7.5　button组件使用SVG图标

button组件使用SVG图标的方法和写入文字效果完全一致，都是通过插槽的方式写入，如代码清单library-07-5所示。

代码清单 library-07-5

```
1.  > examples/src/App.vue: template
2.  <a-button type="danger" size="large">
3.    <a-icon><Delete /></a-icon>
4.    删除
5.    <a-icon><Delete /></a-icon>
6.  </a-button>
7.
8.  > examples/src/App.vue: script
9.  import { Delete } from "@ui-library/icons"
10.
11. > packages/theme/src/button.scss
12. @include b(button) {
13.   ...省略代码
14.   span {
15.     ...省略代码
16.     gap: 0 4px;
17.   }
18. }
```

在 button 组件中，为"删除"文字的左右两侧都添加<a-icon>组件并渲染<Delete />组件，可以看到文字和图标之间没有间距，如图 7-5 左侧所示。根据 UI 组件库设计稿，我们需要为文字和图标添加 4px 的间距，因此可以使用 CSS 的属性 gap 为子元素设置左右两个间距（第 16 行），渲染效果如图 7-5 右侧所示。

图 7-5　使用 gap 属性渲染子元素的间距

本章小结

本章针对 icon 组件使用了 iconfont 和 SVG 模式的图标，可通过属性或插槽等方式渲染图标，结合属性 size 和属性 color 配置图标的尺寸和颜色。

需要注意的是，本章在 UI 组件库中内置了一些常用图标，内置图标的方式会使组件库的代码体积增大，对图标的管理也不友好。因此，可以将图标独立为"图标库"，通过库的模式安装图标，并独立管理图标。

图标库的开发模式与组件库一致，可查看目录 ui-library/icons，通过命令行 pnpm run build:icon 打包图标库。

第8章
输入框组件

输入框（input）组件用于用户在界面中输入和提交文本信息。它为用户提供了一种简单且直观的界面元素，使用户与应用程序或网站进行交互。输入框的使用广泛，包括收集用户信息、搜索、填写表单、用户交互以及数据筛选、过滤等。

8.1 渲染 input 组件

渲染 input 组件之前需要分析 UI 组件库设计稿，将可变化的属性提取出来。如图 8-1 所示，"基础 basis"类型存在 3 种状态，分别是默认、鼠标悬停、获取焦点，不同之处在于输入框的边框颜色；"禁用 disabled"类型改变了边框颜色、背景颜色和文字颜色；"图标 icon"类型则根据输入框的尺寸改变图标的大小。

通过上述分析可了解到 input 组件有哪些可变化的属性，因此我们可以根据 button 组件的开发模式为 input 组件定义私有变量，实现可变化的属性。

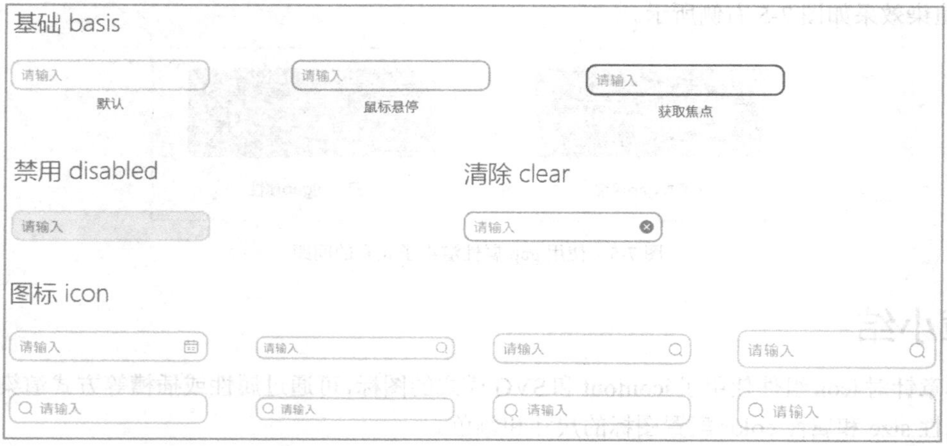

图 8-1 input 组件的 UI 组件库设计稿

8.1.1 构建组件

输入框的标签通常用于接收用户的文本输入、选择框、单选框等简单的交互元素，它的样式和布局比较受限，难以实现复杂的定位、样式装饰或布局要求。因此将 div 标签与输入框的标签相结合，可以更好地控制 input 组件的布局和样式，实现更复杂的布局和外观效果，如代码清单 library-08-1 所示。

代码清单 library-08-1

```
1.  > packages/components/input/src/index.vue
2.  <template>
3.    <div :class="[ns.b()]">
4.      <div :class="[ns.e('wrapper')]">
5.        <input :class="[ns.e('inner')]" placeholder="请输入" />
6.      </div>
7.    </div>
8.  </template>
9.
10. <script>export default { name: "a-input" };</script>
11. <script setup>
12. import { useNamespace } from '@ui-library/hook';
13. const ns = useNamespace("input");
14. </script>
```

根据 3.4 节关于构建组件目录结构的介绍，在 components 目录下新建 input 目录作为 input 组件的结构，并建立 index.vue 文件。在 <template> 模块中，使用 <div> 标签作为组件的根节点，并为属性 class 绑定 ns.b() 方法，生成类名 a-input（第 3 行）；为 <input> 标签的属性 class 绑定 ns.e('inner') 方法，生成类名 a-input__inner（第 5 行）；再为 <input> 标签包裹一层父级 <div> 标签，并绑定 ns.e('wrapper') 方法，生成类名 a-input__wrapper（第 4 行），渲染效果如图 8-2 所示。

图 8-2　为 input 组件生成类名 a-input、a-input__inner 和 a-input__wrapper

8.1.2　渲染组件

input 组件的样式处理方法和 button 组件一致，同样是在 theme/src 目录下新建 input.scss 文件，并引入 index.scss 文件，然后根据 UI 组件库设计稿标注的尺寸，使用 Sass 混合指令书写 CSS 样式，如代码清单 library-08-2 所示，渲染效果如图 8-3 所示。

代码清单 library-08-2

```
1.  > packages/theme/src/input.scss
2.  @use "./common/mixins.scss" as *;
3.  @use "./common/componentVar.scss" as *;
4.  @include b(input) {
5.    display: inline-flex;           // 盒模型，弹性布局
6.    align-items: center;            // 子元素垂直居中
7.    width: 100%;                    // 宽度
```

```
8.      @include e(wrapper) {
9.        display: inline-flex;
10.       align-items: center;
11.       height: 32px;                    // 高度
12.       width: 100%;
13.       border: none;
14.       padding: 0 10px;
15.       box-sizing: border-box;
16.       border: 2px solid #E2E6F1;       // 1 像素，实线边框
17.       border-radius: 12px;             // 圆角
18.       background-color: #fff;          // 默认白色背景
19.       transition: .3s;                 // 300 毫秒过渡
20.       line-height: 32px;               // 行高
21.     }
22.     @include e(inner) {
23.       height: 100%;
24.       width: 100%;
25.       border: none;
26.       padding: 0;
27.       outline: none;
28.       background-color: transparent;
29.       color: #212226;
30.       font-size: 14px;
31.     }
32.   }
33.   input{
34.     &::placeholder { color: #C9CDD9; }                      /* Chrome, Firefox, Opera, Safari 10.1+ */
35.     &::-webkit-input-placeholder { color: #C9CDD9; }  /* WebKit browsers, webkit 内核浏览器 */
36.     &:-moz-placeholder { color: #C9CDD9; }            /* Mozilla Firefox 4 to 18 */
37.     &::-moz-placeholder { color: #C9CDD9; }           /* Mozilla Firefox 19+ */
38.     &:-ms-input-placeholder { color: #C9CDD9; }       /* Internet Explorer 10-11 */
39.     &::-ms-input-placeholder { color: #C9CDD9; }      /* Microsoft Edge */
40.   }
```

图 8-3　渲染 input 组件

8.1.3　样式变量

　　第 6 章使用 Sass 语法生成了 UI 组件库的:root 全局变量，并在 button 组件中使用私有变量，input 组件同样在 componentVar.scss 文件中生成私有变量，然后将变量绑定至 input.scss 指定的样式属性，如代码清单 library-08-3 所示。

代码清单 library-08-3
```
1. > packages/theme/src/common/componentVar.scss
2. @function inputVar(){
```

```scss
3.    $input: (
4.      'text-color': (  // 字体颜色
5.        'default': getVarName('text-color', 'primary')
6.      ),
7.      'border-color': (  // 边框颜色
8.        'default': getVarName('border-color', 'default')
9.      ),
10.     'bg-color': (  // 背景颜色
11.       'default': getVarName('color', 'white')
12.     ),
13.     'size': (  // 组件尺寸
14.       'default': getVarName('component-size', 'default')
15.     ),
16.     'round': (  // 组件圆角
17.       'default': getVarName('component-round', 'default')
18.     ),
19.     'font-size': (  // 文字大小
20.       'default': getVarName('font-size', 'default')
21.     )
22.   );
23.   @return $input;
24. }
25.
26. > packages/theme/src/input.scss
27. @use "./common/mixins.scss" as *;
28. @use "./common/componentVar.scss" as *;
29. @include b(input) {
30.   @include set-component-var('input', inputVar());  // 生成input组件的私有变量
31.   @include e(wrapper) {
32.     height: var(#{getVarName('input', 'size')});
33.     border: 2px solid var(#{getVarName('input', 'border-color')});
34.     line-height: var(#{getVarName('input', 'size')});
35.     border-radius: var(#{getVarName('input', 'round')});
36.     background-color: background-color: var(#{getVarName('input', 'bg-color')});
37.   }
38.   &:hover {
39.     #{getVarName('input', 'border-color')}: var(#{getVarName('border-color', 'hover')});
40.   }
41.   @include s((focus)) {
42.     #{getVarName('input', 'border-color')}: var(#{getVarName('color', 'primary')});
43.   }
44.   @include e(inner) {
45.     color: var(#{getVarName('input', 'text-color')});
46.     font-size: var(#{getVarName('input', 'font-size')});
47.   }
48. }
49. input{
```

```
50.     &::placeholder { color: var(#{getVarName('text-color', 'placeholder')}); }
/* Chrome, Firefox, Opera, Safari 10.1+ */
51.     &::-webkit-input-placeholder { color: var(#{getVarName('text-color',
'placeholder')}); }       /* WebKit browsers，webkit 内核浏览器 */
52.     &:-moz-placeholder { color: var(#{getVarName('text-color', 'placeholder')}); }
/* Mozilla Firefox 4 to 18 */
53.     &::-moz-placeholder { color: var(#{getVarName('text-color', 'placeholder')}); }
/* Mozilla Firefox 19+ */
54.     &:-ms-input-placeholder { color: var(#{getVarName('text-color',
'placeholder')}); }            /* Internet Explorer 10-11 */
55.     &::-ms-input-placeholder { color: var(#{getVarName('text-color',
'placeholder')}); }            /* Microsoft Edge */
56.   }
```

在 componentVar.scss 文件中定义 inputVar()方法返回变量$input（第 2~24 行）。在变量$input 中定义 input 组件可变化的属性 text-color、border-color、bg-color、size、font-size、round，然后在 input.scss 文件中调用 set-component-var 指令生成 input 组件的私有变量（第 30 行）。最后将变量绑定至 input 组件可变化的属性上，如图 8-4 所示。

```
.a-input__wrapper {
    display: inline-flex;
    align-items: center;
    padding: ▶ 0 10px;
    height: var(--a-input-size);
    width: 100%;
    border: ▶ 2px solid  var(--a-input-border-color);
    border-radius: ▶ var(--a-input-round);
    background-color:  var(--a-input-bg-color);
    box-sizing: border-box;
    transition: ▶ 0.3s;
    line-height: var(--a-input-size);
}
```

图 8-4　生成 input 组件的私有变量

8.2　disabled：禁用

禁用输入框的处理方法和 button 组件类似，都是调用 ns.is()方法传入 disabled，但绑定的标签不同，如代码清单 library-08-4 所示。

代码清单 library-08-4
```
1. > packages/components/input/src/index.vue: template
2. <div :class="[ns.b(), ns.is('disabled', disabled)]">
3.   <div :class="[ns.e('wrapper')]">
4.     <input :disabled="disabled" :class="[ns.e('inner')]" placeholder="请输入" />
5.   </div>
6. </div>
7.
8. > packages/components/input/src/index.vue: script
9. /** props */
```

```
10.   const props = defineProps({
11.     disabled: Boolean, // 布尔类型，默认为 false
12.   });
13.
14. > packages/theme/src/input.scss
15.   @include b(input) {
16.     ...省略代码
17.     @include s((disabled)) {
18.       #{getVarName('input', 'border-color')}: var(#{getVarName('border-color', 'disabled')});
19.       #{getVarName('input', 'bg-color')}: var(#{getVarName('color', 'disabled-bg')});
20.       #{getVarName('input', 'text-color')}: var(#{getVarName('text-color', 'placeholder')});
21.     }
22.   }
23.
24. > examples/src/App.vue
25. <a-input disabled></a-input>
```

为 input 组件的根标签绑定的 ns.is 方法传入 disabled，生成类名 .is-disabled（第 2 行），同时为 <input> 标签的自身属性 disabled 绑定 props 对象的 disabled 属性（第 4 行）。因为 ns.b 方法生成的类名 .a-input 和 .is-disabled 同级，因此在 CSS 上需使用"复合选择器"重写边框颜色、背景颜色和文字颜色的私有变量（第 17~21 行），渲染效果如图 8-5 所示。

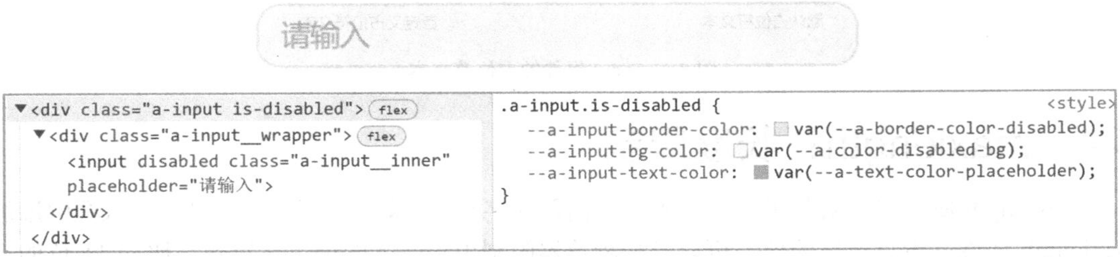

图 8-5 为禁用输入框重写私有变量

8.3 placeholder：占位符

占位符（placeholder）属性是 <input> 标签自带的属性，用于显示在用户未输入任何文本时的占位符文本，为用户提供关于预期输入的提示或说明。输入框的作用就是输入文本，因此可以直接显示占位符文本为"请输入"。在实际业务中，有些输入框可能是"邮箱"或"手机号"类型的，那么可以显示"请输入邮箱"或"请输入 11 位的手机号"，以及其他的占位符文本。使用属性 placeholder 可以实现自定义文本的功能，如代码清单 library-08-5 所示。

代码清单 library-08-5
```
1. > packages/components/input/src/index.vue: template
2. <div :class="[ns.b(), ns.is('disabled', disabled)]">
```

```
3.    <div :class="[ns.e('wrapper')]">
4.      <input ...省略代码 :placeholder="placeholder" />
5.    </div>
6.  </div>
7.
8.  > packages/components/input/src/index.vue: script
9.  /** props */
10. const props = defineProps({
11.   placeholder: {
12.     type: String,
13.     default: '请输入'
14.   },
15.   disabled: Boolean, // 布尔类型，默认为 false
16. });
17.
18. > examples/src/App.vue
19. <a-input placeholder="请输入 11 位的手机号"></a-input>
```

在 props 对象中定义属性 placeholder 为字符串类型，默认值为"请输入"（第 11~14 行），然后将属性 placeholder 绑定至<input>标签的 placeholder 属性（第 4 行），此时的占位符文本会默认显示"请输入"。如果想自定义占位符文本，那么只需要在调用 input 组件时定义 placeholder 属性（第 19 行），渲染效果如图 8-6 所示。

图 8-6　input 组件的占位符文本

8.4　maxlength：长度限制

maxlength 属性用于限制用户在输入框中输入的字符数，即字符长度。通过在输入框中添加 maxlength 属性并设置一定的值，可以告知浏览器限制用户输入的最大字符数，比如 11 位的手机号，确保用户在提交表单时输入的内容不超出预期范围，如代码清单 library-08-6 所示。

代码清单 library-08-6

```
1.  > packages/components/input/src/index.vue: template
2.  <div :class="[ns.b(), ns.is('disabled', disabled)]">
3.    <div :class="[ns.e('wrapper')]">
4.      <input ...省略代码 :maxlength="maxlength" />
5.    </div>
6.  </div>
7.
8.  > packages/components/input/src/index.vue: script
9.  /** props */
10. const props = defineProps({
11.   ...省略代码
12.   maxlength: {
```

```
13.     type:[Number, String],
14.     default: ''
15.   }
16. });
17.
18. > examples/src/App.vue
19. <a-input placeholder="请输入11位的手机号" maxlength="11"></a-input>
```

为 defineProps 对象定义属性 maxlength 为数字类型或字符串类型，默认值为空（第 12~15 行）。然后将 maxlength 属性绑定至<input>标签的 maxlength 属性（第 4 行），在引用 input 组件时，定义属性 maxlength 的值 11（第 19 行），此时只能输入 11 位长度的字符，如图 8-7 所示。

图 8-7 为 input 组件配置 maxlength 属性

8.5 size：尺寸

尺寸可以根据 button 组件的模式进行开发，只是生成"可变化"的私有变量名称不同，如代码清单 library-08-7 所示。

代码清单 library-08-7
```
1.  > packages/components/input/src/index.vue: template
2.  <div :class="[ns.b(), ns.m('size', size)]">
3.    <div :class="[ns.e('wrapper')]">
4.      <input ...省略代码 :maxlength="maxlength" />
5.    </div>
6.  </div>
7.
8.  > packages/components/input/src/index.vue: script
9.  /** props */
10. const props = defineProps({
11.   ...省略代码
12.   size: {
13.     type: String,
14.     default: "default",
15.   },
16. });
17.
18. > packages/theme/src/input.scss
19. @include b(input) {
20.   ...省略代码
21.   @include m(size, small){
22.     #{getVarName('input', 'font-size')}: var(#{getVarName('font-size', 'small')});
23.     #{getVarName('input', 'size')}: var(#{getVarName('component-size', 'small')});
```

```
24.     #{getVarName('input', 'round')}: var(#{getVarName('component-round', 'small')});
25.   }
26.   @include m(size, large){
27.     #{getVarName('input', 'font-size')}: var(#{getVarName('font-size', 'large')});
28.     #{getVarName('input', 'size')}: var(#{getVarName('component-size', 'large')});
29.   }
30. }
31.
32. > examples/src/App.vue
33. <a-input placeholder="请输入 11 位的手机号" maxlength="11"></a-input>
```

在 button 组件中已经配置了 size 的私有变量，与 input 组件使用的私有变量的逻辑一致。为 defineProps 对象定义属性 size 为字符串类型，默认值为空（第 12~15 行），然后调用 ns.m 方法将 size 属性绑定至 input 组件的根元素并生成类名（第 2 行）。对于 CSS 的处理，则引用 @include m 修饰器生成 small 和 large 两种尺寸（第 21~29 行），渲染效果如图 8-8 所示。

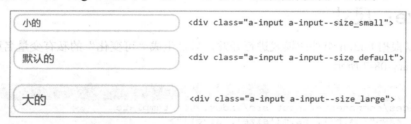

图 8-8　为 input 组件配置不同尺寸

8.6　round：圆角

圆角的逻辑与 UI 组件库是一致的，通过 ns.is('round', round) 生成类名 is-round，并将私有变量 input-round 的值更新为 UI 组件库全局变量的值，如代码清单 library-08-8 所示，渲染效果如图 8-9 所示。

代码清单 library-08-8
```
1.  > packages/components/input/src/index.vue: template
2.  <div :class="[ns.b(), ns.is('round', round)]">...省略代码</div>
3.
4.  > packages/components/input/src/index.vue: script
5.  /** props */
6.  const props = defineProps({
7.    ...省略代码
8.    round: Boolean, // 圆角
9.  });
10.
11. > packages/theme/src/input.scss
12. @include b(input) {
```

```
13.    ...省略代码
14.    @include s((round)) {
15.      #{getVarName('input', 'round')}: var(#{getVarName('component-round',
'round')});
16.    }
17.  }
18.
19.  > examples/src/App.vue
20.  <a-input round></a-input>
```

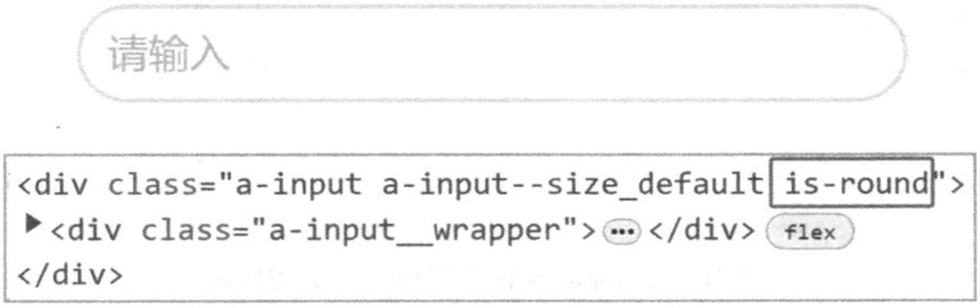

图 8-9 为 input 组件配置圆角

8.7 icon：图标

图标通常用于用户界面的特定描述，简洁地传达输入框的用途和功能，避免过于复杂的描述，以便用户快速理解当前输入框的作用，如搜索、设置、删除等。

根据 UI 组件库设计稿中的"图标 icon"，图标输入框分为"前缀"和"后缀"图标类型，也就是图标位于输入框内部的左侧或右侧，此外，"密码""清除""长度统计"等类型也属于后缀。当出现后缀元素时，需要考虑哪些后缀元素可以同时出现，哪些只能独立出现。

在 UI 组件库设计稿中还有"前置"和"后置"的效果，也就是图标位于输入框外部的左侧或右侧，因此在接下来的开发过程中，分别对"前缀""后缀""前置""后置"4 种情况进行逐步处理，如代码清单 library-08-9 所示。

代码清单 library-08-9
```
1.  > packages/components/input/src/index.vue
2.  <div :class="[ns.b(), ...省略代码">
3.    <div :class="[ns.e('wrapper')]">
4.      <div :class="[ns.e('fix-wrapper')]">前缀区域</div>
5.      <input :disabled="disabled" ...省略代码 />
6.      <div :class="[ns.e('fix-wrapper')]">后缀区域</div>
7.    </div>
8.  </div>
9.
10. > packages/theme/src/input.scss
11. @include b(input) {
12.   ...省略代码
13.   @include e(fix-wrapper) {
14.     display: inline-flex;
```

```
15.         align-items: center;
16.         justify-content: center;
17.         height: 100%;
18.     }
19. }
```

渲染输入框元素（第 5 行），使用<div>标签预留前缀和后缀的渲染区域（第 4、6 行），在<div>标签中调用 ns.e('fix-wrapper')方法生成类名.a-input__fix-wrapper，并设置相应的 CSS 样式（第 13~18 行），渲染效果如图 8-10 所示。

```
<div class="a-input__wrapper"> flex
  <div class="a-input__fix-wrapper">前缀区域</div>
  <input class="a-input__inner" placeholder="请输入" maxlength>
  <div class="a-input__fix-wrapper">后缀区域</div>
</div>
```

图 8-10　为 input 组件设置前缀、后缀区域

8.7.1　渲染 UI 组件库内置的 SVG 图标

SVG 图标是 7.4 节实现的 icon 组件，也是 UI 组件库内置默认提供给用户使用的图标，因此可以结合 icon 组件实现图标渲染，如代码清单 library-08-10 所示。

代码清单 library-08-10
```
1.  > packages/components/input/src/index.vue: template
2.  <div :class="[ns.b(), ...省略代码]">
3.    <div :class="[ns.e('wrapper')]">
4.      <div v-if="isPrefix" :class="[ns.e('fix-wrapper')]">
5.        <div :class="[ns.e('fix'), ns.e('prefix')]">
6.          <a-icon><component :is="prefixIcon" /></a-icon>
7.        </div>
8.      </div>
9.      <input :disabled="disabled" ...省略代码 />
10.     <div v-if="isSuffix" :class="[ns.e('fix-wrapper')]">
11.       <div :class="[ns.e('fix'), ns.e('suffix')]">
12.         <a-icon><component :is="suffixIcon" /></a-icon>
13.       </div>
14.     </div>
15.   </div>
16. </div>
17.
18. > packages/components/input/src/index.vue: script
19. import { computed } from "vue"
20. import { AIcon } from "@ui-library/components"
21. /** props */
22. const props = defineProps({
23.   ...省略代码
24.   prefixIcon: {
25.     type: [String, Object],
```

```
26.        default: ''
27.      },
28.      suffixIcon: {
29.        type: [String, Object],
30.        default: ''
31.      }
32. });
33. // 是否存在前缀
34. const isPrefix = computed(() => {
35.   return props.prefixIcon
36. })
37. // 是否存在后缀
38. const isSuffix = computed(() => {
39.   return props.suffixIcon
40. })
41.
42. > packages/theme/src/input.scss
43. @include b(input) {
44.   ...省略代码
45.   @include e(fix) {
46.     display: flex;
47.     align-items: center;
48.     color: var(#{getVarName('text-color', 'placeholder')});
49.     font-size: var(#{getVarName('input', 'font-size')});
50.     border-radius: var(#{getVarName('input', 'round')});        // 圆角
51.     box-sizing: border-box;
52.     transition: .3s;
53.     background-color: var(#{getVarName('input', 'bg-color')});
54.     height: 100%;
55.   }
56.   @include e(prefix) {
57.     padding-right: var(#{getVarName('input', 'padding')});
58.   }
59.   @include e(suffix) {
60.     padding-left: var(#{getVarName('input', 'padding')});
61.   }
62. }
63.
64. > examples/src/App.vue: template
65. <a-input :suffix-icon="Add" ></a-input>
66. <a-input :prefix-icon="Search"></a-input>
67.
68. > examples/src/App.vue: script
69. import { Search, Add } from "../../packages/icons"
```

在 defineProps 对象中定义 prefixIcon 和 suffixIcon 两个属性，分别对应前缀和后缀，两个属性为字符串或对象类型（第 24~31 行），引入 icon 组件用于渲染图标（第 20 行）。

前缀和后缀区域是固定的图标渲染区域，添加子级<div>标签单独渲染图标，为前缀、后缀区域子级<div>标签绑定 ns.e('fix')和 ns.e('prefix')两个方法（第 5 行），分别生成类名.a-input__fix

和.a-input__prefix。类名.a-input__fix 是 input 组件的公用类名，为其添加样式（第 45~55 行）；类名.a-input__prefix 用于设置前缀的右边距，使图标和文字之间产生间距（第 56~58 行）。

由于 UI 组件库内置的 SVG 图标都是组件，因此需要使用 Vue.js 3 提供的动态组件<component>结合":is"属性渲染组件，将<component>放入<a-icon>图标组件，作为插槽内容渲染，为 is 属性绑定 defineProps 对象的 prefixIcon（第 6 行）。使用 computed 对象判断是否配置了前缀，如果已配置，则返回 props.prefixIcon（第 34~36 行），然后使用 v-if 条件绑定变量 isPrefix（第 4、34 行）。后缀图标的做法和前缀图标相似，只是将 v-if 条件绑定变量改为 isSuffix（第 10~14 行），渲染效果如图 8-11 所示。

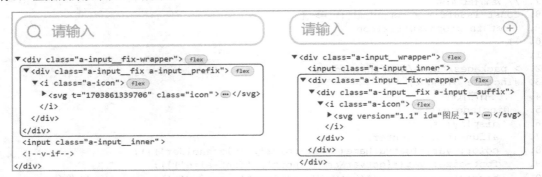

图 8-11　为 input 组件渲染前缀、后缀 SVG 图标

8.7.2　渲染 iconfont 字体图标

在 UI 组件库中，除了使用内置的 SVG 图标，也可以使用阿里图库的 iconfont 字体图标。由于 iconfont 字体图标是通过伪类生成（非组件）的，因此不能使用 Vue.js 3 提供的动态组件，如代码清单 library-08-11 所示。

代码清单　library-08-11
```
1.  > packages/components/input/src/index.vue: template
2.  <div :class="[ns.e('wrapper')]">
3.    <div v-if="isPrefix" :class="[ns.e('fix-wrapper')]">
4.      <div :class="[ns.e('fix'), ns.e('prefix')]">
5.        <a-icon v-if="prefixIcon"><component :is="prefixIcon" /></a-icon>
6.        <a-icon v-if="prefixIconfont">
7.          <i class="iconfont" :class="prefixIconfont"></i>
8.        </a-icon>
9.      </div>
10.   </div>
11.   <input :disabled="disabled" ...省略代码 />
12.   <div v-if="isSuffix" :class="[ns.e('fix-wrapper')]">
13.     <div :class="[ns.e('fix'), ns.e('suffix')]">
14.       <a-icon v-if="suffixIcon"><component :is="suffixIcon" /></a-icon>
15.       <a-icon v-if="suffixIconfont">
16.         <i class="iconfont" :class="suffixIconfont"></i>
```

```
17.         </a-icon>
18.       </div>
19.     </div>
20. </div>
21.
22. > packages/components/input/src/index.vue: script
23. /** props */
24. const props = defineProps({
25.   ...省略代码
26.   prefixIconfont: {
27.     type: String,
28.     default: ''
29.   },
30.   suffixIconfont: {
31.     type: String,
32.     default: ''
33.   }
34. });
35. // 是否存在前缀
36. const isPrefix = computed(() => {
37.   return props.prefixIcon || props.prefixIconfont
38. })
39. // 是否存在后缀
40. const isSuffix = computed(() => {
41.   return props.suffixIcon || props.suffixIconfont
42. })
43.
44. > examples/src/App.vue
45. <a-input suffix-iconfont="icon-del" ></a-input>
46. <a-input prefix-iconfont="icon-search"></a-input>
```

在 defineProps 对象中定义 prefixIconfont 和 suffixIconfont 两个属性，分别对应前缀和后缀，两个属性为字符串类型（第 26~33 行）。在 5.12 节中使用<i>标签渲染 iconfont 字体图标。对于 input 组件也是同样的做法，新增<a-icon>组件并添加<i>标签渲染 iconfont 字体图标，然后在<component>动态组件的<a-icon>组件中使用 v-if 判断是否存在 prefixIcon（第 5 行），存在 prefixIcon 则渲染，反之则不渲染。在<i>标签的<a-icon>组件中使用 v-if 判断是否存在 prefixIconfont（第 6 行），存在 prefixIconfont 则渲染，反之则不渲染，并为<i>标签的 class 属性绑定 prefixIconfont，也就是渲染的图标类名。后缀图标的实现方式和前缀图标一致，只是 v-if 条件和 class 绑定的是 suffixIconfont（14~17 行）。

由于前缀和后缀渲染区域使用 v-if 分别绑定了变量 isPrefix 和 isSuffix，只有变量返回结果时才会渲染，因此需要在 computed 计算属性中添加"或"运算，判断 defineProps 对象的 prefixIconfont 和 suffixIconfont（第 37、41 行），渲染效果如图 8-12 所示。

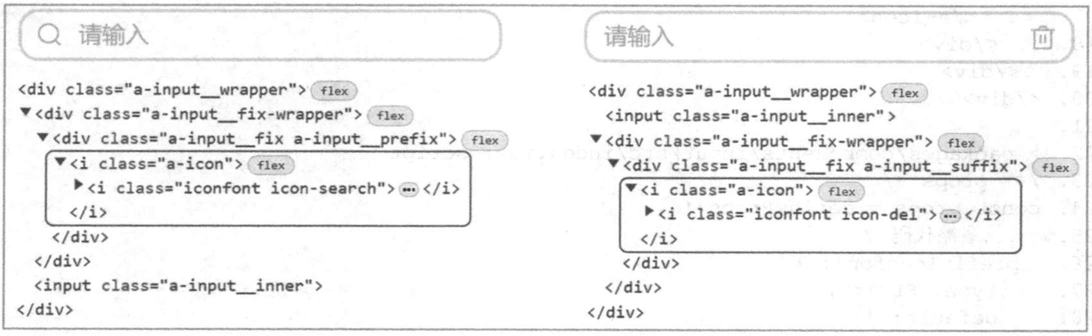

```
<div class="a-input__wrapper"> flex
▼<div class="a-input__fix-wrapper"> flex
  ▼<div class="a-input__fix a-input__prefix"> flex
    ▼<i class="a-icon"> flex
      ▶<i class="iconfont icon-search">…</i>
      </i>
    </div>
  </div>
<input class="a-input__inner">
</div>
```

```
<div class="a-input__wrapper"> flex
  <input class="a-input__inner">
▼<div class="a-input__fix-wrapper"> flex
  ▼<div class="a-input__fix a-input__suffix"> flex
    ▼<i class="a-icon"> flex
      ▶<i class="iconfont icon-del">…</i>
      </i>
    </div>
  </div>
</div>
```

图 8-12　为 input 组件渲染前缀、后缀 iconfont 字体图标

8.8　slot：前置、后置

前置、后置功能是指在输入框外部的左侧或右侧添加一个固定的标签、图标或自定义内容。前置和后置功能可以改善用户体验，使用户更清晰地了解需要输入的信息。在 UI 组件库中，input 组件的前置、后置功能的实现方式与前缀、后缀一致，首先确定前置和后置的渲染区域，如代码清单 library-08-12 所示。

代码清单　library-08-12

```
1.  > packages/components/input/src/index.vue: template
2.  <div :class="[ns.b(), ...省略代码]">
3.    <div :class="[ns.e('aside-wrapper')]">前置内容</div>
4.    <div :class="[ns.e('wrapper')]">...省略代码</div>
5.    <div :class="[ns.e('aside-wrapper')]">后置内容</div>
6.  </div>
7.
8.  > packages/theme/src/input.scss
9.  @include b(input) {
10.   ...省略代码
11.   @include e(aside-wrapper) {
12.     display: inline-flex;
13.     align-items: center;
14.     justify-content: center;
15.     height: var(#{getVarName('input', 'size')});
16.   }
17. }
```

根据 UI 组件库设计稿，前置和后置位于 <input> 标签的外围，因此要新增两个 <div> 标签（第 3、5 行），它们是与 ns.e('wrapper') 同级的兄弟节点关系，然后调用 ns.e('aside-wrapper') 生成类名 .a-input__aside-wrapper，再设置样式（第 11~16 行），如图 8-13 所示。

```
<div class="a-input a-input--size_default">  (flex)
  <div class="a-input__aside-wrapper">前置内容</div>  (flex)
▶ <div class="a-input__wrapper"> ⋯ </div>  (flex)
  <div class="a-input__aside-wrapper">后置内容</div>  (flex)
</div>
```

图 8-13　为 input 组件设置前置、后置区域

8.8.1　渲染前置、后置组件

以渲染前置的 icon 组件和后置的 button 组件为例，既可以渲染 HTML 元素的组件，也可以渲染自定义的任何内容，因此可采用<slot />插槽模式进行渲染，如代码清单 library-08-13 所示。

代码清单 library-08-13
```
1.  > packages/components/input/src/index.vue: template
2.  <div :class="[ns.b(), ...省略代码]">
3.    <!-- 前置 -->
4.    <div v-if="isPrepend" :class="[ns.e('aside-wrapper')]">
5.      <div :class="[ns.e('prepend')]">
6.        <slot v-if="$slots.prepend" name="prepend" />
7.      </div>
8.    </div>
9.    <div :class="[ns.e('wrapper'), ns.is('aside-prepend', isPrepend), ,
ns.is('aside-append', isAppend)]">...省略代码</div>
10.   <!-- 后置 -->
11.   <div v-if="isAppend" :class="[ns.e('aside-wrapper')]">
12.     <div :class="[ns.e('append')]">
13.       <slot v-if="$slots.append" name="append" />
14.     </div>
15.   </div>
16. </div>
17.
18. > packages/components/input/src/index.vue: script
19. import { computed, useSlots } from "vue"
20. const slots = useSlots()
21. // 前置内容
22. const isPrepend = computed(() => {
23.   return slots.prepend
24. })
25. // 后置内容
26. const isAppend = computed(() => {
27.   return slots.append
28. })
29.
30. > packages/theme/src/input.scss
31. @include b(input) {
32.     @include e(wrapper) {
33.         @include s(aside-prepend) {
34.             border-top-left-radius: 0;
```

```scss
35.       border-bottom-left-radius: 0;
36.     }
37.     @include s(aside-append) {
38.       border-top-right-radius: 0;
39.       border-bottom-right-radius: 0;
40.     }
41.   }
42.   ...省略代码
43.   @include e(prepend) {
44.     @include d(button) {
45.       border-top-right-radius: 0;
46.       border-bottom-right-radius: 0;
47.       border-right: none;
48.       min-width: auto;
49.       padding: 0 var(#{getVarName('input', 'padding')});
50.     }
51.   }
52.   @include e(append) {
53.     @include d(button) {
54.       border-top-left-radius: 0;
55.       border-bottom-left-radius: 0;
56.       border-left: none;
57.       min-width: auto;
58.       padding: 0 var(#{getVarName('input', 'padding')});
59.     }
60.   }
61.   @include d(button--default) {
62.     color: var(#{getVarName('input', 'text-color-aside')});
63.   }
64. }
65.
66. > examples/src/App.vue
67. <a-input>
68.   <template #prepend>
69.     <a-button :icon="Search"></a-button>
70.   </template>
71.   <template #append>
72.     <a-button type="success">这是一个按钮</a-button>
73.   </template>
74. </a-input>
```

为前置区域添加子级<div>标签，并绑定 ns.e('prepend')方法，生成类名.a-input__perpend（第5 行）。在该<div>标签中使用<slot />元素，并定义插槽的名称为 prepend，<slot />插槽使用 v-if 判断$slots.prepend 插槽名称 prepend 是否被使用，如果已被使用，则渲染，反之则不渲染（第 6 行）。

在实际应用中，input 组件不一定会使用前置或后置元素，因此在前置区域使用 v-if 条件绑定变量 isPrepend（第 4 行），isPrepend 变量使用 computed 计算属性返回结果 true 或 false（第22~24 行）。需要注意的是，在 computed 计算属性中，slots 对象来自 useSlots（第 19、20、23 行）。

input 组件的 4 个角为圆角,如果使用前置或后置功能,应将其改为直角,因此可在判断 isPrepend 为 true 时生成类名.a-input__aside-prepend,设置左上角、左下角为 0（第 9、33~36 行）；在判断 isAppend 为 true 时生成类名.a-input__aside-append,设置右上角、右下角为 0（第 37~40 行）,渲染效果如图 8-14 所示。

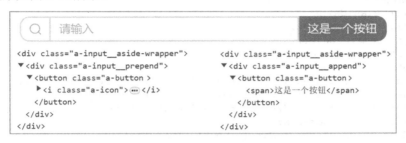

图 8-14　为 input 组件渲染前置、后置组件

8.8.2　渲染前置、后置标识

前置和后置标识属于字符串文本,与<slot />插槽的模式不同,因此可通过属性的方式传入,如代码清单 library-08-14 所示。

代码清单 library-08-14
```
1.  > packages/components/button/src/index.vue: template
2.  <div v-if="isPrepend" :class="[ns.e('aside-wrapper')]">
3.    <div :class="[ns.e('prepend'), (prepend || append) && ns.e('aside')]">
4.      <slot v-if="$slots.prepend" name="prepend" />
5.      <div v-if="prepend">{{ prepend }}</div>
6.    </div>
7.  </div>
8.
9.  > packages/components/button/src/index.vue: script
10. const props = defineProps({
11.   prepend: {
12.     type: String,
13.     default: ''
14.   },
15.   append: {
16.     type: String,
17.     default: ''
18.   }
19. });
20.
21. // 前置内容
22. const isPrepend = computed(() => {
23.   return slots.prepend || props.prepend
24. })
25. // 后置内容
26. const isAppend = computed(() => {
```

```
27.     return slots.append || props.append
28.   })
29.
30. > packages/theme/src/input.scss
31. @include b(input) {
32.   ...省略代码
33.   @include e(aside) {
34.     display: inline-flex;
35.     align-items: center;
36.     justify-content: center;
37.     height: 100%;
38.     border: 1px solid var(#{getVarName('input', 'border-color')});
39.     border-radius: var(#{getVarName('input', 'round')});
40.     background-color: var(#{getVarName('input', 'bg-color-aside')});
41.     box-sizing: border-box;
42.     padding: 0 var(#{getVarName('input', 'padding')});
43.     transition: .3s;
44.     font-size: var(#{getVarName('input', 'font-size')});
45.     user-select: none;
46.     color: var(#{getVarName('input', 'text-color-aside')});
47.   }
48.   @include e(prepend) {
49.     border-top-right-radius: 0;
50.     border-bottom-right-radius: 0;
51.     border-right: none;
52.   }
53.   @include e(append) {
54.     border-top-left-radius: 0;
55.     border-bottom-left-radius: 0;
56.     border-left: none;
57.   }
58. }
59.
60. > examples/src/App.vue
61. <a-input prepend="Http://" append=".com"></a-input>
```

在 defineProps 对象中定义属性 prepend 和 append，分别为前置标识和后置标识（第 10~19 行）。由于渲染的标识是通过属性的方式传入数据，为字符串类型，因此可使用<div>元素进行渲染，并使用 v-if 条件判断 prepend 是否存在，存在则渲染，反之则不渲染（第 5 行）。前置和后置标识的渲染包括背景颜色和边框，可在判断 prepend 或 append 存在时生成类名.a-input__aside（第 3 行），并设置 CSS 样式（第 33~47 行）。

最后通过类名.a-input__prepend 设置前置标识区域的右上角、右下角的圆角为 0，并去除右侧边框（第 48~52 行）；通过类名.a-input__append 设置后置标识区域的左上角、左下角的圆角为 0，并去除左侧边框（第 53~57 行），渲染效果如图 8-15 所示。

图 8-15　为 input 组件渲染前置、后置标识

8.8.3 渲染前缀、后缀标识

前缀、后缀标识与前置、后置标识的不同点在于，前缀和后缀在输入框内部渲染，前置和后置在输入框外部渲染。前缀和后缀的标识在实际业务中也很常见，如前缀的金额标识（¥）、后缀的币种标识（RMB），如代码清单 library-08-15 所示。

代码清单 library-08-15
```
1.  > packages/components/button/src/index.vue: template
2.  <div v-if="isPrefix" :class="[ns.e('fix-wrapper')]"> <!-- 前缀 -->
3.    <div :class="[ns.e('fix'), ns.e('prefix')]">
4.      <span v-if="prefix">{{ prefix }}</span>
5.      <a-icon v-if="prefixIcon"><component :is="prefixIcon" /></a-icon>
6.    </div>
7.  </div>
8.
9.  > packages/components/button/src/index.vue: script
10. const props = defineProps({
11.   prefix: {
12.     type: String,
13.     default: ''
14.   },
15.   suffix: {
16.     type: String,
17.     default: ''
18.   },
19. });
20. const isPrefix = computed(() => {
21.   return ...省略代码 || props.prefix
22. })
23. const isSuffix = computed(() => {
24.   return ...省略代码 || props.suffix
25. })
26.
27. > packages/theme/src/input.scss
28. @include b(input) {
29.   ...省略代码
30.   @include e(fix) {
31.     ...省略代码
32.     > span {
33.       color: var(#{getVarName('input', 'text-color')});
34.       font-size: inherit;
35.     }
36.   }
37. }
38.
39. > examples/src/App.vue
40. <a-input prefix="¥" suffix="RMB"></a-input>
```

在 efineProps 对象中定义属性 prefix 和 suffix，分别为前缀标识和后缀标识（第 10~19 行）。

前缀和后缀的开发方式与前置和后置一致，同样使用 v-if 条件判断 prefix 或 suffix 是否存在，存在则渲染，反之则不渲染（第 4 行）。最后为变量 isPrefix 和 isSuffix 增加新的条件 props.prefix 和 props.suffix（第 21、24 行），渲染效果如图 8-16 所示。

图 8-16　为 input 组件渲染前缀、后缀标识

8.9　password：密码

密码输入框是 input 组件的类型之一，在常规情况下，使用 "•" 或 "*" 隐藏文本，为了能够让用户在输入密码时随时查看自己输入的内容，可在 input 组件后缀添加图标实现按钮，用于切换显示或隐藏密码，如代码清单 library-08-16 所示。

```
代码清单  library-08-16
1.  > packages/components/input/src/index.vue: template
2.  <div :class="[ns.e('wrapper'), ns.is('aside', isAside)]">
3.    <div v-if="isPrefix" :class="[ns.e('fix-wrapper')]">...省略代码</div> <!-- 前缀 -->
4.    <input :type="showPassword ? (passwordVisible ? 'text' : 'password') : type" ...省略代码 />
5.    <!-- 后缀 -->
6.    <div v-if="isSuffix" :class="[ns.e('fix-wrapper')]">
7.      <div :class="[ns.e('fix'), ns.e('suffix')]">
8.        <template v-if="!showPassword">
9.          <span v-if="suffix">{{ suffix }}</span>
10.         <a-icon v-if="prefixIcon"><component :is="prefixIcon" /></a-icon>
11.         <a-icon v-if="prefixIconfont">
12.           <i class="iconfont" :class="prefixIconfont"></i>
13.         </a-icon>
14.       </template>
15.       <a-icon v-if="showPassword" class="pointer" @click="passwordVisible = !passwordVisible">
16.         <component :is="passwordIcon" />
17.       </a-icon>
18.     </div>
19.   </div>
20. </div>
21.
22. > packages/components/input/src/index.vue: script
23. import { Show, Hide } from "@ui-library/icons"
24. const props = defineProps({
25.   type: {
26.     type: String,
27.     default: 'text'
28.   },
29.   showPassword: Boolean, // 是否显示密码图标
```

```
30.   });
31.   // 密码可见，默认 false（不可见）
32.   const passwordVisible = ref(false)
33.   // 密码图标
34.   const passwordIcon = computed(() => {
35.     return passwordVisible.value ? Show : Hide
36.   })
37.   const isSuffix = computed(() => {
38.     return ...省略代码 || props.showPassword
39.   })
40.
41. > packages/theme/src/common.scss
42.   .pointer { cursor: pointer; }
43.
44. > examples/src/App.vue
45.   <a-input type="password" show-password></a-input>
```

将<input />标签的属性值 type 默认为 text，如需设置为密码类型，则将属性值改为 password。在 defineProps 对象中定义属性值为字符串类型，默认值为 text（第 25~28 行），并定义属性 showPassword 为布尔类型，默认值为 false（第 29 行）。接着定义变量 passwordVisible，用于标记显示或隐藏密码文本（第 32 行）。

在<input />标签的属性值中使用三元运算符绑定 showPassword 和 passwordVisible（第 4 行）。如果 showPassword 为 true，则使用第二层三元运算符判断 passwordVisible 是否为 true，如果为 true，那么属性值为 text，反之为 password；如果 showPassword 为 false，则默认 defineProps 对象的属性值为 text。

密码是 input 组件内置的功能，不需要从外部传入图标。但又因后缀可以从外部传入图标，因此可以通过"内置"和"外部传入"两种特性进行区分显示。在后缀渲染区域中添加<template>模板标签，当 showPassword 为 false 时，渲染外部传入的密码图标（第 8~14 行）。接着新增<a-icon>图标组件，当 showPassword 为 true 时，渲染内置的密码图标（第 15~17 行）。在<a-icon>组件内添加<component>动态组件，并为属性 is 绑定变量 passwordIcon（第 16 行）。变量 passwordIcon 调用 computed 计算属性判断 passwordVisible 是否为 true，如果是，则返回 Show，反之则返回 Hide（第 34~36 行）。Show 和 Hide 是来自 icons 图标库的 SVG 图标（第 23 行）。

最后为<a-icon>添加@click 事件，使变量 passwordVisible 在 true 与 false 之间取反（第 15 行），便会触发变量 passwordIcon 的 computed 计算属性更新图标的显示，同时使用<input />标签的属性值在 text 和 password 之间切换（第 4 行），渲染效果如图 8-17 所示。

图 8-17 input 组件的密码类型

8.10 value：数据双向绑定

数据双向绑定是指在前端开发中，在数据模型和视图之间建立双向联系，当数据模型发生变化时，视图也随之更新，反之亦然。通过某种特定的方式将数据模型和界面元素绑定，这些框架和库可以在数据变化时自动更新界面，实现数据双向绑定。在 Vue.js 3 中，可通过 defineModel 对象实现数据双向绑定，如代码清单 library-08-17 所示。

代码清单 library-08-17
```
1.  > packages/components/input/src/index.vue: template
2.  <input @input="handlerInput" :value="modelValue" ...省略代码 />
3.
4.  > packages/components/input/src/index.vue: script
5.  // defineModel
6.  const modelValue = defineModel()
7.  // emits
8.  const emit = defineEmits(['input'])
9.  // 输入事件
10. const handlerInput = (e) => {
11.   const value = e.target.value
12.   modelValue.value = value
13.   // emit input
14.   emit('input', value, e)
15. }
16.
17. > examples/src/App.vue: template
18. <a-input v-model="value" clear @input="handlerInputFun"></a-input>
19. <div>双向绑定数据：{{ value }}</div>
20.
21. > examples/src/App.vue: script
22. import { ref } from "vue"
23. const value = ref('')
24. const handlerInputFun = (val, e) => {
25.   console.log('输入的值：', val)
26.   console.log('输入事件：', e)
27. }
```

在 input 组件中，将 defineModel() 对象赋给变量 modelValue（第 6 行），然后将变量 modelValue 绑定至 <input /> 标签的 value 属性，在 <input /> 标签中添加 @input 输入事件并绑定 handlerInput 方法（第 2 行）。在 handlerInput 方法中获取 input 组件的 value（第 11 行），并将 value 赋给变量 modelValue.value（第 12 行）。

在利用 examples 演示库调试数据时，只需使用 v-model 绑定变量，即可实现数据双向绑定，渲染效果如图 8-18 所示。

图 8-18　使用 defineModel()对象实现数据双向绑定

在实际业务场景中，输入事件具有重要的作用。当用户输入、粘贴或清除文本时，会触发输入事件，这使得应用可以实时捕获用户的输入操作，以便做出相应的处理和改变。因此，input 组件应提供输入事件，确保友好的用户交互体验和实时的数据处理。使用 defineEmits 对象声明可以触发的事件 input 并赋给变量 emit（第 8 行），然后在 handlerInput 方法中调用 emit('input')，并返回当前输入的值 value 以及事件对象（第 14 行）。

在 examples 演示库中添加@input 事件并绑定 handlerInputFun 方法，在 handlerInputFun 方法中便可获取 input 组件输入过程的实时数据（第 18、24~27 行），控制台输出如图 8-19 所示。

图 8-19　使用输入事件获取实时数据

提示：在 Vue.js 3.4 版本中，defineModel()已经是稳定版，可以直接使用。在低于 Vue.js 3.4 版本中使用 defineModel()，会报错"defineModel is not defined"，解决方式是在 vite.config.js 文件中配置 defineModel: true。

8.11　clear：清除数据

input 组件的清除功能主要是为了方便用户快速清除在输入框中输入的内容。清除功能通常以一个清除按钮的形式出现在输入框的右侧，当用户点击清除按钮时，输入框中的文本内容会被清空。既然是在输入框右侧显示按钮，那么我们需要考虑在哪些情况下才会显示清除按钮，如代码清单 library-08-18 所示。

代码清单 library-08-18

```
1.  > packages/components/input/src/index.vue: template
2.  <template v-if="!showPassword || !showClear">...省略代码</template>
3.  <a-icon v-if="showPassword">...省略代码</a-icon>
4.  <a-icon v-if="showClear" class="pointer" @click="handlerClear">
5.    <component :is="ClearFill" />
6.  </a-icon>
7.
8.  > packages/components/input/src/index.vue: script
```

```
9.   import { Show, Hide, ClearFill } from "@ui-library/icons"
10.
11.  // emits
12.  const emit = defineEmits(['input', 'clear'])
13.
14.  // props
15.  const props = defineProps({
16.    ...省略代码
17.    clearance: Boolean,   // 清除
18.  });
19.
20.  // clearIcon
21.  const showClear = computed(() => {
22.    return props.clearance &&    // 配置清除动作
23.      modelValue.value &&         // 存在文本
24.      !props.disabled &&          // 非禁用状态
25.      props.type === 'text'       // type 为文本类型
26.  })
27.
28.  // isSuffix
29.  const isSuffix = computed(() => {
30.    return ...省略代码 || showClear.value
31.  })
32.
33.  // 清除
34.  const handlerClear = () => {
35.    modelValue.value = ''
36.    emit('input', '')
37.    emit('clear')
38.    // focusExpose()   自动聚焦
39.  }
40.
41.  > examples/src/App.vue: template
42.  <a-input v-model="value" clearance @input="handlerInput"
@clear="handlerClear"></a-input>
```

  input 组件的清除按钮通常在用户输入文本时，也就是存在字符时才会显示。由于在输入框中可输入文本（处于非禁用状态），再加上 defineProps 对象的 clearance 属性启动清除功能，因此共有 3 个条件。如果再加上输入框的类型为 text 类型，那么就有 4 个条件。因此，定义变量 showClear 调用 computed 计算属性（第 21 行）。computed 的逻辑为：启用清除动作 props.clearance、存在文件 modelValue.value、非禁用状态!props.disabled 以及 props.type === 'text'（第 22~25 行）。这 4 个条件都成立，便会显示清除按钮。

  清除功能是 input 组件的内置功能，与密码类型类似。单独定义<a-icon>，并使用动态组件<component>为属性 is 绑定 ClearFill 图标（第 5 行）。为 v-if 条件绑定变量 showClear，当 showClear 为 true 时，显示图标（第 4~6 行）；如果变量 showClear 为 false，则渲染外部传入的图标（第 2 行）。

清除功能体现为在点击图标时清除文本，可为图标添加@click 事件，绑定 handlerClear 方法（第 4 行），在方法中重置 modelValue.value 为空，清除文本（第 35 行），并调用 emit 触发 input 和 clear 事件，实现外部回调（第 42 行）。最后为后缀变量 isSuffix 增加变量 showClear.value，判断是否显示后缀渲染区域（第 30 行），渲染效果如图 8-20 所示。

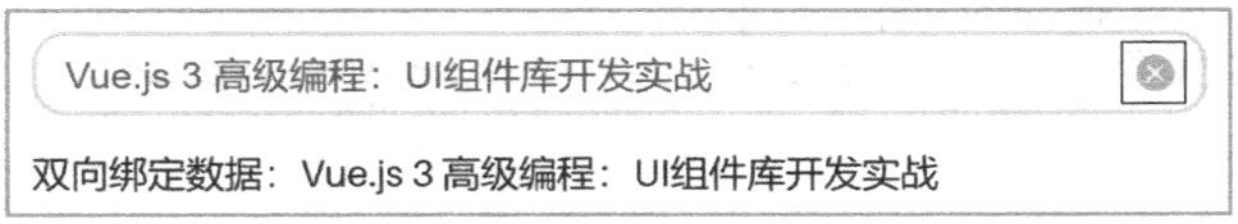

图 8-20　input 组件的清除功能

提示：在点击清除图标时，输入框的光标会消失，因此可以调用 focusExpose 方法（第 38 行），将光标聚焦回输入框，详见 8.15 节。

8.12　count：统计字符数

输入框长度的统计功能用于显示用户输入的文本内容的字符数。这种功能通常用于限制用户输入的文本长度或者提供即时的统计字符数，可确保用户输入符合特定格式要求的文本。例如，在创建密码时限制最大字符数，在社交媒体中发布内容时显示剩余的可用字符数等，如代码清单 library-08-19 所示。

代码清单 library-08-19

```
1.  > packages/components/input/src/index.vue: template
2.  <input :class="[ns.e('inner'), ns.is('color-danger', isColorDanger)]" />
3.  <div v-if="isSuffix" :class="[ns.e('fix-wrapper')]">
4.    <div :class="[ns.e('fix'), ns.e('suffix')]">
5.      <template v-if="!showPassword || !showClear || !showLimit">...省略代码
</template>
6.      <a-icon v-if="showPassword">...省略代码</a-icon>
7.      <a-icon v-if="showClear">...省略代码</a-icon>
8.      <div v-if="showLimit" :class="[ns.e('count')]">
9.        {{ valueLength }} / {{ maxlength }}
10.     </div>
11.   </div>
12. </div>
13.
14. > packages/components/input/src/index.vue: script
15. // props
16. const props = defineProps({
17.   ...省略代码
18.   count: Boolean, // 统计
19. });
20.
21. // showLimit
22. const showLimit = computed(() => props.count && !props.disabled && props.maxlength)
23. // 文本长度
```

```
24.   const valueLength = computed(() => modelValue.value.length)
25.   const isColorDanger = computed(() => {
26.     return props.maxlength && props.count && props.maxlength
27.   })
28.
29.   // isSuffix
30.   const isSuffix = computed(() => {
31.     return ...省略代码 || showClear.value || showLimit.value
32.   })
33.
34.  > packages/theme/src/input.scss
35.  @include b(input) {
36.    ...省略代码
37.    @include e(count) {
38.      font-size: 12px;      // 默认字体大小
39.      margin-left: var(#{getVarName('input', 'padding')});  // 左侧边距
40.      line-height: 1;        // 根据该元素本身的字体大小
41.      white-space: pre;      // 处理元素内的空白符
42.    }
43.    @include e(inner) {
44.      @include s(color-error) {
45.        // 重置输入框文本颜色变量
46.        #{getVarName('input', 'text-color')}: var(#{getVarName('color', 'error')});
47.      }
48.    }
49.  }
50.
51.  > examples/src/App.vue: template
52.  <a-input v-model="value" count :maxlength="30"></a-input>
```

　　输入框的字符数统计通常发生在用户输入文本时，因此输入框处于非禁用状态，并且需要配置 defineProps 对象的属性 count 和 maxlength 都为 true，也就是有 3 个条件。定义变量 showLimit 调用 computed 计算属性，computed 的逻辑是：启用统计 props.count、非禁用状态!props.disabled、长度限制 props.maxlength。这 3 个条件都成立，便会显示字符长度统计（第 22 行）。

　　字符长度统计是 input 组件的内置功能，和密码类型一致。单独定义<div>标签，并使用 v-if 条件绑定变量 showLimit，绑定 ns.e('count')生成类名.a-input__count，设置 CSS 样式（第 8、37~42 行）。为<div>标签分别绑定变量 valueLength 和 maxlength（第 9 行），变量 valueLength 调用 computed 计算属性统计<input />标签属性 value 绑定的变量 modelValue 的长度（第 24 行），直接读取 defineProps 对象属性的 maxlength。

　　根据 UI 组件库设计稿，当文本超出统计字符长度时，输入框的字符会变为"红色"，因此在<input />标签中调用 ns.is('color-error, isColorError)。变量 isColorError 使用 computed 计算属性判断 defineProps 对象的属性 count 和 maxlength 是否为 true，如果都为 true 且输入的文本字符长度大于 maxlength，则返回 true，这样便会追加类名.is-color-error，即可重置 input 组件的私有变量值（第 2、43~48 行）。最后为后缀变量 isSuffix 增加变量 showLimit.value，判断是否显示后

缀渲染区域（第 29 行），渲染效果如图 8-21 所示。

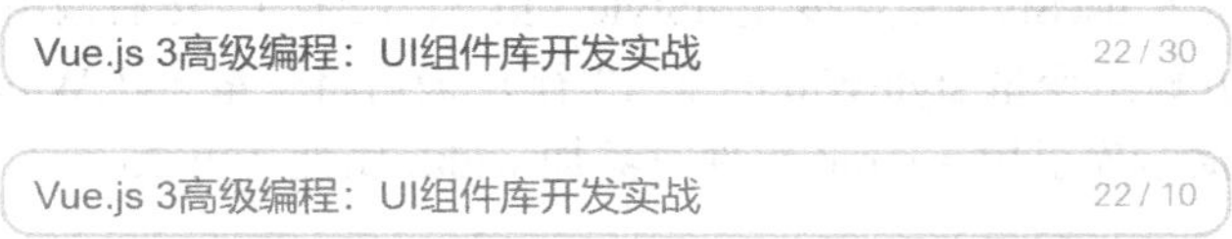

> [Vue.js 3高级编程：UI组件库开发实战   22 / 30]
>
> [Vue.js 3高级编程：UI组件库开发实战   22 / 10]

图 8-21　为 input 组件设置字符统计功能

8.13　width：宽度

input 组件的宽度通常用于控制用户输入框的可见范围，它影响着用户输入内容在界面上的显示方式。输入框的宽度可以是页面布局的一部分，通过控制它，可以平衡界面的美观度和功能性，如代码清单 library-08-20 所示。

代码清单 library-08-20
```
1.  > packages/components/input/src/index.vue: template
2.  <div :style="[styleWidth]" :class="[ns.b(), ...省略代码]"></div>
3.
4.  > packages/components/input/src/index.vue: script
5.  import { useNamespace, useStyle } from '@ui-library/hook';
6.  const props = defineProps({
7.    ...省略代码
8.    width: { type: String, default: "" },
9.  });
10.
11. // style
12. const uStyle = useStyle();
13. const styleWidth = computed(() => uStyle.width(props.width))
14.
15. > packages/hook/use-style/index.js
16. export const useStyle = () => {
17.   // 宽度
18.   const width = (value) => {
19.     return value ? {'width': value} : {}
20.   }
21.   return { fontSize, color, width }
22. }
23.
24. > examples/src/App.vue: template
25. <a-input width="200px" v-model="value"></a-input>
```

input 组件自动适应父元素的宽度，由于父元素是 div，也是"块元素"，所以 input 组件的宽度默认是 100%。如需修改宽度，只需为父元素设置行间样式 width。

在 defineProps 对象中定义属性 width 为字符串类型，默认值为空，然后在 hook/use-style/index.js 中添加 width 方法，并定义参数 value 接收数据（第 18 行）。使用三元运算判断参数 value 是否存在数据，如果存在，则返回 CSS 的 width 属性，反之则返回空的 Object

对象（第 19 行），然后返回 width（第 21 行）。

在 input 组件中引入 useStyle 赋给变量 uStyle（第 5、12 行）。然后调用 uStyle.width 方法传入数据 props.width，并赋给变量 styleWidth（第 13 行）。最后将变量 styleWidth 绑定至父元素（第 2 行），在 examples 演示库引用 a-input 组件时添加 width 属性（第 25 行），渲染效果如图 8-22 所示。

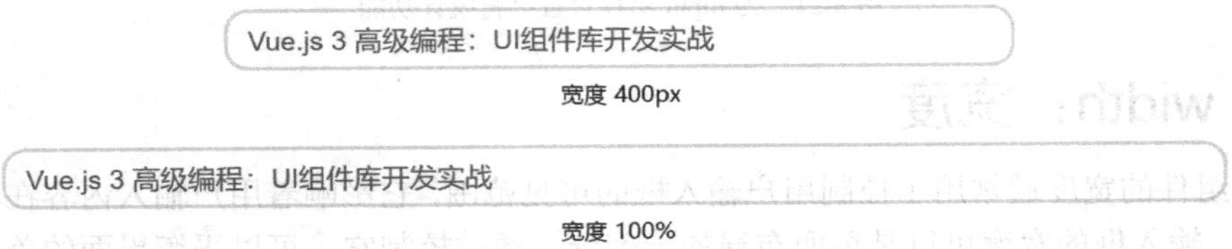

图 8-22　调整 input 组件的宽度

8.14　event：事件

input 组件的事件通常涉及用户与输入框的交互或输入框状态的变化，包括用户单击、获取焦点、失去焦点、鼠标悬停、输入，以及其他与输入框相关的操作。事件描述可以包括用户的动作、触发事件的因素，以及与输入框交互时触发的任何响应或效果。

8.14.1　focus 和 blur

获取焦点（focus）事件是用户将鼠标光标聚焦在输入框时的特定操作，比如改变输入框的样式、显示辅助信息等。失去焦点（blur）事件是用户在鼠标光标离开输入框时的特定操作，比如验证输入内容、隐藏辅助信息等。这些事件可以帮助开发人员实现更好的用户交互体验，提高页面的可用性，如代码清单 library-08-21 所示。

代码清单 library-08-21
```
1.  > packages/components/input/src/index.vue: template
2.  <div :style="[styleWidth]" :class="[ns.b(), ns.is('focus', isFocus), ...省略代码]">
3.    <div :class="[ns.e('wrapper'), ...省略代码]">
4.      ...省略代码
5.      <input
6.        @input="handlerInput"
7.        @focus="focusEvent"     // 绑定获取焦点事件
8.        @blur="blurEvent"       // 绑定失去焦点事件
9.      />
10.   </div>
11. </div>
12.
13. > packages/components/input/src/index.vue: script
14. import { ...省略代码, useEvent } from '@ui-library/hook';
15. // emits
```

```
16.   const emit = defineEmits(['input', 'clear', 'focus', 'blur'])
17.   // event
18.   const { isFocus, focusEvent, blurEvent } = useEvent()
19.
20. > packages/hook/use-event/index.js
21.   import { ref, getCurrentInstance } from "vue"
22.   export const useEvent = () => {
23.     let isFocus = ref(false)
24.     // emit
25.     const { emit } = getCurrentInstance()
26.     // 获取焦点
27.     const focusEvent = (e) => {
28.       isFocus.value = true
29.       emit('focus', e)
30.     }
31.     // 失去焦点
32.     const blurEvent = (e) => {
33.       isFocus.value = false
34.       emit('blur', e)
35.     }
36.     return { focusEvent, blurEvent, isFocus }
37.   }
38.
39. > examples/src/App.vue: template
40. <a-input v-model="value" @focus="handlerFocusFun"
@blur="handlerBlurFun"></a-input>
41.
42. > examples/src/App.vue: script
43. const handlerFocusFun = (e) => {
44.   console.log('获取焦点：', e)
45. }
46. const handlerBlurFun = (e) => {
47.   console.log('失去焦点：', e)
48. };
```

focus 事件和 blur 事件是 Web 应用中的常见交互，并且大多数 HTML 元素都有这两种事件，如 input、select、textarea、button 等。为了使常用事件公共化，可在 hook 目录下新建 use-event.js 文件，用于处理事件对象。

在 use-event.js 文件中定义 useEvent 组合式函数（第 22 行），并定义变量 isFocus、方法 focusEvent 和 blurEvent（第 23、27、32 行），分别为：标记是否已获取焦点、获取焦点事件和失去焦点事件。然后引入 Vue.js 3 的 emit 事件对象（第 25 行），在 focusEvent 方法中触发 emit('focus', e)（第 29 行）、在 blurEvent 方法中触发 emit('blur', e)（第 34 行），最后返回 focusEvent、blurEvent 和 isFocus（第 36 行）。

在 input 组件中引入 useEvent，并解构 focusEvent、blurEvent 和 isFocus（第 14、18 行），在<input />标签中添加@focus 事件绑定 focusEvent 方法，添加@blur 事件绑定 blurEvent 方法（第 7、8 行）。为 isFocus 绑定在 input 组件的根节点生成的类名.is-focus，便可在获取焦点时通过类

名 is-focus 重写 input 组件的私有变量，改变其边框颜色，如图 8-23 所示。

```
<div data-v-7a7a37b1 class="a-input is-focus a-i
nput--size_default" style="width: 400px;"> ⋯
```

图 8-23　改变 input 组件的边框颜色

8.14.2　mouseenter 和 mouselevel

鼠标移入（mouseenter 和 mouseover）事件在鼠标进入元素所占据的空间时触发。mouseenter 事件与 mouseover 事件的不同点在于，mouseenter 事件不会冒泡。鼠标移出（mouselevel）事件在鼠标离开元素所占据的空间时触发，同样不会产生冒泡，如代码清单 library-08-22 所示。

代码清单　library-08-22

```
1.  > packages/components/input/src/index.vue: template
2.  <div :style="[styleWidth]" :class="[ns.b(), ...省略代码]
3.    @mouseenter="mouseenterEvent"
4.    @mouseleave="mouselevelEvent"
5.  ">省略代码</div>
6.
7.  > packages/components/input/src/index.vue: script
8.  import { ...省略代码, useEvent } from '@ui-library/hook';
9.  // emits
10. const emit = defineEmits([...省略代码, 'blur', 'mouseenter', 'mouselevel'])
11. // event
12. const { ...省略代码, isHover, mouseenterEvent, mouselevelEvent } = useEvent()
13.
14. > packages/hook/use-event/index.js
15. import { ref, getCurrentInstance } from "vue"
16. export const useEvent = () => {
17.   let isFocus = ref(false)
18.   let isHover = ref(false)
19.   ...省略代码
20.   / 移入事件
21.   const mouseenterEvent = (e) => {
22.     isHover.value = true
23.     emit('mouseenter', e)
24.   }
25.   // 移出事件
26.   const mouselevelEvent = (e) => {
27.     isHover.value = false
28.     emit('mouselevel', e)
29.   }
30.   return { ...省略代码, isHover, mouseenterEvent, mouselevelEvent }
31. }
32.
```

```
33. > examples/src/App.vue: template
34. <a-input v-model="value" @mouseenter="mouseenterFun"
    @mouseleave="mouselevelFun"></a-input>
35.
36. > examples/src/App.vue: script
37. const mouseenterFun = (e) => {
38.   console.log('移入事件: ', e)
39. }
40. const mouselevelFun = (e) => {
41.   console.log('移出事件: ', e)
42. }
```

mouseenter、mouselevel 事件的开发方式与 focus、blur 事件一致,在 use-event.js 文件中定义 isHover、mouselevelEvent 和 handlerMouseLevel(第 18、21~24、26~29 行),分别为:标记是否移入元素、鼠标移入和鼠标移出。将其引入 input 组件后,再绑定根节点的@mouseenter 事件和@mouseleave 事件(第 3、4 行)。

8.14.3　compositionstart、compositionupdate 和 compositionend

输入法交互事件常用于处理非拉丁语系(如中文、日文等)输入法输入文本的过程。开始输入(compositionstart)事件在输入法输入开始时触发,输入更新(compositionupdate)事件在输入法输入过程中持续触发,结束输入(compositionend)事件在输入法输入结束时触发,如代码清单 library-08-23 所示。

代码清单 library-08-23

```
1.  > packages/components/input/src/index.vue: template
2.  <input ...省略代码
3.    @compositionstart="compositionStartEvent"
4.    @compositionupdate="compositionUpdateEvent"
5.    @compositionend="handlerCompositionEnd"
6.  />
7.
8.  > packages/components/input/src/index.vue: script
9.  import { ...省略代码, useEvent } from '@ui-library/hook';
10. // emits
11. const emit = defineEmits([...省略代码, 'compositionstart', 'compositionupdate',
    'compositionend'])
12. // event
13. const { ...省略代码, isComposition, compositionStartEvent, compositionUpdateEvent,
    compositionEndEvent } = useEvent()
14. // 输入结束
15. const handlerCompositionEnd = (e) => {
16.   compositionEndEvent(e).then(() => {
17.     handlerInput(e)
18.   })
19. }
20.
21. const handlerInput = (e) => {
```

```
22.     if(isComposition.value) { return false } // 输入法输入时，阻止
23.     ...省略代码
24.   }
25.
26. > packages/hook/use-event/index.js
27. export const useEvent = () => {
28.      ...省略代码
29.   let isComposition = ref(false)
30.   /** 输入法事件 */
31.   const compositionStartEvent = (e) => {    // 开始输入
32.     isComposition.value = true
33.     emit('compositionstart', e)
34.   }
35.   const compositionUpdateEvent = (e) => {    // 输入更新
36.     emit('compositionupdate', e)
37.   }
38.   const compositionEndEvent = () => {        // 结束输入
39.     return new Promise((resolve, reject) => {
40.       if(isComposition.value) {
41.         isComposition.value = false
42.         resolve()
43.         return false
44.       }
45.       reject()
46.     })
47.   }
48.   return { ...省略代码, isComposition, compositionStartEvent,
compositionUpdateEvent, compositionEndEvent }
49. }
50.
51. > examples/src/App.vue: template
52. <a-input v-model="value"
53.   @input="henalerInput"
54.   @compositionstart="handlerCompositionStart"
55.   @compositionupdate="handlerCompositionUpdate"
56.   @compositionend="handlerCompositionEnd">
57. </a-input>
58.
59. > examples/src/App.vue: script
60. const henalerInput = (e) => { console.log('输入时：', e) }
61. const handlerCompositionStart = (e) => { console.log('开始：', e) }
62. const handlerCompositionUpdate = (e) => { console.log('更新：', e) }
63. const handlerCompositionEnd = (e) => { console.log('结束：', e) }
```

输入法事件的 3 个过程分别是：开始、持续输入和结束。@input 事件也在输入文本时触发，两者的不同点在于，输入法事件有过程，@input 事件无过程。使用输入法输入文本的效果如图 8-24 所示。

图 8-24　使用输入法输入文本

在图 8-24 中，在使用输入法输入"库"字时，只在输入框中显示了"ylk"，并没有把"库"字写入输入框，拼音、全拼等输入法也同理，只有按下空格键才会将文字真正写入输入框。整个过程包含输入法的开始、持续输入和结束阶段。如果输入数字、英文等，则只触发 @input 事件，不会触发输入法事件。因此，为了区分 @input 事件和输入法的持续输入事件，需要定义变量 isComposition（第 29 行）。

在使用输入法输入内容时，必定会触发 compositionstart 事件绑定的 compositionStartEvent 方法（第 3 行）。可在调用 compositionStartEvent 方法时，设置变量 isComposition 为 true（第 32 行），原因在于，输入框在输入文本时会触发 @input 事件绑定的 handlerInput 方法，因此可在 handlerInput 方法中判断 isComposition 是否为 true，如果为 true，则阻止 handlerInput 方法的逻辑被执行（第 22 行）。

为 compositionupdate 事件直接绑定 compositionUpdateEvent 方法（第 4 行）；compositionend 事件的处理比较特殊，绑定的是 handlerCompositionEnd 方法（第 5 行），在该方法中调用 compositionEndEvent 方法返回 Promise 对象（第 15~19 行）。在 compositionEndEvent 方法中重置变量 isComposition 为 false（第 41 行），并回调 resolve()，此时在 handlerCompositionEnd 方法中便可通过 .then 回调执行 handlerInput 方法（第 16~18 行）。

8.14.4　change、keydown 和 keyup

change 事件在用户完成表单输入后、元素的值改变且失去焦点时触发，主要用在对表单元素的监听中，用于即时处理输入的数据。

键盘（keydown 和 keyup）事件在用户按下或释放按键时触发，通常用于捕获用户对键盘输入的即时反馈或执行特定的操作，如代码清单 library-08-24 所示。

代码清单 library-08-24

```
1.  > packages/components/input/src/index.vue: template
2.  <input ...省略代码
3.    @change="changeEvent"
4.    @keydown="keydownEvent"
5.    @keyup="keyupEvent"
6.  />
7.
8.  > packages/components/input/src/index.vue: script
9.  import { ...省略代码, useEvent } from '@ui-library/hook';
10. // emits
11. const emit = defineEmits([...省略代码, 'change', 'keydown', 'keyup'])
12. // event
13. const { ...省略代码, changeEvent, keydownEvent, keyupEvent } = useEvent()
```

```
14.
15. > packages/hook/use-event/index.js
16. export const useEvent = () => {
17.     ...省略代码
18.     // change
19.     const changeEvent = (e) => { emit('change', e) }
20.     // 键盘事件
21.     const keydownEvent = (e) => { emit('keydown', e) }
22.     const keyupEvent = (e) => { emit('keyup', e) }
23.     return { ...省略代码, changeEvent, keydownEvent, keyupEvent }
24. }
```

8.15 expose：暴露对象

defineExpose 函数是 Vue.js 3 Composition API 中的一个特性，用于将子级组件中的任何内部状态或方法暴露给父级组件，从而使父级组件能够访问子级组件内部的状态和方法。这样做既有助于管理组件之间的数据，也有助于降低组件之间的耦合度，从而使组件通信和应用逻辑的组织方式更灵活，如代码清单 library-08-25 所示。

代码清单 library-08-25

```
1.  > packages/components/input/src/index.vue: template
2.  <input ref="_ref" />
3.
4.  > packages/components/input/src/index.vue: script
5.  import { ref, computed, useSlots, shallowRef } from "vue";
6.  import { ...省略代码, useExpose } from '@ui-library/hook';
7.  // input
8.  const _ref = shallowRef(null)
9.  const { focusExpose, blurExpose, selectExpose } = useExpose(_ref)
10. ...省略代码
11. defineExpose({
12.     ref: _ref, // 输入框对象
13.     focus: focusExpose,
14.     blur: blurExpose,
15.     select: selectExpose,
16.     clear: handlerClear
17. })
18.
19. > packages/hook/use-expose/index.js
20. import { nextTick } from "vue"
21. export const useExpose = (elem) => {
22.     const _ref = elem
23.     // 自动聚集
24.     const focusExpose = async () => {
25.         await nextTick()
26.         _ref.value?.focus()
27.     }
28.     // 失焦
```

```
29.   const blurExpose = async () => {
30.     await nextTick()
31.     _ref.value?.blur()
32.   }
33.   // 全选文本
34.   const selectExpose = () => {
35.     _ref.value?.select()
36.   }
37.   return { focusExpose, blurExpose, selectExpose }
38. }
39.
40. > examples/src/App.vue: template
41. <a-input ref="input" v-model="value"></a-input>
42.
43. > examples/src/App.vue: script
44. import { onMounted, ref } from "vue"
45. const value = ref('Vue.js 3 高级编程：UI 组件库开发实战')
46. const input = ref(null)
47. onMounted(() => {
48.   input.value.select() // 调用子级组件暴露的 select()方法，默认选中输入框文本
49. })
```

在 hook 目录下新 use-expose 文件夹，在 index.js 文件中定义需要暴露的方法 focusExpose、blurExpose 和 selectExpose 并返回（21~38 行）。引入 input 组件并解构 3 个方法（第 6、9 行），将需要暴露给父级组件的方法放置在 defineExpose 对象中（第 11~17 行）。最后在 examples 演示库中调用子级组件暴露的 select()方法选中输入框文本（第 44~49 行），如图 8-25 所示。

图 8-25　调用子级组件暴露的 select()方法选中输入框文本

8.16　textarea：文本域

文本域（textarea）组件类似于 input 组件，不同的是，textarea 允许用户输入多行文本，并且可以启用文本域的上、下、左、右拖曳或缩放功能。textarea 的开发和 input 组件类似，只是去除了前缀、后缀、前置和后置功能，如代码清单 library-08-26 所示。

代码清单 library-08-26

```
1. > packages/components/textarea/src/index.vue: template
2. <div :style="[styleWidth]" :class="[ns.b()]" @mouseenter="mouseenterEvent"
@mouseleave="mouselevelEvent">
3.   <textarea
4.     ref="_ref"
5.     @input="handlerInput"
6.     @focus="focusEvent"
7.     @blur="blurEvent"
8.     @compositionstart="compositionStartEvent"
```

```
9.      @compositionupdate="compositionUpdateEvent"
10.     @compositionend="handlerCompositionEnd"
11.     @change="changeEvent"
12.     @keydown="keydownEvent"
13.     @keyup="keyupEvent"
14.     :rows="rows"
15.     :placeholder="placeholder"
16.     :value="modelValue"
17.     :disabled="disabled"
18.     :maxlength="maxlength"
19.     :class="[ns.e('inner'), ns.is('color-danger', isColorDanger), ns.is('focus', isFocus), ns.is('disabled', disabled)]"
20.   />
21.   <div v-if="showLimit" :class="[ns.e('count')]">{{ valueLength }} / {{ maxlength }}</div>
22. </div>
23.
24. > packages/components/textarea/src/index.vue: script
25. /** props */
26. const props = defineProps({
27.   ...省略代码,
28.   rows: {
29.     type: [String, Number],
30.     default: ''
31.   }
32. });
```

将 input 组件所在的文件夹复制一份,将文件名改为 textarea,然后将 index.vue 文件中的前缀、后缀、前置和后置功能删除,仅保留"统计"功能。

在 defineProps 对象中新增 rows 属性,并绑定至 textarea 的属性 rows(第 28、14 行)。rows 属性用于指定文本区域的显示行数,通过设置 rows,可以控制文本域的初始高度,使用户看到多行的文本域。最后将"统计"功能的字符计算定位在右下角,如图 8-26 所示。

图 8-26   渲染 textarea 组件

本章小结

本章开发的 input 组件比其他组件更复杂,input 组件需要处理各种用户交互事件,比如 focus、blur、change 等。这些事件能提供即时反馈,提升用户体验。通过设置长度限制、图标、字符统计,以及 input 组件暴露内部的方法供父级组件使用,开发人员可以更灵活地实现业务交互。

第 9 章 布局组件

布局（layout）组件是一种用于在网页或应用程序中管理和组织页面布局的组件。通过使用 layout 组件，开发人员可以轻松地划分页面的各个区域，比如头部、侧边栏、主体内容区、列数等分等，并在这些区域中放置其他组件或元素，快速搭建出符合设计要求的页面布局，提供良好的用户体验。

9.1 grid：栅格分栏

栅格分栏（grid）布局通常是指使用网格系统实现响应式布局。通常情况下，栅格分栏布局支持设置列和行，以便进行元素布局。这使得开发人员能够轻松地定位和放置元素，创建各种适应不同屏幕尺寸和设备的布局，如图 9-1 所示。

图 9-1　栅格分栏布局

9.1.1 渲染 grid 组件

出于灵活性和可定制性的考量，第三方 UI 组件库（如 Ant Design、Element Plus 等）中的栅格分栏布局通常选择将页面的宽度划分为 24 份。这样可以提供非常细致的布局控制，使开发人员可以更精确地定义各个元素在页面上的位置和宽度。因此，我们在开发 grid 组件时也遵循 24 份原则，如代码清单 library-09-1 所示。

```
代码清单 library-09-1
1.  > packages/components/row/src/index.vue: template
2.  <component :is="tag" :class="[ns.b()]"><slot /></component>
3.
4.  > packages/components/row/src/index.vue: script
5.  import { useNamespace } from '@ui-library/hook';
```

```
6.   const ns = useNamespace("row");
7.   /** props */
8.   const props = defineProps({
9.     tag: {
10.      type: String,
11.      default: "div",
12.    }
13.  });
14.
15.  > packages/components/row/col/index.vue: template
16.  <component :is="tag" :class="[ns.b(), classCol]"><slot /></component>
17.
18.  > packages/components/row/col/index.vue: script
19.  import { computed } from "vue"
20.  import { useNamespace } from '@ui-library/hook';
21.  const ns = useNamespace("col");
22.  /** props */
23.  const props = defineProps({
24.    tag: {
25.      type: String,
26.      default: "div",
27.    },
28.    span: {
29.      type: Number,
30.      default: 0
31.    }
32.  });
33.
34.  const classCol = computed(() => {
35.    return props.span ? ns.b(props.span) : ''
36.  })
37.
38.  > examples/src/App.vue
39.  <template>
40.    <a-row>
41.      <a-col :span="24">gird 栅栏布局</a-col>
42.    </a-row>
43.  </template>
```

将"行"分成多个"列",在代码上的表现就是用一个"父级"包裹多个"子级",使子级形成"一排",如果子级宽度总和大于父级宽度,则子级会自动换行。因此,我们需要"父级"和"子级"两个组件,可分别将其定义为 row 和 col。由于栅格分栏只是分配不同的等分区域,实际的元素渲染在组件中进行,因此可采用<slot />插槽模式。

在 components 目录下新建 row 和 col 文件夹,作为 grid 组件的父级和子级。row 和 col 两个组件均采用 Vue.js 3 提供的动态组件<component>,并为 is 属性绑定 defineProps 对象的 tag,默认渲染为<div>元素(第 2、9~12、16、24~27 行)。

col 组件的等分逻辑是:检测 defineProps 对象的 span 属性是否存在值(第 28~31 行),如果存在,则变量 classCol 返回结果(第 34~36 行),生成等分的类名(第 16 行),渲染效果如图

9-2 所示。

```
<div class="a-row">
  <div class="a-col a-col-24">gird栅栏布局</div>
</div>
```

图 9-2　为 row、col 组件渲染栅格分栏布局

9.1.2　CSS 弹性布局

栅格分栏布局需使子级组件 col "并排" 排列，为了更好、更方便地实现左对齐、右对齐、两侧分布等效果，需要采用弹性（flex）布局。flex 布局大幅简化了网页布局的复杂性，允许开发人员通过灵活的属性控制，轻松实现各种常见的布局需求，如代码清单 library-09-2 所示。

代码清单 library-09-2
```
1.  > packages/theme/src/row.scss
2.  @use "./common/mixins.scss" as *;
3.  @include b(row) {
4.    display: flex;         // 弹性盒模型
5.    flex-wrap: wrap;       // 允许换行
6.    box-sizing: border-box;
7.  }
8.
9.  > packages/theme/src/col.scss
10. @use "sass:math";
11. @use "./common/mixins.scss" as *;
12. @use "./common/config.scss" as *;
13. @include b(col) {
14.   display: block;
15.   min-height: 1px;
16. }
17.
18. @for $i from 1 through 24 {
19.   .#{$namespace}-col-#{$i} {
20.     $value: math.div(1, 24) * $i * 100 * 1%;
21.     min-width: $value;
22.     flex: 0 0 $value
23.   }
24. }
25.
26. > examples/src/App.vue: template
27. <a-row>
28.   <a-col :span="6"><div class="bg bg-odd"></div></a-col>
29.   <a-col :span="6"><div class="bg bg-even"></div></a-col>
30.   <a-col :span="6"><div class="bg bg-odd"></div></a-col>
31.   <a-col :span="6"><div class="bg bg-even"></div></a-col>
32. </a-row>
33.
```

```
34. > examples/src/App.vue: style
35. .bg { width: 100%; height: 32px; }
36. .bg-odd { background-color: #CCD1DD; }
37. .bg-even { background-color: #E5E7EE; }
```

在9.1.1节中，col是子级组件，row是父级组件。row组件需要使子级组件形成"并排"排列效果，并且能自动换行，因此可将row组件定义为弹性盒模型，并添加属性flex-wrap，允许子级组件自动换行（第3~7行）。在图9-2中，col子级组件生成的等分类名是.a-col-24，其中的.a-col固定不变，数字24是通过span属性配置生成的，24是最大等分数，宽度为100%。由于等分的数量是1~24，因此可采用Sass的@for循环，生成1~24不同等分的百分比（第18~24行）。

变量$namespace是config.scss文件的配置项，先组合固定的字符"-icon"，再组合@for循环的变量$i，即可生成指定的等分类名.a-col-1 ~ .a-col-24（第19行）。然后调用Sass的内置方法math.div，计算出1除以24的结果，再乘以变量$i、100和1%，就可以得到不同等分的百分比宽度（第20行）。最后为属性min-width和flex赋值（第21、22行），渲染效果如图9-3所示。

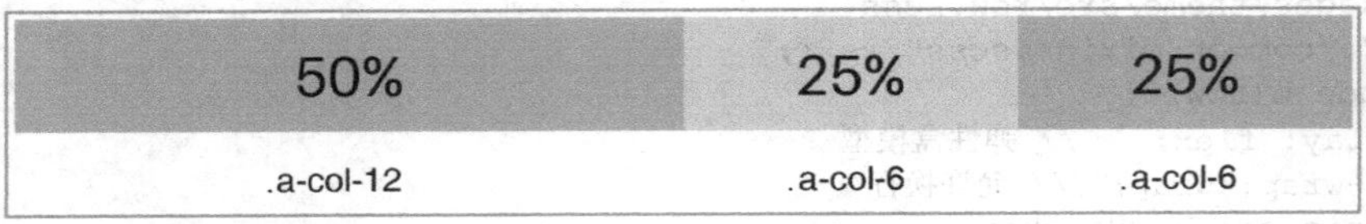

50%	25%	25%
.a-col-12	.a-col-6	.a-col-6

图9-3　为row、col组件渲染不等分宽度

9.1.3　gutter：间距分隔

间距分隔（gutter）属性用于定义栅格单元之间的空白间隔。通过调整gutter属性，开发人员可以灵活地改变栅格单元之间的距离，以满足页面布局的需求。一般来说，gutter属性可以用来设定栅格单元之间的外边距或内边距，使页面具有更合理的布局和更好的美观性，如代码清单library-09-3所示。

```
代码清单 library-09-3
1.  > packages/components/row/src/index.vue: template
2.  <component ...省略代码 :style="[styleMargin]"><slot /></component>
3.
4.  > packages/components/row/src/index.vue: script
5.  import { computed } from "vue"
6.  /** props */
7.  const props = defineProps({
8.    ...省略代码
9.    gutter: { type: Number, default: 0 }
10. });
11. const styleMargin = computed(() => {
12.   const gutter = props.gutter
13.   const value = gutter ? -gutter / 2 + 'px' : null
14.   return value ? { 'marginLeft': value, 'marginRight': value } : {}
15. })
```

```
16.
17. > packages/components/col/src/index.vue: template
18. <component ...省略代码 :style="[styleGutter]"><slot /></component>
19.
20. > packages/components/col/src/index.vue: script
21. import { computed } from "vue"
22. import { useNamespace, useParent } from '@ui-library/hook';
23. const uParent = useParent("row")
24. // 计算间距
25. const styleGutter = computed(() => {
26.   const gutter = uParent.props('gutter')
27.   const value = gutter ? gutter / 2 + 'px' : null
28.   return value ? { 'paddingLeft': value, 'paddingRight': value } : {}
29. })
30.
31. > examples/src/App.vue: template
32. <a-row :gutter="30">
33.   <a-col :span="6"><div class="bg bg-odd"></div></a-col>
34.   <a-col :span="6"><div class="bg bg-even"></div></a-col>
35.   <a-col :span="6"><div class="bg bg-odd"></div></a-col>
36.   <a-col :span="6"><div class="bg bg-even"></div></a-col>
37. </a-row>
```

间距使 col 子级组件之间产生空隙，因此应将 gutter 属性定义添加到 row 组件中。为 defineProps 对象添加 gutter 属性，并设置其为数字类型，默认值为 0（第 9 行）。为 col 子级组件引入 useParent，并调用 props 方法获取父级组件 gutter 属性的值赋给变量 gutter（第 23、26 行），间距的空隙占位是左右各取一半，所以使用三元运算判断变量 gutter 是否存在值，若存在，则先除以 2，再组合单位 px 赋给变量 value；若不存在，则为 null（第 27 行）。最后使用三元运算判断变量 value 存在值时，生成 paddingLeft 左内边距和 paddingRight 右内边距的样式赋给变量 styleGutter（第 25 行），并将变量 styleGutter 绑定到 col 组件上（第 18 行），如图 9-4 所示。

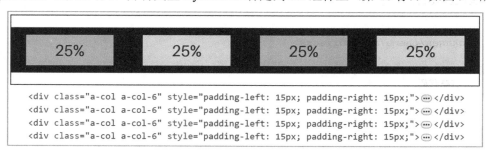

图 9-4　为 row、col 组件渲染间距

从图 9-4 中可以看到，col 组件之间已经生成了 padding-left 和 padding-right 属性，使元素之间产生了间距。但最左边和最右边的元素并没有完全贴边，也就是黑色背景的元素 row 组件没有使子级组件左右两边紧贴左侧和右侧边框。为了使子级组件左右两侧贴边，需要让 row 组件左右两侧各往外偏移 15px，也就是设置 margin-left 和 margin-right 为负数。具体的实现方式

与变量 styleGutter 相似，只是将生成的 CSS 样式改为 marginLeft 和 marginRight（第 11~15、2 行），如图 9-5 所示。

图 9-5　使 col 组件左右贴边

9.1.4　offset：偏移

偏移（offset）属性通常是以栅格单元的单位进行定义的，通过设置 offset 属性将栅格单元向右移动指定的等份。这使得页面布局能够更加丰富多样，满足各种设计需求。通常偏移属性可以与栅格系统的其他属性配合使用，例如与宽度属性结合，实现更灵活的布局效果，如代码清单 library-09-4 所示。

代码清单 library-09-4

```
1.  > packages/components/col/src/index.vue: template
2.  <component :is="tag" :class="[ns.b(), classCol]" :style="[styleGutter]">
3.
4.  > packages/components/col/src/index.vue: script
5.  /** props */
6.  const props = defineProps({
7.    ...省略代码
8.    offset: { type: Number, default: 0 }
9.  });
10.
11. const classOffset = computed(() => {
12.   const offset = props.offset ? ns.b(`offset-${props.offset}`) : {}
13.   return offset
14. })
15.
16. > packages/theme/src/col.scss
17.
18. @for $i from 1 through 24 {
19.   ...省略代码
20.   // offset
21.   .#{$namespace}-col-offset-#{$i} {
22.     margin-left: math.div(1, 24) * $i * 100 * 1%;
23.   }
24. }
25.
26. > examples/src/App.vue: template
27. <a-row :gutter="10">
28.   <a-col :span="6" :offset="6"><div class="bg bg-odd"></div></a-col>
29.   <a-col :span="6"><div class="bg bg-even"></div></a-col>
30.   <a-col :span="6"><div class="bg bg-odd"></div></a-col>
```

```
31.    <a-col :span="6"><div class="bg bg-even"></div></a-col>
32. </a-row>
```

offset 偏移同样为 24 等分，使用指定的等分数便能使元素向右侧偏移相应的百分比。因此，在 col.scss 文件的 @for 指令中循环生成不同等分的百分比的类名 .a-col-offset-1 ~ .a-col-offset-24（第 18~20 行）。为 col.vue 组件的 defineProps 对象定义 offset 属性并设置为数字类型，默认值为 0（第 8 行）。然后定义变量 classOffset，在 computed 对象中使用三元运算判断 offset 是否存在值，存在则返回类名，否则不返回数据（第 11~14 行）。最后将变量 classOffset 绑定到根元素（第 2 行），渲染效果如图 9-6 所示。

图 9-6　使 col 组件产生偏移

9.1.5　justif：对齐

栅格分栏布局的对齐属性用于控制子元素在布局容器中的对齐方式。通常使用 flex 布局实现响应式布局和灵活布局。在实际应用中，根据设计和交互需求，应选择合适的对齐属性来布局页面，如代码清单 library-09-5 所示。

代码清单 library-09-5
```
1.  > packages/components/row/src/index.vue: template
2.  <component :class="[ns.b(), ns.is(`justify-${justify}`, !!justify)]"><slot /></component>
3.
4.  > packages/components/row/src/index.vue: script
5.  /** props */
6.  const props = defineProps({
7.    ...省略代码,
8.    justify: {
9.      type: String,
10.     default: ''
11.   }
12. });
13.
14. > packages/theme/src/row.scss
15. @include b(row) {
16.   display: flex;
17.   flex-wrap: wrap;
18.   box-sizing: border-box;
19.   @include s(justify-center) { justify-content: center; }
20.   @include s(justify-end) { justify-content: flex-end; }
21.   @include s(justify-start) { justify-content: flex-start; }
22.   @include s(justify-space-around) { justify-content: space-around; }
23.   @include s(justify-space-between) { justify-content: space-between; }
```

```
24.      @include s(justify-space-evenly) { justify-content: space-evenly; }
25.  }
26.
27.  > examples/src/App.vue: template
28.  <a-row :gutter="10" :gutter="center">
29.    <a-col :span="3"><div class="bg bg-odd"></div></a-col>
30.    <a-col :span="3"><div class="bg bg-even"></div></a-col>
31.    <a-col :span="3"><div class="bg bg-odd"></div></a-col>
32.    <a-col :span="3"><div class="bg bg-even"></div></a-col>
33.  </a-row>
```

在 flex 模式中，要实现子元素的布局，只需使用属性 justify-content，常用的布局有 start（起始对齐）、center（居中对齐）、end（结束对齐）、space-between（两端对齐）、space-around（均分布）和 space-evenly（等距离分布）等，可根据 6 种对齐方式定义对应的类名（第 19~24 行）。

在 defineProps 对象中定义属性 justify 为字符串类型，默认值为空字符，在根元素中调用 nb.is 方法，当属性 justify 存在值时生成类名（第 2 行）。

9.1.6　gap：行间距

栅格分栏布局使用内边 paddingLeft 和 paddingRight 在水平方向产生间距。如果要产生垂直方向的行间距，可以使用 CSS 3 的 gap 属性，如代码清单 library-09-6 所示。

代码清单 library-09-6

```
1.  > packages/components/row/src/index.vue: template
2.  <component ...省略代码 :style="[styleMargin, styleRowGap]"><slot /></component>
3.
4.  > packages/components/row/src/index.vue: script
5.  /** props */
6.  const props = defineProps({
7.    ...省略代码
8.    gap: {
9.      type: Number,
10.     default: 0
11.   }
12. });
13.
14. const styleRowGap = computed(() => {
15.   return props.gap ? { row-gap: `${props.gap}px` } : {}
16. })
17.
18. > examples/src/App.vue: template
19. <a-row :gutter="10" :gap="10">
20.   <a-col :span="3"><div class="bg bg-odd"></div></a-col>
21.   <a-col :span="3"><div class="bg bg-even"></div></a-col>
22.   <a-col :span="3"><div class="bg bg-odd"></div></a-col>
23.   <a-col :span="3"><div class="bg bg-even"></div></a-col>
24. </a-row>
```

gap 属性可以使父元素的子元素之间产生行间距。在 row 组件的 defineProps 对象中定义属性 gap 为数字类型，并设置默认值为 0（第 8~11 行）。使用 computed 计算属性结合三元运算判断 props.gap 的值是否存在，若存在，则返回 CSS 的 row-gap 属性设置行间距，反之返回空对象（第 15 行），将结果赋给变量 styleRowGap（第 14 行），最后将变量 styleRowGap 绑定到 style（第 2 行），渲染效果如图 9-7 所示。

图 9-7 为子元素设置行间距

9.2 container：容器组件

容器（container）组件通常用于组织和管理页面布局。这些容器可以是栅格分栏、弹性布局等，开发人员可以使用这些组件更轻松地设计和布局页面的外观和结构，如图 9-8 所示。

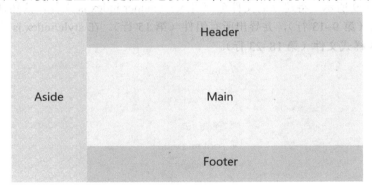

图 9-8 容器组件的布局效果

9.2.1 容器组件的结构

container 组件的结构也可以理解为框架组件结构。container 组件专注于定义应用的整体布局结构，通常由侧边栏（Aside）、头部（Header）、主体（Main）和脚部（Footer）等 4 个部分构成，以确保界面元素的正确和协调排列。在构建 container 组件结构时，将 4 个部分放在同一个文件夹中，如代码清单 library-09-7 所示。

代码清单 library-09-7
```
1.  > packages/components/container/index.js
2.  import { componentInstall } from "@ui-library/utils"
3.  import Container from "./src/container.vue"
4.  import Aside from "./src/aside.vue"
```

```
5.  import Header from "./src/header.vue"
6.  import Footer from "./src/footer.vue"
7.  import Main from "./src/main.vue"
8.  // 提供按需加载的方式
9.  export const AContainer = componentInstall(Container)
10. export const AAside = componentInstall(Aside)
11. export const AHeader = componentInstall(Header)
12. export const AFooter = componentInstall(Footer)
13. export const AMain = componentInstall(Main)
14. // 导出组件
15. export default { AContainer, AAside, AHeader, AFooter, AMain }
16.
17. > packages/components/container/src/style/index.js
18. import "@ui-library/theme/src/initRoot.scss"
19. import "@ui-library/theme/src/container.scss"
20. import "@ui-library/theme/src/aside.scss"
21. import "@ui-library/theme/src/header.scss"
22. import "@ui-library/theme/src/footer.scss"
23. import "@ui-library/theme/src/main.scss"
```

在 container/src 目录下建立 aside.vue、container.vue、footer.vue、header.vue 和 main.vue 共 5 个文件，如图 9-9 所示。将建立的文件全部引入 container/index.js 文件，依然使每个文件提供按需加载的方式（第 9~13 行），并导出所有组件（第 15 行）。在 style/index.js 文件中引入所有子级容器的 CSS 样式文件（第 18~23 行）。

图 9-9　container 组件的文件结构

9.2.2　容器外层

容器外层是指可以包裹子级容器的父级容器，也就是 src 目录下的 container.vue 文件。容器外层可以手动或自动使子级容器形成横向、纵向布局，如图 9-10 所示。

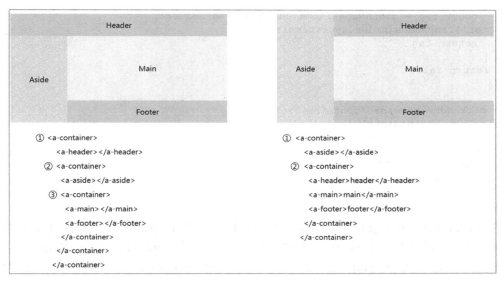

图 9-10 子级容器的布局

在图 9-10 中，容器结构由 Header、Aside、Main、Footer 等组成。其中，序号①②③都是 container 组件，并且都包裹着子级容器。container 组件包裹的子级容器有横向布局和纵向布局。比较特殊的是纵向布局，container 组件只要检测到包裹的子级容器中存在 header 容器或 footer 容器，就会使子级容器形成纵向布局，如代码清单 library-09-8 所示。

代码清单 library-09-8
```
1.  > packages/components/container/src/container.vue: template
2.  <section :class="[ns.b(), ns.is('vertical', isVertical)]"><slot /></section>
3.
4.  > packages/components/container/src/container.vue: script
5.  import { computed, useSlots } from "vue"
6.  // slots
7.  const slots = useSlots()
8.  /** props */
9.  const props = defineProps({
10.    direction: {
11.      type: String,
12.      default: ''
13.    }
14. });
15.
16. const isVertical = computed(() => {
17.   if (props.direction === 'vertical') { return true }
18.   if (props.direction === 'horizontal') { return false }
19.   if(slots && slots.default) {
20.     const slotsNodes = slots.default()
```

```
21.     const tag = slotsNodes.some(nodes => ['a-header',
'a-footer'].includes(nodes.type.name))
22.     return tag
23.   }
24.   return false
25. })
26.
27. > packages/theme/src/container.scss
28. @include b(container) {
29.   display: flex;
30.   flex: 1;
31.   @include s(vertical) {
32.     flex-direction: column;
33.   }
34. }
```

在 container 组件中，要使子级容器呈纵向布局，既可以手动配置，也可以通过程序自动处理。如果采用手动配置，则可以在 defineProps 对象添加属性 direction（第 9~14 行）。然后在 computed 计算属性中判断 props.direction 的值是否为 vertical，如果是，则返回 true，表示纵向布局（第 17 行）；如果 props.direction 的值为 horizontal，则返回 false，表示横向布局，也就是默认布局（第 18 行）。

如果使程序自动处理，则获取 slots 插槽的内容，在确认存在 slots 插槽时执行（第 19 行）。将获取的插槽对象赋给变量 slotsNodes（第 20 行），接着使用数组的 some 方法遍历变量 slotsNodes 集合的 name 属性，也就是子级容器 name 属性的值，使用 includes 判断是否包含 a-header 或 a-footer，如果是，则返回 true（第 21 行）。然后将 computed 计算属性的结果赋给变量 isVertical，调用 ns.is 方法绑定变量 isVertical 并生成类名 is-vertical（第 2 行）。最后通过生成的类名设置属性 flex-direction（第 31~33 行），如图 9-11 所示。

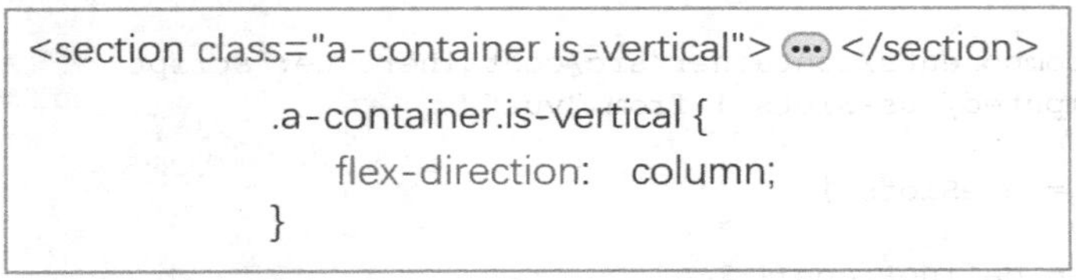

```
<section class="a-container is-vertical"> ⋯ </section>
            .a-container.is-vertical {
                flex-direction:  column;
            }
```

图 9-11　为 container 组件生成类名 is-vertical

本章小结

layout 组件是一种可以包含和组织其他 UI 组件的容器，通常用于划分页面的不同区域和排列页面元素。gird 组件的开发使用了 1~24 等分的计算方式，读者应了解列与列之间的计算方式，以及间距、偏移、行间距等的计算方式。

第 10 章 多选框组件

多选框（checkbox）组件是一种常见的用户界面元素，通常用于让用户选择一个或多个选项。它通常由一个小方框和一个标签组成，用户通过点击方框来选择或取消选择对应的选项。checkbox 组件在网页设计和应用程序开发中被广泛使用，可以用于表单、设置界面、筛选器等场景。

10.1 渲染 checkbox 组件

checkbox 组件的 UI 渲染方式和 input 组件一致，如图 10-1 所示，分别有默认、鼠标悬停、勾选等状态，可变化的属性有边框、背景颜色、文字颜色、尺寸、圆角等。根据不同的状态及可变化属性，我们优先渲染 checkbox 组件的"基础 basis" UI 效果。

图 10-1　checkbox 组件的 UI 渲染

10.1.1 构建组件

checkbox 组件实际上是 HTML 中的 input 元素，其属性为 checkbox。由于现代浏览器的厂商较多，如 Chrome、Firefox、Safari、Edge 等，对 input 元素默认样式的处理方式、渲染引擎有所不同，因此 CSS 的一些特性也存在一定的差异，导致相同的代码在不同浏览器中显示的外观和交互行为不一致。为了解决上述问题，并更好地还原 UI 组件库设计稿的效果，提升 checkbox 组件的交互动画效果，以及用户对选中与非选中的感知和反馈，我们将采用 CSS 样式来定制 checkbox 组件的外观和交互行为，如代码清单 library-10-1 所示。

```
代码清单 library-10-1
1.  > packages/components/checkbox/src/index.vue: template
2.  <component :is="tag" :class="[ns.b()]">
3.    <span :class="[ns.e('input-wrapper')]">
4.      <input :class="[ns.e('input')]" type="checkbox" />
5.      <span :class="[ns.e('inner')]">
6.        <a-icon size="11"><Check /></a-icon>
7.      </span>
```

```
8.        </span>
9.        <span :class="[ns.e('label')]"><slot /></span>
10.   </component>
11.
12. > packages/components/checkbox/src/index.vue: script
13. import Aicon from "@ui-library/components/icon"
14. import { useNamespace, useParent } from '@ui-library/hook';
15. import { Check } from "@ui-library/icons"
16. const ns = useNamespace("checkbox");
17. /** props */
18. const props = defineProps({
19.   tag: { type: String, default: "label" }
20. });
```

HTML 的 label 标签可用于标记表单控件，使表单控件形成关联关系，当点击 label 标签时，触发与 label 标签关联的控件。在定义 checkbox 组件时，使用 Vue.js 3 的动态组件 component，并使属性 is 与变量 tag 绑定，由于变量 tag 的值是 label，因此 checkbox 的根元素会被渲染为 label 标签（第 2 行）。

为 label 标签添加两个子级 span 标签，第一个为多选框左侧的勾选区域，生成类名 a-checkbox__wrapper（第 3~8 行）；第二个为多选框右侧的文本区域，生成类名 a-checkbox__label（第 9 行）。

在类名为 a-checkbox__wrapper 的 span 标签下分别添加子元素 input 和 span（第 4~7 行），分别生成类名 a-checkbox__input 和 a-checkbox__inner。其中，input 元素会被隐藏，而子级的 span 则会模拟多选框的状态效果。

类名为 a-checkbox__label 的 span 标签为 slot 插槽，为渲染的组件传入文本内容，checkbox 组件如图 10-2 所示。

```
□✓广东省
<label class="a-checkbox">
  ▼<span class="a-checkbox__wrapper">
      <input class="a-checkbox__input" type="checkbox">
    ▼<span class="a-checkbox__inner">
        ▶ <i class="a-icon" style="font-size: 11px;">…</i>  flex
      </span>
    </span>
    <span class="a-checkbox__label">广东省</span>
</label>
```

图 10-2 构建 checkbox 组件

提示：checkbox 组件可以使 label 标签包裹 input 组件，形成关联关系。label 标签也可能通过原生属性 for 的值与原生属性 id 的值进行关联。

10.1.2 渲染组件

在 theme/src 目录下新建 checkbox.scss 文件并引入 index.scss 文件，同时引入 checkbox 组件的 style/index.js 文件。根据 UI 组件库设计稿的效果，使用 Sass 封装的 BEM 命名规则书写 CSS 样式，如代码清单 library-10-2 所示。

代码清单 library-10-2

```
> packages/components/checkbox/src/style/index.js
import "@ui-library/theme/src/initRoot.scss"
import "@ui-library/theme/src/checkbox.scss"

> packages/theme/src/checkbox.scss
@include b(checkbox) {
  display: inline-flex;
  align-items: center;
  height: 32px;
  margin-right: 24px;
  cursor: pointer;
  @include e(wrapper) { position: relative; }
  @include e(input) {
    position: absolute;
    margin: 0;
    padding: 0;
    outline: none;
    border: none;
    background: none;
    z-index: 1;
    // opacity: 0;
    // height: 0;
    // width: 0;
  }
  @include e(inner) {
    position: relative;
    display: flex;
    align-items: center;
    justify-content: center;
    height: 20px;
    width: 20px;
    border-radius: 8px;
    border: 2px solid #E2E6F1;
    background-color: #fff;
    color: #fff;
    box-sizing: border-box;
    transition: var(--a-transition);
    @include d(icon) {
      transform: scale(0);
      transition: var(--a-transition);
    }
  }
  @include e(label) {
```

```
44.        margin-left: 8px;
45.        line-height: 1;
46.        color: #4d5059;
47.        font-size: 14px;
48.     }
49. }
```

在 CSS 样式中，第 13~24 行是原生 checkbox 选择框的实现代码，其中"注释掉"了第 21~23 行，这是为了查看原生选择框的效果。第 25~42 行是通过 span 标签模拟实现选择框的代码，其中第 38~41 行是"勾选"图标的实现代码，使用 transform 属性设置缩放比例 scale 为 0，待勾选后将图标的缩放比例设置为 1，就会产生由小到大的放大效果，渲染效果如图 10-3 所示。

图 10-3　渲染 checkbox 组件

10.1.3　样式变量

checkbox 组件的样式变量是私有变量，需要在 componentVar.scss 文件中定义。在该文件中定义 checkboxVar()方法，返回 checkbox 组件的样式变量，然后在 checkbox.scss 文件中调用 set-component-var 混合指令，生成 checkbox 组件私有变量，将生成的变量绑定至 checkbox 组件可变化的属性中，如代码清单 library-10-3 所示，生成的样式变量如图 10-4 所示。

代码清单 library-10-3
```
1.  > packages/theme/src/common/componentVar.scss
2.  @function checkboxVar($type: ''){
3.    $checkbox: (
4.      /** 默认状态 *******************************/
5.      'text-color': ('default': getVarName('text-color', 'primary')),  // label 文本颜色
6.      'border-color': ('default': getVarName('border-color', 'default')), // 边框颜色
7.      'bg-color': ('default': getVarName('color', 'white')),      // 背景颜色
8.      'icon-color': ('default': getVarName('color', 'white')),    // 图标颜色
9.      /** 禁用状态 *******************************/
10.     'disabled-text-color': ('default': getVarName('text-color', 'disabled')),
11.     'disabled-border-color': ('default': getVarName('border-color', 'disabled')),
12.     'disabled-bg-color': ('default': getVarName('color', 'disabled-bg')),
13.     'disabled-icon-color': ('default': getVarName('border-color', 'disabled')),
14.     /** 勾选状态 *******************************/
15.     'checked-text-color': ('default': getVarName('text-color', 'primary')),
16.     'checked-border-color': ('default': getVarName('color', 'primary')),
17.     'checked-bg-color': ('default': getVarName('color', 'primary')),
18.     'checked-icon-color': ('default': getVarName('color', 'white')),
19.     // 组件尺寸
```

```
20.         'size': ('default': getVarName('component-size', 'default')),
21.         // label 文字尺寸
22.         'font-size': ('default': getVarName('font-size', 'default')),
23.     );
24.     @return $checkbox;
25. }
26.
27. @include b(checkbox) {
28.     @include set-component-var('checkbox', checkboxVar());
29.     height: var(#{getVarName('checkbox', 'size')});
30.     &:hover {
31.         #{getVarName('checkbox', 'border-color')}: var(#{getVarName('border-color', 'hover')});
32.     }
33.     @include e(inner) {
34.         border: 2px solid var(#{getVarName('checkbox', 'border-color')});
35.         background-color: var(#{getVarName('checkbox', 'bg-color')});
36.         color: var(#{getVarName('checkbox', 'icon-color')});
37.     }
38.     @include e(label) {
39.         color: var(#{getVarName('checkbox', 'text-color')});
40.         font-size: var(#{getVarName('checkbox', 'font-size')});
41.     }
42. }
```

```
.a-checkbox {
    --a-checkbox-text-color:           ■ var(--a-text-color-primary);
    --a-checkbox-border-color:         □ var(--a-border-color-default);      } 默认状态变量
    --a-checkbox-bg-color:             □ var(--a-color-white);
    --a-checkbox-check-color:          □ var(--a-color-white);

    --a-checkbox-disabled-text-color:      ■ var(--a-text-color-disabled);
    --a-checkbox-disabled-border-color:    □ var(--a-border-color-disabled);  } 禁用状态变量
    --a-checkbox-disabled-bg-color:        □ var(--a-color-disabled-bg);
    --a-checkbox-disabled-check-color:     □ var(--a-border-color-disabled);

    --a-checkbox-checked-text-color:       ■ var(--a-text-color-primary);
    --a-checkbox-checked-border-color:     ■ var(--a-color-primary);          } 勾选状态变量
    --a-checkbox-checked-bg-color:         ■ var(--a-color-primary);
    --a-checkbox-checked-icon-color:       □ var(--a-color-white);

    --a-checkbox-size:       var(--a-component-size-default);
    --a-checkbox-font-size:  var(--a-font-size-default);
}
```

图 10-4　checkbox 组件的样式变量

10.2　theme：主题

主题是 UI 组件库的全局规范，涉及主题渲染的任何组件都是从变量$types 中读取的。6.1 节已经定义了变量$types 的值为 primary、success、warning、error，因此通过循环$types 属性可以自动生成主题变量。

10.2.1 生成主题变量

在 componentVar.scss 文件的 checkboxVar 方法中接收参数$type,并为可变化的属性添加 type,所有 type 均使用 getVarName 方法获取 UI 组件库的全局变量名称,如代码清单 library-10-4 所示。

代码清单 library-10-4
```scss
1.  > packages/theme/src/common/componentVar.scss
2.  @function checkboxVar($type: ''){
3.    $checkbox: (
4.      /** 默认状态 ******************************/
5.      'text-color': ('type': getVarName('color', $type)),
6.      'border-color': ('type': getVarName('color', $type)),
7.      'bg-color': ('type': getVarName('color', $type)),
8.      'icon-color': ('type': getVarName('color', 'white')),
9.      /** 禁用状态 ******************************/
10.     'disabled-text-color': ('type': getVarName('color', $type, 'light-5')),
11.     'disabled-border-color': ('type': getVarName('color', $type, 'light-5')),
12.     'disabled-bg-color': ('type': getVarName('color', $type, 'light-5')),
13.     'disabled-icon-color': ('type': getVarName('color', 'white')),
14.     /** 勾选状态 ******************************/
15.     'checked-text-color': ('type': getVarName('color', $type)),
16.     'checked-border-color': ('type': getVarName('color', $type)),
17.     'checked-bg-color': ('type': getVarName('color', $type)),
18.     'checked-icon-color': ('type': getVarName('color', 'white')),
19.   );
20.   @return $checkbox;
21. }
```

在 checkbox 组件"基础 basis"类型的 UI 效果中,可变化的属性有 text-color、border-color、bg-color、icon-color,会发生变化的状态有默认、禁用、勾选状态。因此,需要为这几个变量配置 type 属性,并使用 getVarName 混合指令获取:root 全局变量。

10.2.2 渲染主题

6.2 节中已经定义了属性$types 并将其作为 UI 组件库主题色的变量,因此在渲染 checkbox 组件的主题时,可通过循环属性$types 的值生成 checkbox 组件的主题色,如代码清单 library-10-5 所示。

代码清单 library-10-5
```scss
1.  > packages/theme/src/checkbox.scss
2.  @include b(checkbox) {...省略代码}
3.  // 生成主题变量
4.  @each $type in $types {
5.    $className: '.a-checkbox--' + $type;
6.    #{$className} {
7.      @include s((checked)) {
8.        @include set-component-var('checkbox', checkboxVar($type), 'type');
```

```
 9.     }
10.   };
11. }
12.
13. > packages/components/checkbox/src/checkbox.vue: template
14. <component:class="[ns.b(), ns.m(type), ns.is('checked', true)]"> ...省略代码
15.
16. > packages/components/checkbox/src/checkbox.vue: script
17. const props = defineProps({
18.   ...省略代码,
19.   type: { type: String, default: "primary" },
20. });
```

为 checkbox 组件生成主题色时需要注意，checkbox 组件在"已勾选"状态下才会渲染主题色，并且"已勾选"状态下的默认颜色是 primary。因此，还需要为 checkbox 组件增加 is-checked 类名来标记 checkbox 组件已被勾选（第 19 行）。

在 checkbox.scss 文件中使用 @each 循环变量 $types（第 4 行），完成类名拼接后赋给 $className（第 5 行），再调用 mixins.scss 的混合指令 "s((checked))" 生成主题色的变量（第 7 行）。最后在 checkbox.vue 文件中调用 ns.m 和 ns.is 两个方法生成类名（第 14 行），渲染效果如图 10-5 所示。

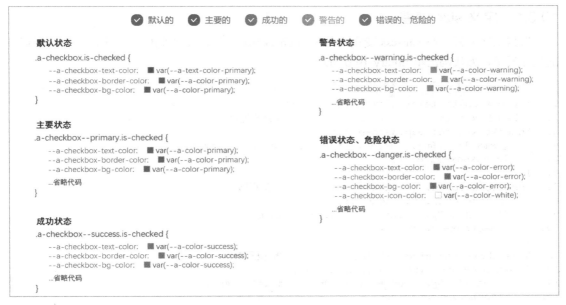

图 10-5 为 checkbox 组件生成主题色

从图 10-5 中可以看到，在 checkbox 组件被勾选时，每种主题色的状态都添加了类名 is-checked，通过类名 is-checked 可读取对应主题色的变量，再对 checkbox 组件可变化的私有变量进行覆盖，即可修改主题色。

10.3 size：尺寸

checkbox 组件的 size 属性和其他组件的 size 属性存在一定的区别，如图 10-6 所示，其中标注了"多选框尺寸"、"图标"和"文字"3 种类型。"文字"是 UI 组件库的全局变量$font-size，"多选框尺寸"和"图标"则是 checkbox 组件独有的属性尺寸，包含"宽度"、"圆角"和"图标"。因此，需要额外定义属于 checkbox 组件的属性变量。

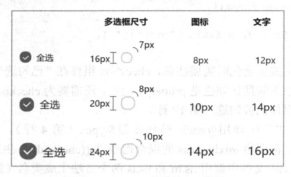

图 10-6　checkbox 组件的 size 属性

10.3.1　定义 size 变量

根据上述分析，在 var.scss 变量文件中为 checkbox 组件定义"宽度"、"圆角"和"图标"的尺寸规范，如代码清单 library-10-6 所示。

```scss
代码清单 library-10-6
> packages\theme\src\common\var.scss
$checkbox-size: () !default;          // 多选框宽度
$checkbox-size: map.deep-merge(
    ('small': 16px, 'default': 20px, 'large': 24px ),
    $checkbox-size
);
$checkbox-round: () !default;         // 多选框圆角
$checkbox-round: map.deep-merge(
    ('small': 7px, 'default': 8.5px, 'large': 10.5px),
    $checkbox-round
);
$checkbox-icon-size: () !default;     // 多选框图标
$checkbox-icon-size: map.deep-merge(
    ('small': 8px, 'default': 10px, 'large': 14px),
    $checkbox-icon-size
);

/** 全局配置 */
$global: (
    ...省略代码
    'checkbox-size': $checkbox-size,
```

```
22.        'checkbox-round': $checkbox-round,
23.        'checkbox-icon-size': $checkbox-icon-size,
24.    );
25.
26.    > packages/theme/src/common/componentVar.scss
27.    @function checkboxVar($type: ''){
28.      $checkbox: (
29.        'check-size': ('default': getVarName('checkbox-size', 'default')),
30.        'check-round': ('default': getVarName('checkbox-round', 'default')),
31.        'check-icon-size': ('default': getVarName('checkbox-icon-size', 'default')),
32.      );
33.      @return $checkbox;
34.    }
35.
36.    > packages/theme/src/checkbox.scss
37.    @include e(inner) {
38.      height: var(#{getVarName('checkbox', 'check-size')});
39.      width: var(#{getVarName('checkbox', 'check-size')});
40.      border-radius: var(#{getVarName('checkbox', 'check-round')});
41.      @include d(icon) {
42.        font-size: var(#{getVarName('checkbox', 'check-icon-size')});
43.      }
44.    ...省略代码
45.    @each $s in $sizes {
46.      @include m(size, $s){
47.        #{getVarName('checkbox', 'font-size')}: var(#{getVarName('font-size', $s)});
48.        #{getVarName('checkbox', 'size')}: var(#{getVarName('component-size', $s)});
49.        #{getVarName('checkbox', 'checkbox-size')}: var(#{getVarName('checkbox-size', $s)});
50.        #{getVarName('checkbox', 'checkbox-round')}: var(#{getVarName('checkbox-round', $s)});
51.        @include d(icon) {
52.          #{getVarName('checkbox', 'checkbox-icon-size')}: var(#{getVarName('checkbox-icon-size', $s)});
53.        }
54.      }
55.    }
56. }
```

在 var.scss 全局变量文件中定义变量$checkbox-size、$checkbox-round、$checkbox- icon-size，分别对应多选框的尺寸、圆角、图标尺寸（第 2~16 行），然后将其添加到变量$global 中（第 21~23 行）。在 componentVar.scss 文件中定义 checkbox 组件的私有变量 check-size、check-round、check-icon-size，分别调用 getVarName 混合指令读取 var.scss 的全局变量（第 29~31 行）。最后将私有变量添加到可变化的属性 height、width、border-radius 和 font-size 中（第 38~43 行）。

checkbox 组件尺寸的变化方法与 button 组件一致，使用 mixins.scss 文件的修饰器方法 "m" 生成类名 a-checkbox--size_small、a-checkbox--size_default 和 a-checkbox--size_large，再通过类

名读取 var.scss 文件的全局变量，覆盖私有变量的值（第 45~55 行）。

10.3.2 配置 size 变量

在为 checkbox 组件生成 size 私有变量后，接下来应在 defineProps 对象中定义 size 属性接收父级组件传入的值，并更新到 ns.m 方法中，以改变 checkbox 组件的尺寸，如代码清单 library-10-7 所示，渲染效果如图 10-7 所示。

```
代码清单 library-10-7
1.  > packages/components/checkbox/src/checkbox.vue: template
2.  <component :is="tag" :class="[...省略代码, ns.m('size', size),]">
3.
4.  > packages/components/checkbox/src/checkbox.vue: script
5.  const props = defineProps({
6.    ...省略代码,
7.    size: {type: String, default: "" },
8.  });
9.
10. > examples/src/App.vue
11. <a-checkbox type="primary" size="small">小的</a-checkbox>
12. <a-checkbox type="success">默认的</a-checkbox>
13. <a-checkbox type="danger" size="large">大的</a-checkbox>
```

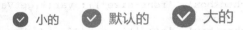

<label class="...省略代码 is-checked a-checkbox--size_small"> ⋯ </label>
<label class="...省略代码 is-checked a-checkbox--size_default"> ⋯ </label>
<label class="...省略代码 is-checked a-checkbox--size_large"> ⋯ </label>

图 10-7 改变 checkbox 组件的尺寸

10.4 composables：组合式函数

composables 是 Vue.js 3 的组合式 API，用来封装和复用有状态逻辑的函数。composable 是一种设计模式，它允许开发人员将可复用的逻辑抽象成单独的函数，这些函数可以在组件之间共享。

为什么要在 checkbox 组件部分学习 composables 组合式函数呢？回看前文中的 button、icon、input、layout 等组件的开发过程，基本上都是通过父级组件传入属性来改变 UI 组件的渲染效果的，并没有太多数据上的交互变化。而 checkbox 组件与此有明显的不同之处，从 UI 组件库设计稿中可以看到 checkbox 组件具有"全选"类型，如图 10-8 所示。

图 10-8 checkbox 组件的"全选"类型

"全选"多选框的状态会随着第二行中多个选项数据勾选状态的变化而变化。此外，UI 组件库设计稿中还有"禁用"和"异步交互"等交互效果。如果把这些交互效果都写到 checkbox.vue 文件中，那么不但会导致 checkbox.vue 文件的代码量越来越大，而且会使后期维护难上加难。为了使代码更有逻辑性，更好地维护代码，下面以 checkbox 组件的 size 属性为例，使用 composables 组合式函数的方式来开发。

10.4.1 定义状态模块

composables 组合式函数，简单理解就是将各种不同的逻辑模块拆分成独立的函数，然后在需要使用这些模块的函数时，将其引入即可。例如，本节要开发的 checkbox 组件的 size 属性属于一种交互状态，因此可以建立 use-checkbox-state.js 文件单独处理这些状态的逻辑，如代码清单 library-10-8 所示。

代码清单 library-10-8
```
1. > packages/components/checkbox/src/composables/use-checkbox-state.js: js
2. import { computed } from "vue"
3. export function useCheckboxState({ props }){
4.   const checkboxSize = computed(() => props.size)
5.   return { checkboxSize }
6. }
```

在 src 目录下新建 composables 目录，在该目录下新建 use-checkbox-state.js 文件。在文件中定义 useCheckboxState 函数（第 3 行），在函数中定义变量 checkboxSize，并使用 computed 计算属性返回 props.size（第 4 行），然后返回变量 checkboxSize（第 5 行），最后使用 export 关键字导出 useCheckboxState 函数。

10.4.2 应用状态模块

composables 组合式函数的应用很简单，通过使用 import 关键字将函数导入文件，再解构出函数内部定义的变量，如下代码所示。

代码
```
1. > packages/components/checkbox/src/checkbox.vue: template
2. <component :is="tag" :class="[...省略代码, ns.m('size', checkboxSize),]">
3.
4. > packages/components/checkbox/src/checkbox.vue: script
5. import { useCheckboxState } from "./composables/use-checkbox-state"
6. const { checkboxSize } = useCheckboxState({props})
```

在上述代码清单中，将 use-checkbox-state.js 的 useCheckboxState 组合式函数导入 checkbox.vue 文件（第 5 行），在调用 useCheckboxState 函数时，以 JSON 对象格式传入 props 对象，然后解构出变量 checkboxSize（第 6 行），再将其绑定到组件的 ns.m 方法中（第 2 行）。如果有多个模块文件都使用 import 关键字导入，则再解构出函数中的变量即可。

在项目开发中，上述方式是最常见的，如果在项目中使用了很多依赖包，那么可以使用 import 关键字将需要用到的包一个个导入后再进行使用，当然，这样做的前提是每个包都是独立的。如果不同模块之间有一定的联系，那么笔者更喜欢另一种导入方式，如图 10-9 所示。

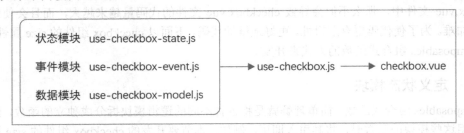

图 10-9　导入多个 composables 组合式函数

图 10-9 中定义了状态模块、事件模块和数据模块，将这 3 个模块统一导入 use-checkbox.js 文件，然后将 use-checkbox.js 文件中的所有函数和变量导入 checkbox.vue 文件。对于 checkbox.vue 文件而言，这种方式只有一个数据来源，无须考虑多个模块的数据来源，如代码清单 library-10-9 所示。

```
代码清单 library-10-9
1.  > packages/components/checkbox/src/composables/use-checkbox.js
2.  import { useCheckboxState } from "./use-checkbox-state"
3.  export function useCheckbox({props}){
4.      const { checkboxSize } = useCheckboxState({props})
5.      return { checkboxSize }
6.  }
7.
8.  > packages/components/checkbox/src/checkbox.vue: script
9.  import { useCheckbox } from "./composables/use-checkbox"
10. const { checkboxSize } = useCheckbox({props})
```

在 composables 目录下新建 use-checkbox.js 文件，use-checkbox.js 同样使用 export 关键字导出 useCheckbox 函数。然后将状态模块文件 use-checkbox-state.js 导入 use-checkbox.js 文件（第 2 行），在 useCheckbox 组合式函数中解构变量 checkboxSize 并返回（第 4、5 行）。最后在 checkbox.vue 文件中引入 use-checkbox.js 文件并解构变量 checkboxSize（第 9、10 行）。

10.5　disabled：禁用

禁用（disabled）交互也属于一种状态，可以在 use-checkbox-state.js 文件中定义禁用的变量，接收 props.disabled 的值，返回交互结果，如代码清单 library-10-10 所示。

代码清单 library-10-10

```js
1.  > packages/components/checkbox/src/composables/use-checkbox-state.js
2.  export function useCheckboxState({ props }){
3.      const checkboxSize = computed(() => props.size)
4.      const isDisabled = computed(() => props.disabled)
5.      return { checkboxSize, isDisabled }
6.  }
7.
8.  > packages/components/checkbox/src/composables/use-checkbox.js
9.  import { useCheckboxState } from "./use-checkbox-state"
10. export function useCheckbox({ props }){
11.     // useCheckboxState
12.     const { checkboxSize, isDisabled } = useCheckboxState({ props })
13.     return { checkboxSize, isDisabled }
14. }
15.
16. > packages/components/checkbox/src/checkbox.vue: template
17. <component :is="tag" :class="[...省略代码, ns.is('disabled', isDisabled)]">
18.   <span :class="[ns.e('wrapper')]">
19.      <input :class="[ns.e('input')]" type="checkbox" :disabled="isDisabled" />
20.   </span>
21. </component>
22.
23. > packages/components/checkbox/src/checkbox.vue: script
24. /** props */
25. const props = defineProps({
26.    ...省略代码
27.    disabled: Boolean
28. });
29. const { checkboxSize, isDisabled } = useCheckbox({props})
30.
31. > packages/theme/src/checkbox.scss
32. @each $type in $types {
33.    $className: '.a-checkbox--' + $type;
34.    #{$className} {
35.      @include s((checked)) {...省略代码 }
36.      @include s((disabled)) {
37.         #{getVarName('checkbox', 'border-color')}: var(#{getVarName('checkbox', 'disabled-border-color')});
38.         #{getVarName('checkbox', 'bg-color')}: var(#{getVarName('checkbox', 'disabled-bg-color')});
39.         #{getVarName('checkbox', 'text-color')}: var(#{getVarName('checkbox', 'disabled-text-color')});
40.      }
41.    };
42. }
```

在 use-checkbox-state.js 文件中定义变量 isDisabled，使用 computed 计算属性返回

props.disabled 属性的值（第 4、5 行）。然后将变量 isDisabled 解构至 useCheckbox.js 主入口文件（第 12、13 行），接着在 checkbox.vue 文件的 useCheckbox 函数中解构变量 isDisabled（第 29 行），最后使用 ns.is 方法绑定变量 isDisabled（第 17 行），同时为 input 标签中的 disabled 属性绑定变量 isDisabled（第 19 行）。

对于 checkbox 组件的样式，需要在循环主题时调用 mixins.scss 文件的混合指令"s"生成禁用样式的效果（第 36~40 行）。这是因为在已勾选状态下的多选框会生成 is-checked 类名的变量值，需要使用相同权重类名的样式对 is-checked 类名的变量值进行覆盖，从而重写变量的颜色值。勾选与未勾选状态的 checkbox 组件的禁用效果如图 10-10 所示。

图 10-10　勾选与未勾选状态的 checkbox 组件的禁用效果

10.6　group：多选框组

多选框组允许用户选择一个或多个选项，并且每个多选框都是独立的，用户可以单独选择或取消选择某个选项。这种界面元素通常在表单或设置界面中使用，以便用户可以在特定主题或配置场景下进行多项选择。

checkbox 组件多选框组的开发方式和 button 组件类似，需要增加一个 checkboxGroup.vue 文件，并以插槽的方式渲染 checkbox 组件，如代码清单 library-10-11 所示。

```
代码清单 library-10-11
1.  > packages/components/checkbox/src/checkboxGroup.vue
2.  <template>
3.    <div :class="[ns.b()]">
4.      <slot />
5.    </div>
6.  </template>
7.  <script>export default { name: "a-checkbox-group" };</script>
8.  <script setup>
9.  import { useNamespace } from '@ui-library/hook';
10. const ns = useNamespace("checkbox-group");
11. </script>
12.
13. > packages/components/checkbox/index.js
14. import CheckboxGroup from "./src/checkboxGroup.vue"
15. export const ACheckboxGroup = componentInstall(CheckboxGroup)
16. // 导出组件
17. export default { Checkbox, CheckboxGroup }
18.
19. > packages/components.js
```

```
20.    ...省略代码
21.    import { ACheckbox, ACheckboxGroup } from "./components/checkbox/index.js"
22.    export default [...省略代码, ACheckbox, ACheckboxGroup]
23.
24.    > examples/src/App.vue
25.    <a-checkbox-group>
26.      <a-checkbox disabled>默认的</a-checkbox>
27.      ...省略代码
28.    </a-checkbox-group>
```

checkboxGroup.vue 多选框组的组件文件与 checkbox.vue 文件同级，同样是引入 index.js，并使用 componentInstall 函数按需加载组件（第 15 行），然后在 packages/components.js 文件中导出该组件，将其作为全局注册组件使用（第 21、22 行）。最后在 examples 演示包中使用 <a-checkbox-group>（第 26~28 行），渲染效果如图 10-11 所示。

```
<div class="a-checkbox-group">
  ▶ <label class="a-checkbox a-checkbox--default ...省略代码"> ⋯ </label> flex
  ▶ <label class="a-checkbox a-checkbox--primary ...省略代码"> ⋯ </label> flex
  ▶ <label class="a-checkbox a-checkbox--success ...省略代码"> ⋯ </label> flex
  ▶ <label class="a-checkbox a-checkbox--warning ...省略代码"> ⋯ </label> flex
  ▶ <label class="a-checkbox a-checkbox--danger ...省略代码"> ⋯ </label> flex
</div>
```

图 10-11　checkbox 组件多选框组

10.6.1　provide、inject：通信

Vue.js 3 中的 provide 和 inject 是一对用于父级组件向其所有子级组件注入依赖的高级选项。provide 和 inject 选项允许用户在祖先组件中提供数据，并在子孙组件中接收数据。这样可以避免通过 props 层层传递数据，而是可以直接在组件树中的任何地方访问所需的数据。

checkboxGroup 组件和 buttonGroup 组件都涉及父级和子级组件的概念，button 组件与 buttonGroup 组件通过 useParent 函数递归获取父级组件的数据实现通信，递归的方式会消耗比较多的性能，并且使用方式复杂。因此，在 checkboxGroup 组件中将使用 provide 和 inject 实现父级和子级组件的数据通信，如代码清单 library-10-12 所示。

代码清单 library-10-12
```
1.  > packages/components/checkbox/src/checkboxGroup.vue: script
2.  import { provide } from "vue";
3.  provide('checkboxGroupKey', {size: 'large'})
4.
5.  > packages/components/checkbox/src/checkbox.vue: script
6.  import { inject } from "vue"
7.  const checkboxGroupKey = inject('checkboxGroupKey', undefined)
8.  console.log('checkboxGroupKey', checkboxGroupKey)
```

provide 和 inject 的用法很简单，在 checkboxGroup 组件中使用 provide 对象注入数据（第 3

行），第一个参数是自定义 key，第二个参数是需要注入的数据，可以是任意类型的。

在 checkbox 组件中使用 inject 接收数据（第 7 行），第一个参数是 provide 对象注入的 key，第二个参数是默认值，可设置为 undefined（可选），然后赋给变量 checkboxGroupKey（第 7 行），控制台的输出结果如图 10-12 所示。

checkboxGroupKey ▶ {size: 'large'}

图 10-12 inject 接收 provide 的数据

10.6.2 size：尺寸

checkboxGroup 组件的父级组件和子级组件配置属性的处理逻辑与 button 组件类似。如果在 checkboxGroup 组件中设置 size 属性，那么所有 checkbox 组件的尺寸都会随着父级组件尺寸的变化而变化，如代码清单 library-10-13 所示。

代码清单 library-10-13

```
1.  > packages/components/checkbox/src/constant.js
2.  export const CHECKBOX_GROUP_KEY = Symbol('checkboxGroupKey')
3.
4.  > packages/components/checkbox/src/checkboxGroup.vue
5.  import { provide, toRefs } from "vue";
6.  import { CHECKBOX_GROUP_KEY } from "../constant"
7.  const props = defineProps({
8.    size: { type: String, default: 'default' }
9.  })
10. provide(CHECKBOX_GROUP_KEY, {...toRefs(props)})
11.
12. > packages/components/checkbox/src/composables/use-checkbox-group.js
13. import { inject } from "vue"
14. import { CHECKBOX_GROUP_KEY } from "../constant"
15. export function useCheckboxGroup(){
16.   const checkboxGroupKey = inject(CHECKBOX_GROUP_KEY, undefined)  // checkboxGroup 数据
17.   const isGroup = checkboxGroupKey !== undefined    // checkboxGroup 是否存在
18.   return { checkboxGroupKey, isGroup }
19. }
20.
21. > packages/components/checkbox/src/composables/use-checkbox.js
22. import { useCheckboxGroup } from "./use-checkbox-group"
23. export function useCheckbox({ props }){
24.   const { checkboxGroupKey, isGroup } = useCheckboxGroup()
25.   const { checkboxSize, isDisabled } = useCheckboxState({ props, checkboxGroupKey, isGroup })
26. }
27.
28. > packages/components/checkbox/src/composables/use-checkbox-state.js
29. export function useCheckboxState({ props, checkboxGroupKey, isGroup }){
```

```
30.    const checkboxSize = computed(() => {
31.      return props.size || (isGroup ? checkboxGroupKey?.size.value : props.size)
32.    })
33.  }
```

provide 和 inject 对象使用相同的 key 注入和接收数据，因此我们可以定义 constant.js 常量文件，预先定义两者使用的 key（第 2 行），然后将其引入 checkboxGroup 组件（第 6、10 行）。checkboxGroup 组件使用 provide 对象注入数据时使用了 Vue.js 3 的 toRefs 对象，原因在于 provide 和 inject 不具有响应式的特性，需要使用 toRefs 对象解决数据响应式问题（第 10 行）。

为了判断是否存在 checkboxGroup 组件，可以在 composables 目录下新建 use-checkbox-group.js 模块文件，作为父级"组"的判断条件。同样引入 inject 对象和常量 CHECKBOX_GROUP_KEY（第 13、14 行），接着定义组合式函数 useCheckboxGroup，并使用 inject 对象接收数据，设置默认值为 undefined 再赋给变量 checkboxGroupKey（第 16 行），然后判断 checkboxGroupKey 是否等于 undefined，若不等于，则表示存在父级"组"，接着赋给变量 isGroup（第 17 行），最后返回 checkboxGroupKey 和 isGroup（第 18 行）。

在 use-checkbox.js 主入口文件中引入 use-checkbox-group（第 22 行），并解构 checkboxGroupKey 和 isGroup（第 24 行），然后将两者传入 useCheckboxState 函数（第 25 行）。useCheckboxState 组合式函数的变量 checkboxSize 使用"或"运算判断 props.size 是否存在，如果 props.size 不存在，再使用三元运算判断 isGroup 是否存在，若存在，则使用父级"组"的 size 属性，否则仍然使用 props.size（第 31 行）。

10.7　model：数据绑定

checkbox 组件和 checkboxGroup 组件的数据绑定都使用组合式函数的方式处理，但两者有一定的区别。checkbox 组件绑定的数据类型为基础数据类型，如 String、Number、Boolean。checkboxGroup 组件绑定的数据类型为对象数据类型 Array，如代码清单 library-10-14 所示。

代码清单 library-10-14
```
1.  > packages/components/checkbox/src/checkbox.vue
2.  const checkboxModel = defineModel({type: [String, Number, Boolean], default: ''})
3.
4.  > packages/components/checkbox/src/checkboxGroup.vue
5.  const groupModel = defineModel({type: Array, default: () => ([])})
6.  provide(CHECKBOX_GROUP_KEY, {...toRefs(props), groupModel})
```

10.7.1　数据模块定义

数据绑定属于独立的逻辑，同样在 composables 目录下建立文件和处理数据，如代码清单 library-10-15 所示。

代码清单 library-10-15
```
1.  > packages/components/checkbox/src/checkbox.vue: template
```

```
2.    <input :class="[ns.e('input')]" v-model="model" type="checkbox" />
3.
4.  > packages/components/checkbox/src/checkbox.vue: script
5.    const { checkboxSize, isDisabled, model } = useCheckbox({ props, checkboxModel })
6.
7.  > packages/components/checkbox/src/composables/use-checkbox.js
8.  import { useCheckboxModel } from "./use-checkbox-model"
9.  export function useCheckbox({ props, checkboxModel }){
10.    const { model } = useCheckboxModel({ props, checkboxModel, checkboxGroupKey, isGroup })
11.    return {...省略代码, model }
12. }
13.
14. > packages/components/checkbox/src/composables/use-checkbox-model.js
15. import { computed } from "vue"
16. export function useCheckboxModel({ props, checkboxModel, checkboxGroupKey, isGroup }){
17.    const model = computed({
18.      get(){
19.        return isGroup ? checkboxGroupKey.groupModel.value : checkboxModel.value
20.      }
21.    })
22.    return { model }
23. }
```

为 checkbox 组件定义双向绑定数据变量 checkboxModel，将该变量传入 useCheckbox 函数（第 5 行）。在 composables 目录下新建 use-checkbox-model.js 文件，定义 useCheckboxModel 组合式函数，并接收 props、checkboxModel、checkboxGroupKey 和 isGroup 这 4 个参数（第 16 行）。组合式函数中的变量 model 使用三元运算判断 isGroup 是否存在，若存在，则使用父级"组"双向绑定数组的对象 groupModel，反之使用 checkboxModel（第 18~20 行），最后返回变量 model（第 22 行）。

在 use-checkbox.js 主入口文件中引入 use-checkbox-model.js 文件解构并返回 model（第 8、10、11 行），在 checkbox.vue 文件中也可以解构 model，并在 input 标签中使用 v-model 绑定 model（第 2、5 行）。

10.7.2 数据测试

由于 checkbox 和 checkboxGroup 两个组件绑定的数据类型分别为"基础类型"和"对象类型"，因此需要测试两个组件绑定数据的正确性，如代码清单 library-10-16 所示。

代码清单 library-10-16

```
1.  > examples/src/App.vue: template
2.  <!-- 多选框组 -->
3.  <a-checkbox-group v-model="value1">
4.    <a-checkbox>广东省</a-checkbox>
5.    <a-checkbox>广西省</a-checkbox>
6.  </a-checkbox-group>
```

```
7.
8.   <!-- 多选框   -->
9.   <a-checkbox v-model="value2">广东省</a-checkbox>
10.
11.  > examples/src/App.vue: script
12.  const value1 = ref(['guangdong', 'guangxi'])
13.  const value2 = "guangdong"
14.
15.  packages/components/checkbox/src/composables/use-checkbox-model.js
16.  const model = computed(...省略代码)
17.  console.log('model', model.value)
```

在 examples 演示库中渲染"多选框组"和"多选框"(第 3~6、9 行),并使用 v-model 绑定变量 value1 和 value2(第 12、13 行)。在 use-checkbox-model.js 文件中打印输出结果,如图 10-13 所示。

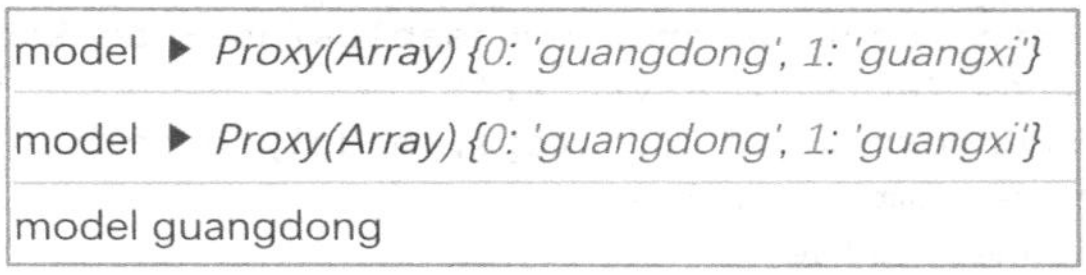

| model ▶ *Proxy(Array) {0: 'guangdong', 1: 'guangxi'}* |
| model ▶ *Proxy(Array) {0: 'guangdong', 1: 'guangxi'}* |
| model guangdong |

图 10-13 测试绑定的数据

10.8 checked:勾选

勾选(checked)是用户在点击多选框组时产生的选中与非选中的交互。在 10.1 节中,渲染 checkbox 组件时调用了 ns.is('checked', true)方法默认将多选框设置为勾选状态,现在通过绑定的数据来实现是否勾选状态。由于多选框组件分为 checkbox 组件和 checkboxGroup 组件,两个组件的数据类型不一样,因此需要分开处理。

勾选与非勾选是一种状态变更,可以在 use-checkbox-state.js 文件中处理,如代码清单 library-10-17 所示。

代码清单 library-10-17
```
1.   > packages/components/checkbox/src/composables/use-checkbox-state.js
2.   const isDisabled = ...省略代码
3.   /** checked */
4.   const isChecked = computed(() => true)
5.   return {checkboxSize, isDisabled, isChecked }
6.
7.   > packages/components/checkbox/src/composables/use-checkbox.js
8.   const { checkboxSize, isDisabled, isChecked } = useCheckboxState({ ...省略代码 })
9.   return { ...省略代码, isChecked }
10.
11.  > packages/components/checkbox/src/checkbox.vue
12.  <component :is="tag" :class="[...省略代码, ns.is('checked', isChecked)]">
13.  const { ...省略代码, isChecked } = useCheckbox({ props, checkboxModel })
```

在 use-checkbox-state.js 文件中定义变量 isChecked 并返回（第 4、5 行），在 use-checkbox.js 主入口文件中解构 isChecked 并返回（第 8、9 行）。最后在 checkbox.vue 文件中将 isChecked 绑定至 ns.is('checked', isChecked)方法（第 12、13 行），这样就可以通过 use-checkbox-state.js 文件来实现后续的业务逻辑。

10.8.1 checkbox 组件

当 input 标签的 type 属性的值为 checkbox 时，则为一个多选框。多选框的勾选与非勾选状态与 input 标签的 checked 属性有关，当 checked 的值为 true 时，表示勾选状态，反之表示非勾选状态。因此，我们可以根据 v-model 传入的数据类型判断 checkbox 组件是否为勾选状态，如代码清单 library-10-18 所示。

代码清单 library-10-18

```
1.  > packages/components/checkbox/src/composables/use-checkbox-state.js
2.  import { types } from "@ui-library/utils"
3.  export function useCheckboxState({ props, model, checkboxGroupKey, isGroup }){
4.    const isChecked = computed(() => {
5.      const value = model.value
6.      // Boolean
7.      if(types().isBoolean(value)) {
8.        return value
9.      }
10.     return false
11.   })
12. }
13.
14. packages/components/checkbox/src/composables/use-checkbox-model.js
15. export function useCheckboxModel({ ...省略代码 }){
16.   const model = computed({
17.     get(){
18.       return isGroup ? checkboxGroupKey.groupModel.value : checkboxModel?.value
19.     },
20.     set(val){
21.       checkboxModel.value = val
22.     }
23.   })
24. }
25.
26. examples/src/App.vue
27. <a-checkbox v-model="value2">广东省</a-checkbox>
28. const value2 = ref(true)
```

在 isChecked 对象中获取 model.value，也就是 v-model 绑定的数据，将其赋给变量 value（第 5 行）。然后使用 types()的 isBoolean 方法判断数据是否为布尔类型，如果是，则返回变量 value（第 7~9 行），否则默认返回 false（第 10 行）。

在 use-checkbox-model.js 文件的 model 的 computed 计算属性中添加 set 对象（第 20~22 行）。

如果 v-model 的值发生变化，则进入 set 对象，将新的值赋给 checkboxModel（第 21 行）。由于 model 被更新，即被 isChecked 的 computed 计算属性监听到，因此会重新返回最新的结果。

在使用 examples 演示库测试时，可以默认传入 true，也就是默认勾选状态，再次点击时即可实现勾选与非勾选状态的切换，如图 10-14 所示。

图 10-14　checkbox 组件的勾选与非勾选状态

提示：types()方法属于工具类，本书并未详细讲解，读者如需了解详情，可在 utils 目录中查看 types.js 文件。

10.8.2　checkboxGroup 组件

checkboxGroup 组件是多个 checkbox 组件的集合，每个 checkbox 组件都有自身绑定的值，而 v-model 则绑定 checkboxGroup 组件，并且是数组类型的。因此，在判断勾选状态时，可以采用数组的 includes 方法，如代码清单 library-10-19 所示。

```
代码清单 library-10-19
1.  > packages/components/checkbox/src/checkbox.vue: template
2.  <input ...省略代码 type="checkbox" v-model="model" :value="value" />
3.
4.  > packages/components/checkbox/src/checkbox.vue: script
5.  const props = defineProps({
6.    ...省略代码
7.    value: { type: [String, Number, Boolean], default: undefined },
8.  })
9.
10. packages/components/checkbox/src/composables/use-checkbox-model.js
11. export function useCheckboxModel({ ...省略代码 }){
12.   const model = computed({
13.     get(){...省略代码},
14.     set(val){
15.       if (isGroup && Array.isArray(val)) {
16.         checkboxGroupKey?.changeEvent?.(val)
17.       } else{
18.         checkboxModel.value = val
19.       }
20.     }
21.   })
22. }
23.
24. > packages/components/checkbox/src/composables/use-checkbox-state.js
25. export function useCheckboxState({ props, model, checkboxGroupKey, isGroup }){
26.   const isChecked = computed(() => {
```

```
27.     const value = model.value
28.     if(types().isBoolean(value)) {...省略代码}
29.     // Array
30.     if(types().isArray(value)){  // 或者使用 Array.isArray(value)
31.       return value.includes(props.value)
32.     }
33.     return false
34.   })
35. }
36.
37. > packages/components/checkbox/src/checkboxGroup.vue
38. const changeEvent = async (value) => {
39.   groupModel.value = value
40. }
41. provide(CHECKBOX_GROUP_KEY, {...toRefs(props), groupModel, changeEvent})
```

当 input 标签的 type 属性为 checkbox 类型且绑定的数据为数组类型时，那么勾选状态和 input 标签的 value 属性之间存在关联关系。如果 value 属性的值包含在 v-model 绑定的数据中，则表示已勾选，如果不包含，则表示未勾选。因此，应在 checkbox 组件的 prosp 对象中定义 value 属性接收数据，并将数据绑定到 input 标签的 value 属性中（第 2、7 行）。

当点击多选框选项使 v-model 绑定的值发生变化时，就会触发 use-checkbox-model.js 文件变量 model 的 computed 计算属性的 set 对象（第 14~19 行）。当 set 对象中的 isGroup 为"true"且更新的数据为数组类型时，调用 checkboxGrouop 组件的 changeEvent 方法执行双向绑定数据的更新（第 15~17 行）。此时更新的是 checkboxGrouop 组件 v-model 双向绑定的数据，而不是 checkbox 组件双向绑定的数据。也就是说，需要从 checkboxGrouop 组件的 provide 对象注入自定义方法 changeEvent，再进行调用（第 41 行），在 changeEvent 方法中更新双向绑定的数据（第 39 行）。

当双向绑定的数据被更新时，use-checkbox-state.js 中的 model 会监听到数据的变化，接着在确认双向绑定的数据为数组类型后（第 30 行），调用数组的 includes 方法判断双向绑定的数据是否包含 props.value 的值，如果包含，则返回 true，否则返回 false，从而实现 checkboxGroup 组件的勾选与非勾选状态，如图 10-15 所示。

图 10-15　checkboxGroup 组件的勾选与非勾选状态

10.9　event：事件

事件（event）的处理方式与 8.14 节中 input 输入框处理事件的方式是一致的，当 v-model 双向绑定的数据发生变化时，则提供额外的方法让用户处理其他的业务，如代码清单 library-10-20 所示。

代码清单 library-10-20

```
1.  > packages/components/checkbox/src/composables/use-checkbox-event.js
2.  import { getCurrentInstance } from "vue"
3.  export function useCheckboxEvent(){
4.    const { emit } = getCurrentInstance()
5.    const changeEvent =  (e) => emit('change', e.target.checked, e)
6.    return { changeEvent }
7.  }
8.
9.  > packages/components/checkbox/src/composables/use-checkbox.js
10. import { useCheckboxEvent } from "./use-checkbox-event"
11. export function useCheckbox({ props, checkboxModel }){
12.   const { changeEvent } = useCheckboxEvent()
13.   return {...省略代码, changeEvent}
14. }
15.
16. > packages/components/checkbox/src/checkbox.vue: template
17. <input ...省略代码 type="checkbox" @change="changeEvent" />
18.
19. > packages/components/checkbox/src/checkbox.vue: script
20. const { ...省略代码, changeEvent } = useCheckbox({ props, checkboxModel }))
21.
22. > packages/components/checkbox/src/checkboxGroup.vue
23. const changeEvent = async (value) => {
24.   groupModel.value = value
25.   emit('change', value)
26. }
27. provide(CHECKBOX_GROUP_KEY, {...省略代码})
28.
29. > examples/src/App.vue
30. <template>
31.   <a-checkbox-group v-model="value1" @change="handleChangeboxGroup">...省略代码
32.   <a-checkbox  v-mode="value2" @change="handleChangebox">广东省</a-checkbox>
33. </template>
```

对事件的处理依然采用组合式函数，在 composables 目录下新建 use-checkbox-event.js 文件，并定义 useCheckboxEvent 函数（第 3 行），在该函数中定义 changeEvent 方法并调用 Vue.js 3 的 emit 对象回调 change 方法（第 5 行），然后返回 changeEvent（第 6 行）。

为 use-checkbox.js 主入口文件引入 useCheckboxEvent 函数并解构 changeEvent 方法（第 10、12、13 行）。当 v-model 双向绑定数据发生变化时，默认触发 input 标签的 onChange 方法，因此在该方法中绑定 changeEvent 即可（第 17、20 行）。

对于 checkboxGroup 组件，只需要在 changeEvent 方法中调用 emit 对象回调 change 方法（第 25 行），用户在使用组件时即可使用@change 触发 change 方法（第 31、32 行），如图 10-16 所示。

checkboxGroup的值：	▶ (2) ['guangdong', 'guangxi']
checkbox的值：	true

图 10-16 emit 对象回调 change 方法

10.10 async：异步

异步（async）是一种人机交互动作，是对用户点击多选框选项与服务端产生数据交互后返回结果这一过程的体现，也可以理解为加载（loading）提示，与 5.13 节 button 组件的加载按钮类似，不同点在于 button 组件不需要处理数据，而 checkbox 组件需要处理数据。在 UI 组件库设计稿中，checkbox 组件的加载效果如图 10-17 所示。

图 10-17 checkbox 组件的加载效果

10.10.1 渲染加载效果

从 UI 组件库设计稿中可以看到，加载效果和禁用效果的渲染方法相似，不同点在于加载效果显示为一个 loading 图标，与 button 组件的 loading 按钮一致，如代码清单 library-10-21 所示。

```
代码清单 library-10-21
1.  > packages/components/checkbox/src/checkbox.vue: template
2.  <component :is="tag" :class="[
3.    ...省略代码
4.    ns.is('loading', isLoading),
5.    ns.is('disabled', isDisabled || isLoading)
6.  ]">
7.    <span :class="[ns.e('wrapper')]">
8.      <input :class="[ns.e('input')]" type="checkbox" ...省略代码/>
9.      <span :class="[ns.e('inner')]">
10.       <a-icon>
11.         <Loading v-if="isLoading" :class="[`${ns.is('loading-transition', isLoading)}`]" />
12.         <Check v-else />
13.       </a-icon>
14.     </span>
15.   </span>
16. </component>
17.
18. > packages/components/checkbox/src/composables/use-checkbox-state.js
19. export function useCheckboxState({ props, model, checkboxGroupKey, isGroup }){
20.   const isLoading = ref(true)
21.   return {...省略代码, isLoading }
```

```
22.    }
23.
24.    > packages/theme/src/checkbox.scss
25.    @include b(checkbox) {
26.      @include s((checked, loading)) { ...省略代码 }
27.    }
28.    @each $type in $types {
29.      $className: '.a-checkbox--' + $type;
30.      #{$className} {
31.        @include s((checked, loading)) { ...省略代码 }
32.      };
33.    }
```

在 use-checkbox-state.js 文件中定义变量 isLoading 并返回（第 20、21 行），在 checkbox 组件中引入 loading 图标，然后使用 v-if 判断 isLoading 是否为 true，如果是，则渲染 loading 图标，否则渲染 Check 图标（第 11、12 行）。由于加载效果和禁用效果一致，所以只需要在混合指令上添加 loading 即可（第 26、31 行），渲染效果如图 10-18 所示。

图 10-18 checkbox 组件的加载效果

10.10.2 事件交互

我们在构建 checkbox 组件时，已经了解到 label 标签嵌套 input 标签会使两者自动形成关联关系，无论是点击"多选框"还是"文本区域"，都会触发 input 组件，因此无法实现异步交互。为了实现异步交互，可以在 label 标签中绑定点击事件，通过阻止默认行为的方式实现交互效果，如代码清单 library-10-22 所示。

代码清单 library-10-22
```
1.  > packages/components/checkbox/src/composables/use-checkbox-event.js
2.  export function useCheckboxEvent({ props, model, isDisabled, isGroup, isLoading }){
3.    const clickEvent = (e) => {
4.      console.log('clickEvent')
5.    }
6.    return { changeEvent, clickEvent }
7.  }
8.
9.  > packages/components/checkbox/src/checkbox.vue: template
10. <component :is="tag" :class="[...省略代码]" @click="clickEvent">...省略代码</component>
```

1. 阻止默认行为

默认行为一般是指 HTML 元素自带的行为。例如，点击 a 标签的 href 属性，发生自动跳转；点击 form 标签的 action 属性，提交表单内容要带有地址；label 标签与 input 控件关联，触发多选框。因此，我们需要阻止默认行为，使其不被执行，如代码清单 library-10-23 所示。

代码清单 library-10-23
```
1.  > packages/components/checkbox/src/composables/use-checkbox-event.js
2.  export function useCheckboxEvent({ props, model, checkboxGroupKey, isDisabled,
    isGroup, isLoading }){
3.    const beforeChange = computed(() => isGroup ?
    checkboxGroupKey?.beforeChange?.value : props.beforeChange)
4.    const isBeforeChange = computed(() => types().isFunction(beforeChange.value))
5.    const clickEvent = (e) => {
6.      if(isBeforeChange.value || isDisabled.value || isLoading.value) {
7.        const ev = e || window.event
8.        ev.preventDefault();
9.      }
10.     console.log('clickEvent')
11.   }
12. }
13.
14. > examples/src/App.vue
15. <a-checkbox-group v-model="value1" :before-change="asyncChange">...省略代码
    </a-checkbox-group>
16. <a-checkbox v-model="value2" :before-change="asyncChange">广东省</a-checkbox>
17. const asyncChange = () => {...省略代码}
```

在 use-checkbox-event.js 文件中使用三元运算判断 igGroup 是否存在，若存在，则获取父级"组"的 beforeChange，否则获取 props.beforeChange（第 3 行）。然后判断 beforeChange 是否是一个函数，将判断结果赋给变量 isBeforeChange（第 4 行）。

label 和 input 标签并不是在所有情况下都阻止默认行为，而是在异步、禁用和加载这 3 种情况下才阻止。因此，在 clickEvent 事件中使用"或"运算判断是否满足 isBeforeChange.value、isDisabled.value、isLoading.value 时（第 6 行），只要其中任意一个条件成立，就可以执行 ev.preventDefault()阻止默认行为（第 8 行）。此时再为 checkbox 组件或 checkboxGroup 组件添加 beforeChange 属性，多选框无法实现交互（第 15~17 行）。

2. 事件冒泡

事件冒泡是 JavaScript 中的一种事件传播机制。当一个元素触发了某个事件时，这个事件会在该元素上被处理，并且随着时间的推移，逐级传递给它的父元素，一直传递到文档的根元素，这种传播过程被称为事件冒泡。

label 绑定单击（click）事件时会出现事件冒泡的问题，将 checkbox 组件上的 beforeChange 属性去除后，再点击多选框选项，并在 clickEvent 中输入内容，在控制台中会看到 clickEvent 被执行了两次，如图 10-19 所示。

图 10-19 clickEvent 的加载效果

clickEvent 被执行两次的原因在于,当 input 的 type 属性类型为 checkbox 或 radio 时,会默认存在 click 事件。所以在每次点击多选框选项时,input 的默认 click 事件都被执行了一次,label 标签绑定的 click 事件也被执行了一次。解决方式是用 Vue.js 3 的修饰符.stop 阻止事件冒泡,如代码清单 library-10-24 所示。

代码清单 library-10-24

```
1. > packages/components/checkbox/src/checkbox.vue
2. <input :class="[ns.e('input')]" type="checkbox" ...省略代码 @click.stop/>
```

10.10.3 数据交互

checkbox 组件和 checkboxGroup 组件在启用异步功能后,多选框会被默认为无法自动形成勾选与非勾选状态的切换交互。Vue.js 3 是一种用于构建用户界面的渐进式 JavaScript 框架,它的核心原理是通过数据驱动视图,实现响应式的 UI 更新。因此,在异步功能下,我们需要修改 v-model 绑定的数据,使 UI 视图产生变化,如代码清单 library-10-25 所示。

代码清单 library-10-25

```
1.  > packages/components/checkbox/src/composables/use-checkbox-event.js
2.  export function useCheckboxEvent({ props, model, checkboxGroupKey, isDisabled, isGroup, isLoading }){
3.    const clickEvent = (e) => {
4.      if(isBeforeChange.value || isDisabled.value || isLoading.value) {
5.        const ev = e || window.event
6.        ev.preventDefault();
7.        if(isBeforeChange.value && !isDisabled.value && !isLoading.value) {
8.          isLoading.value = true
9.          beforeChange.value(props?.params).then(() => {
10.           updateData()
11.           isLoading.value = false
12.         }).catch(() => {
13.           isLoading.value = false
14.         })
15.       }
16.     }
17.   }
18.
19.   const updateData = () => {
20.     if(isGroup){
21.       const index = model.value.findIndex(v => v === props.value)
22.       (index !== -1) ? model.value.splice(index, 1) : model.value.push(props.value)
23.     }else{
24.       model.value = !model.value
25.     }
26.   }
27. }
```

为所有的多选框均绑定 click 事件,触发 clickEvent 方法,但只有使用了 beforeChange 属性才会触发异步逻辑，因此需要排除"禁用"和"加载"两种状态的情况（第 7 行）。beforeChange 返回的是 Promise 对象，只有在 Promise 对象为 resolve 时才需要改变数据，所以应在回调.then 中执行 updateDate 方法（第 10 行），并设置 isLoading 为 false，取消加载状态。在回调.catch 时同样取消加载状态。

updateData 函数需要区分 checkbox 组件和 checkboxGroup 组件两种数据。如果 isGroup 存在，则数据为数组类型（第 20 行），优先使用数组的 findIndex 方法获取当前选项的索引位置（第 21 行），再判断索引是否为-1，若不为-1，则索引存在，表示用户在做取消勾选动作，接着应使用 splice 方法删除指定索引的数据；反之表示用户在做勾选动作，则应对数据进行 push 操作（第 22 行）。如果 isGroup 不存在，表示单独使用 checkbox 组件，那么可以直接对数据进行取反操作（第 24 行）。

10.11 all：全选

全选是一种常见的用户界面功能，通常呈现为多选框或开关按钮的形式，出现在与多个选项或条目相关的界面中，以便用户一次性选择所有选项，而无须逐一点击。全选功能可以简化用户的选择过程，特别是当列表中包含大量选项时。全选功能的 UI 效果如图 10-20 所示。

图 10-20　全选功能的 UI 效果

10.11.1 渲染全选组件

checkboxAll 组件和 checkboxGroup 组件一样，均属于父级"组"组件。checkboxAll 组件由 checkbox 组件和 checkboxGroup 组件组成，如代码清单 library-10-26 所示。

```
代码清单 library-10-26
1.  > packages/components/checkbox/src/checkboxAll.vue: template
2.  <div :class="[ns.b()]">
3.    <div :class="ns.e('label')">
4.      <Checkbox all v-model="checkAll" :size="size">全选</Checkbox>
5.    </div>
6.    <div :class="ns.e('wrapper')">
7.      <CheckboxGroup v-model="allModel" :size="size">
8.        <slot />
9.      </CheckboxGroup>
10.   </div>
11. </div>
12.
13. > packages/components/checkbox/src/checkboxAll.vue: script
```

```
14.    ...省略代码
15.    import Checkbox from "./checkbox.vue"
16.    import CheckboxGroup from "./checkboxGroup.vue"
17.    import { CHECKBOX_ALL_KEY } from "./constant"
18.    const ns = useNamespace("checkbox-all");
19.    // defineModel
20.    const allModel = defineModel({ type: Array, default: () => ([]) })
21.    const checkAll = ref(false)
22.    /** emit */
23.    const emit = defineEmits(['change'])
24.    /** props */
25.    const props = defineProps({
26.      size: {
27.        type: String,
28.        default: "default",
29.      }
30.    });
31.
32.    const changeEvent = async (value) => {
33.      allModel.value = value
34.      emit('change', value)
35.    }
36.
37.    provide(CHECKBOX_ALL_KEY, {
38.      ...toRefs(props),
39.      allModel,
40.      changeEvent
41.    })
```

checkboxAll 组件的构建方式与 checkbox 组件类似，都需要添加到 components/index.js 和 packages/components.js 两个文件中。相同点是 changeEvent 方法和 provide 对象（第 32~35、37~41 行），不同点在于 provide 对象使用的常量 key（第 17、37 行），checkboxAll 组件的渲染效果如图 10-21 所示。

图 10-21 checkboxAll 组件的渲染效果

10.11.2 渲染部分选中状态

checkboxAll 组件的部分选中状态是指选项没有全部被选中，如图 10-22 所示。此时需要用一个部分选中状态的符号或图标来标记多选框的状态，如代码清单 library-10-27 所示。

图 10-22 checkboxAll 组件的部分选中状态

代码清单 library-10-27

```
1.  > packages/components/checkbox/src/checkbox.vue: template
2.  <component :is="tag" :class="[
3.      ...省略代码
4.      ns.is('checked', isChecked || indeterminate)
5.    ]" >
6.    <span :class="[ns.e('wrapper')]">
7.      <input :class="[ns.e('input')]" ...省略代码/>
8.      <span :class="[ns.e('inner'), ns.is('indeterminate', indeterminate)]">
9.        <template v-if="indeterminate">
10.         <i :class="[ns.e('indeterminate', indeterminate)]"></i>
11.       </template>
12.       <a-icon v-else>...省略代码</a-icon>
13.     </span>
14.   </span>
15. </component>
16.
17. > packages/components/checkbox/src/checkbox.vue: script
18. const props = defineProps({ ...省略代码, indeterminate: Boolean });
19.
20. packages/theme/src/common/var.scss
21. ...省略代码
22. $checkbox-indeterminate-size: () !default;              // 多选框宽度、高度尺寸
23. $checkbox-indeterminate-size: map.deep-merge(
24.     ('small': 8px, 'default': 10px, 'large': 12px),
25.     $checkbox-indeterminate-size
26. );
27. $checkbox-indeterminate-round: () !default;             // 多选框圆角
28. $checkbox-indeterminate-round: map.deep-merge(
29.     ('small': 3px, 'default': 4px, 'large': 5px),
30.     $checkbox-indeterminate-round
31. );
32. $global: (
33.     ...省略代码
34.     'checkbox-indeterminate-size': $checkbox-indeterminate-size,
35.     'checkbox-indeterminate-round': $checkbox-indeterminate-round,
36. );
37.
38. > packages/theme/src/checkbox.scss
39. @include b(checkbox) {
40.     ...省略代码
41.     @include e(inner) {
42.       @include d(icon) { ...省略代码 }
43.       @include s((indeterminate)) { background: #fff; }
44.     }
45.     @include e(indeterminate) {
46.       background-color: var(#{getVarName('checkbox', 'bg-color')});
47.       border-radius: var(#{getVarName('checkbox', 'check-indeterminate-round')});
48.       display: block;
49.       width: var(#{getVarName('checkbox', 'check-indeterminate-size')});
```

```
50.     height: var(#{getVarName('checkbox', 'check-indeterminate-size')});
51.     box-sizing: border-box;
52.     transition: var(--a-transition);
53.   }
54.   @include e(label) {...省略代码}
55.   @each $s in $sizes {
56.     @include m(size, $s){
57.       ...省略代码
58.       #{getVarName('checkbox', 'check-indeterminate-size')}: var(#{getVarName('checkbox-indeterminate-size', $s)});
59.       #{getVarName('checkbox', 'check-indeterminate-round')}: var(#{getVarName('checkbox-indeterminate-round', $s)});
60.       @include d(icon) {
61.         #{getVarName('checkbox', 'checkbox-icon-size')}: var(#{getVarName('checkbox-icon-size', $s)});
62.       }
63.     }
64.   }
65. }
66.
67. > packages/components/checkbox/src/checkboxAll.vue
68. <Checkbox all v-model="checkAll" :indeterminate="indeterminate">全选</Checkbox>
```

在图 10-22 中，多选框的部分选中状态有"宽高"和"圆角"两种尺寸，分为 small、default 和 large 共 3 种类型。根据图中标注的尺寸数据，在 var.scss 文件中配置全局变量（第 22~31 行），然后在 checkbox.scss 文件中绑定变量（第 45~64 行）。

在 checkbox 组件的 props 对象中定义属性 indeterminate，为布尔类型（第 18 行）。多选框仅有"图标"和"部分选中状态"两种元素，使用 v-if 判断属性 indeterminate 是否为 true，若是，则显示部分选中状态（第 9~11 行），否则显示图标（第 12 行）。最后在 checkboxAll 组件中传入 indeterminate 状态（第 68 行），如图 10-23 所示。

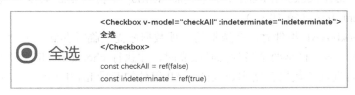

图 10-23　checkAll 和 indeterminate 均为 true 时，显示部分选中状态

10.11.3　存储选项数据

全选交互的逻辑以用户勾选的个数与选项总和为判断条件，如果勾选的个数与选项总和相等，则表示全选，否则表示部分勾选或未勾选。因此，在处理交互过程前，应先获取所有选项，将其存储为临时数据，如代码清单 library-10-28 所示。

代码清单 library-10-28
```
1. > packages/components/checkbox/src/composables/use-checkbox-group.js
```

```
2.  import { CHECKBOX_GROUP_KEY, CHECKBOX_ALL_KEY } from "../constant"
3.  export function useCheckboxGroup(){
4.    const checkboxAllKey = inject(CHECKBOX_ALL_KEY, undefined)
5.    const isAll = checkboxAllKey !== undefined
6.    return {...省略代码, checkboxAllKey, isAll}
7.  }
8.
9.  > packages/components/checkbox/src/composables/use-checkbox.js
10. const { model } = useCheckboxModel({...省略代码, checkboxAllKey, isAll})
11.
12. > packages/components/checkbox/src/composables/use-checkbox-model.js
13. export function useCheckboxModel({ ...省略代码, checkboxAllKey, isAll }){
14.   const model = computed({
15.     get(){...省略代码},
16.     set(val){
17.       if (isGroup && Array.isArray(val)) {
18.         checkboxGroupKey?.changeEvent?.(val)
19.         checkboxAllKey?.changeEvent?.(val)
20.       }
21.     }
22.   })
23.   // 临时存储
24.   isAll && !props.all && checkboxAllKey?.setValuesEvent?.(props.value)
25. }
26.
27. > packages/components/checkbox/src/checkboxAll.vue
28. const list = ref([])
29. const setValuesEvent = (val) => list.value.push(val)
```

checkboxAll 是一个父级"组"组件，use-checkbox-group.js 文件使用同样的方式返回 checkboxAllKey 和 isAll（第 3~6 行），再从 use-checkbox.js 主入口文件传入 useCheckboxModel 函数（第 10 行）。在 useCheckboxModel 函数中判断 isGroup 是否为 true，若是，则调用 checkboxAll 组件的 changeEvent 方法（第 19 行）。

只有针对 checkboxAll 组件时，多选框的选项数据才需要临时存储。因此，只有当 isAll 为 true 且不是"全选"多选框这两个条件同时成立时，才执行 checkboxAll 组件的 setValuesEvent 方法（第 24 行），将选项的数据存储到 checkboxAll 组件的变量 list 中（第 28、29 行），如图 10-24 所示。

> Proxy(Array) {0: 'shenzhen', 1: 'guangzhou', 2: 'huizhou', 3: 'shanwei'}

图 10-24 临时存储的选项数据

10.11.4 全选交互

checkbox 组件根据属性 indeterminate 完成对部分勾选状态的 UI 渲染，在交互过程中，只需要考虑用户勾选的选项个数与临时存储数据的总和，如代码清单 library-10-29 所示。

代码清单 library-10-29

```
1.  > packages/components/checkbox/src/checkboxAll.vue: template
2.  <Checkbox :indeterminate="indeterminate" @change="handleAll">全选</Checkbox>
3.
4.  > packages/components/checkbox/src/checkboxAll.vue: script
5.  const changeEvent = async (val) => {
6.    allModel.value = val
7.    changeAllEvent(val)
8.    emit('change', val)
9.  }
10.
11. const handleAll = (val) => {
12.   allModel.value = val ? list.value : []
13.   indeterminate.value = false
14. }
15.
16. const changeAllEvent = (val) => {
17.   const checkedCount = val.length
18.   checkAll.value = checkedCount === list.value.length
19.   indeterminate.value = checkedCount > 0 && checkedCount !== list.value.length
20. }
```

在 checkboxAll 组件中，用户点击多选框选项会触发 changeEvent 方法，因此可以在该方法中调用 changeAllEvent 方法（第 7 行）。在 changeAllEvent 方法中，将已勾选的选项长度存储在变量 checkedCount 中（第 17 行），如果变量 checkedCount 的长度等于变量 list 的长度，则表示全选，变量 checkAll 的值为 true，"全选"多选框显示为勾选状态，否则显示未非勾选状态（第 18 行）。如果用户勾选了 1 个或多个选项，那么 checkedCount 的总和肯定大于 0，但勾选的选项个数不等于变量 list 的长度，因此判定为部分选中状态，此时变量 indeterminate 的值为 true（第 19 行）。

若用户点击的是"全选"多选框，即只有全选和非全选两种状态，那么可以直接设置变量 indeterminate 为 false（第 13 行）。如果 handleAll 方法的返回值为 true，则将变量 list 的数据直接赋给 allModel，否则赋值为空数组（第 12 行）。

本章小结

本章主要介绍了多选框组件的开发过程，以便于读者进一步掌握 Scss 全局变量和私有变量的应用，并深入学习 Vue.js 3 的组合式函数的应用，实现解耦业务模块，再重新汇集使用，体验开发业务、降低耦合度的逻辑思想。

本章完成了多选框组件核心功能的开发，当然，组件的开发并不是完美的，读者可以根据自己的业务需求，不断扩展组件的功能。

单选框（radio）组件与多选框组件有近 90%的功能和 UI 是一致的，不同点在于 radio 组件只有单选的概念，如果你能完整地开发出多选框组件，那么想必开发 radio 组件也是信手拈来。因此，本书不再介绍 radio 组件，如需学习 radio 组件的源码，可查看本书配套代码中的分支 library-radio。

第 11 章
开关组件

开关（switch）组件是一种常用的用户界面控件，用于在应用程序或网页中切换功能、选项或状态。在界面设计中，switch 组件通常用于表示一个二进制选择，例如打开/关闭、启用/禁用等。

在 UI 组件库设计稿中，switch 组件是以"开关"模式设计的，包含"基础""主题""文字""图标""加载""尺寸"等类型，并且需要实现开关切换过程的过渡动画效果。

11.1 渲染 switch 组件

switch 组件的效果如图 11-1 所示，其中有打开、关闭和禁用 3 种状态，可变化的属性有背景颜色、文字颜色、宽度、高度等。根据这些状态及可变化的属性，我们优先渲染 switch 组件的"基础 basis" UI 效果。

图 11-1 switch 组件的效果

11.1.1 构建组件

根据第 2 章中的组件结构和样式规范来构建 switch 组件，首先定义组件名称为 ASwitch，并引入 components/index.js 和 packages/index.js 文件，然后根据 UI 组件库设计稿定义 switch 组件的基础结构，如代码清单 library-11-1 所示。

```
代码清单 library-11-1
1.  > packages/components/switch/src/index.vue: template
2.  <div :class="[ns.b()]">
3.    <div :class="[ns.e('input')]">
4.      <input type="checkbox" />
5.    <span :class="ns.e('handel')">
6.      <button type="button" :class="[ns.e('button')]">按钮</button>
7.      <span :class="[ns.e('inner')]">
8.        <span :class="[ns.e('inner-checked')]">开</span>
9.        <span :class="[ns.e('inner-unchecked')]">关</span>
10.     </span>
```

```
11.         </span>
12.       </div>
13.    </div>
```

上述代码清单实现了 switch 组件的基本结构，首先生成一个 button 按钮（第 6 行），它是 switch 组件中间的"白色圆"；然后生成左右两边的文本和图标渲染区域（第 7~10 行）。其中要注意的是 input 标签（第 4 行），我们将 v-model 绑定在 input 上，当值发生变化时，会触发 input 的 change 方法，可以通过该方法处理额外的业务，如 form 表单校验。由于 switch 组件只有打开和关闭两种状态，因此可设置 input 的 type 属性类型为 checkbox。switch 组件的基础结构如图 11-2 所示。

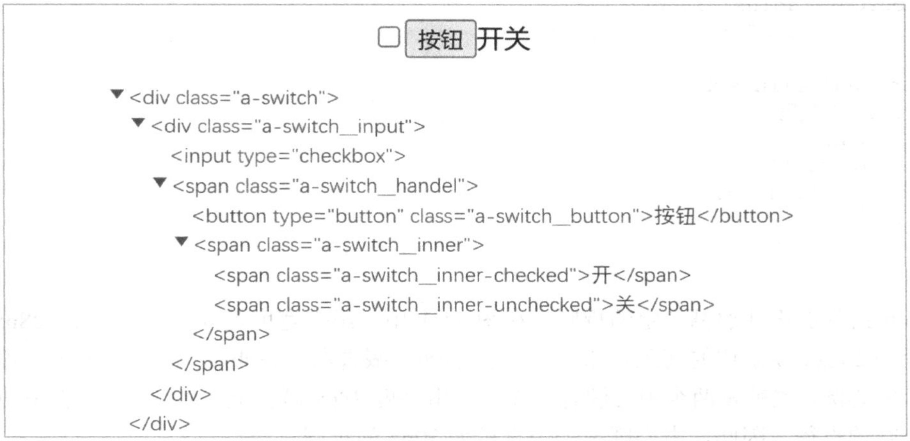

```
□ 按钮 开关
▼<div class="a-switch">
  ▼<div class="a-switch__input">
     <input type="checkbox">
    ▼<span class="a-switch__handel">
       <button type="button" class="a-switch__button">按钮</button>
      ▼<span class="a-switch__inner">
         <span class="a-switch__inner-checked">开</span>
         <span class="a-switch__inner-unchecked">关</span>
       </span>
     </span>
   </div>
 </div>
```

图 11-2 switch 组件的基础结构

11.1.2 渲染组件

在 theme/src 目录下新建 switch.scss 文件并引入 index.scss 文件，同时引入 switch 组件的 style/index.js 文件。根据 UI 组件库设计稿的效果，使用 Sass 封装的 BEM 命名规则书写 CSS 样式，如代码清单 library-11-2 所示。

代码清单　library-11-2
```
1.   > packages/components/switch/src/index.vue: template
2.   @include b(switch) {
3.     ...省略代码
4.     input { display: none; }
5.     @include e(input) {
6.       height: 28px;
7.       width: 48px;
8.       border-radius: 100px;
9.       ...省略代码
10.    }
11.    @include e(handel) {
```

```
12.     ...省略代码
13.     padding: 4px;
14.   }
15.   @include e(button) {
16.     ...省略代码
17.     position: absolute;
18.     top: 4px;
19.     left: 4px;
20.     z-index: 10;
21.     width: 20px;
22.     height: 20px;
23.     background-color: #fff;
24.     border-radius: 100px;
25.     > i { transform: scale(.8); }
26.   }
27.   @include e(inner) {
28.     ...省略代码
29.     > span {
30.       width: 50%;
31.       height: 100%;
32.     }
33.   }
34. }
```

　　switch 组件的设计遵循一定的规则，在图 11-1 中，组件宽度是 48px、高度是 28px、中心圆的直径是 20px、4 条边的间距是 4px。用宽度 48px 减去左右各 4px，也就是 40px，刚好可以放下两个中心圆，也就是两个中心圆各占 50%。用高度 28px 减去上下各 4px，也就是 20px，正好是中心圆的直径。因此，中心圆是一个直径为 20px 的正圆。

　　根据上述理解，switch 组件的样式就清晰了。switch 组件的宽和高由 e(inner) 生成（第 5~10 行），同时设置属性 border-radius 的值为 100px，用于生成圆角。switch 组件的中心圆由 e(button) 生成（第 15~26 行），其中设置属性 position 的值为 absolute，用于绝对定位，这样便可以挡住下层的文字和图标。switch 组件在打开和关闭状态下，左右两侧的文字各占一半，也就是宽度占 50%（第 30 行）。

　　需要注意的是，e(handel) 生成的类名包裹了中心圆和文字区域，并设置内边距为 4px。这里的内边距便是上文提到的间距，通过去除间距得到两个中心圆的位置，也就是每个中心圆各占 50%，如图 11-3 所示。

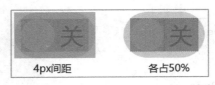

图 11-3　switch 组件的 4px 间距占位

11.1.3 私有全局变量

在前几章的组件开发过程中，组件的变量都定义在 var.scss 全局文件中，var.scss 文件提供了整个项目中所有组件共用的变量。但有些组件并不完全跟随全局变量，例如 checkbox 组件和 radio 组件的多选框和单选框的尺寸、圆角等，都是独有的，其他的组件基本用不到。考虑到这两个组件在尺寸、交互方面是一致的，因此把它们都放在了 va.scss 全局文件中。下面要开发的 switch 组件的尺寸也不完全遵循 UI 组件库的设计规范，switch 组件自身的尺寸也不与其他组件共用。在这种情况下，我们可以将 switch 组件的变量定义在 switch.scss 文件中，作为私有全局变量处理，如代码清单 library-11-3 所示。

代码清单 library-11-3

```
1.  > packages/theme/src/common/mixins.scss
2.  @mixin set-global-var(){...省略代码}
3.  @mixin set-private-var($block, $data){
4.    @each $key, $value in $data {
5.      #{createVarName(($block, $key))}: #{$value};
6.    }
7.  }
8.
9.  > packages/theme/src/switch.scss
10. @use "sass:map";
11. @use "./common/mixins.scss" as *;
12. @use "./common/componentVar.scss" as *;
13. $switch-size: () !default;
14. $switch-size: map.deep-merge(
15.   ('small': 24px, 'default': 28px, 'large': 32px),
16.   $switch-size
17. );
18. $switch-width: () !default;
19. $switch-width: map.deep-merge(
20.   ('small': 40px, 'default': 48px, 'large': 56px),
21.   $switch-width
22. );
23. :root {
24.   @include set-private-var('switch-size', $switch-size);
25.   @include set-private-var('switch-width', $switch-width);
26. }
27. @include b(switch) {...省略代码}
```

在 mixins.scss 文件中，定义 set-private-var 指令接收$block 和$data 参数（第 3~7 行），set-private-var 指令的逻辑和 set-global-var 基本一致，只是少了一层 @each 循环。

在 switch.scss 文件中定义可变化的属性 switch-size 和 switch-width（第 13~22 行），在语法上与 var.scss 定义全局变量完全一致，然后在:root 中使用指令 @include set-private-var，并传入生成变量的名称 switch-size 及数据，即可生成 switch 组件的私有全局变量（第 24、25 行），如图 11-4 所示。

```
:root {
    --a-switch-height-small:   24px;
    --a-switch-height-default: 28px;
    --a-switch-height-large:   32px;
    --a-switch-width-small:    40px;
    --a-switch-width-default:  48px;
    --a-switch-width-large:    56px;
}
```

图 11-4 switch 组件的私有全局变量

11.1.4 私有样式变量

switch 组件的样式变量与 checkbox 组件的样式变量有很多相似之处，如关闭状态（默认状态）、打开状态（选中状态）、禁用状态。可变化的属性包括背景颜色、宽度、高度、字体大小等。因此，可以将 checkbox 组件的私有变量复制一份，将其更新为 switch 组件的私有样式变量，如代码清单 library-11-4 所示。

代码清单 library-11-4

```
1.  > packages/theme/src/common/componentVar.scss
2.  @function switchVar($type: ''){
3.    $switch: (
4.      'text-color': ('default': getVarName('text-color', 'primary')),
5.      'bg-color': ('default': getVarName('neutral-color', 'light-6')),
6.      'disabled-text-color': ('default': getVarName('text-color', 'disabled')),
7.      'disabled-bg-color': ('default': getVarName('color', 'disabled-bg')),
8.      'checked-text-color': ('default': getVarName('color', 'white')),
9.      'checked-bg-color': ('default': getVarName('color', 'primary')),
10.     'width': ('default': getVarName('switch-width', 'default')),
11.     'height': ('default': getVarName('switch-height', 'default')),
12.     'font-size': ('default': getVarName('font-size', 'default')),
13.   );
14.   @return $switch;
15. }
16.
17.
18. @include b(switch) {
19.   @include set-component-var('switch', switchVar());
20.   ...省略代码
21.   @include e(input) {
22.     height: var(#{getVarName('switch', 'height')});
23.     width: var(#{getVarName('switch', 'width')});
24.     background-color: var(#{getVarName('switch', 'bg-color')});
25.   }
26.   @include e(button) {
27.     width: calc(var(#{getVarName('switch', 'height')}) - 8px);
28.     height: calc(var(#{getVarName('switch', 'height')}) - 8px);
29.     font-size: var(#{getVarName('switch', 'font-size')});
```

```
30.      color: var(#{getVarName('text-color', 'primary')});
31.    }
32.    @include e(inner) {
33.      color: var(#{getVarName('switch', 'text-color')});
34.      > span {
35.        font-size: var(#{getVarName('switch', 'font-size')});
36.      }
37.    }
38.  }
```

通过将混合指令 set-component-var 生成的变量绑定到可变化的属性上，可以绑定 switch 组件的私有样式变量，如图 11-5 所示。其中要注意的是中心圆的绑定方式，这里使用 calc 函数计算，将 switch 组件的高度减去上、下两个边距的 8px，得到中心圆的宽度和高度（第 27、28 行）。

```
.a-switch {
  --a-switch-text-color:         var(--a-text-color-primary);      } 默认状态
  --a-switch-bg-color:           var(--a-neutral-color-light-6);

  --a-switch-disabled-text-color: var(--a-text-color-disabled);    } 禁用状态
  --a-switch-disabled-bg-color:   var(--a-color-disabled-bg);

  --a-switch-checked-text-color:  var(--a-color-white);            } 打开状态
  --a-switch-checked-bg-color:    var(--a-color-primary);

  --a-switch-width:   var(--a-switch-width-default);               } 宽高尺寸
  --a-switch-height:  var(--a-switch-height-default);

  --a-switch-font-size:  var(--a-font-size-default);               } 字体尺寸
}
```

图 11-5　switch 组件的私有样式变量

11.2　theme：主题

switch 组件的主题（theme）渲染与 checkbox 组件完全一致，在 componentVar.scss 文件的 switchVar 方法中接收参数 $type，并为可变化的属性添加 type 属性，所有的 type 均使用 getVarName 方法获取 UI 组件库的全局变量名称，如代码清单 library-11-5 所示。

代码清单 library-11-5

```
1.  > packages/theme/src/common/componentVar.scss
2.  @function switchVar($type: ''){
3.    $switch: (
4.      'text-color': ('default':, ...省略代码, 'type': getVarName('color', 'white')),
5.      'bg-color': ('default':, ...省略代码, 'type': getVarName('color', $type)),
6.      'disabled-text-color': ('default': , ...省略代码, 'type': getVarName('color', 'white')),
7.      'disabled-bg-color': ('default': , ...省略代码, 'type': getVarName('color', $type, 'light-5')),
8.      'checked-text-color': ('default': , ...省略代码, 'type': getVarName('color', 'white')),
```

```
9.         'checked-bg-color': ('default': , ...省略代码, 'type': getVarName('color',
$type)),
10.     );
11.   }
12. 
13. > packages/theme/src/switch.scss
14. @include e(button) {
15.   ...省略代码
16.   > i { transform: scale(.8); }
17.   @include s((checked)) { left: 50%; }
18. }
19. @each $type in $types {
20.   $className: '.a-switch--' + $type;
21.   #{$className} {
22.     @include s((checked)) {
23.       @include set-component-var('switch', switchVar($type), 'type');
24.     }
25.   };
26. }
27. 
28. > packages/components/switch/src/index.vue: template
29. <div :class="[ns.e('input'), ns.m(type), ns.is('checked', isChecked)]">
30. <button type="button" :class="[ns.e('button'), ns.is('checked',
isChecked)]"></button>
31. 
32. > packages/components/switch/src/index.vue: template
33. const props = defineProps({
34.   type: {type: String, default: 'primary' }
35. })
36. /** 选中 */
37. const isChecked = computed(() => true)
```

在 componentVar.scss 文件中为可变化的属性添加 type 属性，自动生成对应主题色的变量（第 2~11 行），然后在 switch.scss 文件中使用@each 循环全变量$type（第 19 行），并且在 s((checked)) 指令下生成变量（第 22~25 行），这是因为 switch 组件只有在打开的状态下才会发生主题色的变化。

在 switch 组件的 index.vue 文件中使用 defineProps 对象定义 type 属性，设置默认值为 primary，再使用 ns.m 绑定生成类名（第 33~35、29 行）。使用 computed 计算属性返回 true，将其赋给变量 isChecked（第 37 行），然后分别与 ns.e('input')和 ns.e('button')绑定（第 29、30 行）。要使中心圆靠右，需要在样式的 e(button)中添加 s((checked))，设置属性 left 的值为 50%（第 17 行），渲染效果如图 11-6 所示。

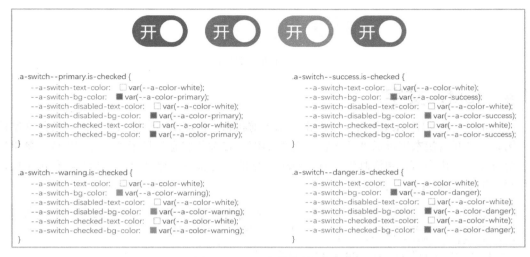

图 11-6　switch 组件的主题渲染

11.3　size：尺寸

switch 组件的尺寸（size）遵循全局变量$sizes，与 checkbox 组件一致，使用@each 循环全局变量$sizes，修改 switch 组件的私有变量，如代码清单 library-11-6 所示。

```
代码清单 library-11-6
1.  > packages/theme/src/switch.scss
2.  @include b(switch) {...省略代码}
3.  @each $s in $sizes {
4.    @include m(size, $s){
5.      #{getVarName('switch', 'font-size')}: var(#{getVarName('font-size', $s)});
6.      #{getVarName('switch', 'height')}: var(#{getVarName('switch-height', $s)});
7.      #{getVarName('switch', 'width')}: var(#{getVarName('switch-width', $s)});
8.    }
9.  }
10.
11. > packages/components/switch/src/index.vue: template
12. <div :class="[ns.e('input'), ...省略代码, ns.m('size', size)]"> ...省略代码</div>
13.
14. > packages/components/switch/src/index.vue: script
15. const props = defineProps({
16.   ...省略代码
17.   size: {type: String, default: 'default'}
18. })
```

在 switch.scss 文件中使用@each 循环全局变量$sizes，然后覆盖 switch 组件可变化的变量 switch-font-size、switch-width 和 switch-height（第 3~9 行）。在 index.vue 文件的 defineProps 对象中定义 size 属性（第 17 行），然后将其绑定到 switch 组件，渲染效果如图 11-7 所示。

图 11-7 switch 组件的尺寸

11.4 text：文字

switch 组件的文字（text）既可以用来显示当前状态，也可以用于提供与状态相关的文字信息。例如，在开启状态下显示"已开启"或"启用"，在关闭状态下显示"已关闭"或"禁用"。通过配置文字，用户可以更清晰地了解 switch 组件的当前状态，提升用户体验和组件可用性，如代码清单 library-11-7 所示。

代码清单 library-11-7
```
1.  > packages/components/switch/src/index.vue: template
2.  <span :class="[ns.e('inner')]">
3.    <span :class="[ns.e('inner-checked')]">
4.      <span v-if="checkedText">{{ firstChatAt(checkedText) }}</span>
5.    </span>
6.    <span :class="[ns.e('inner-unchecked')]">
7.      <span v-if="uncheckedText">{{ firstChatAt(uncheckedText) }}</span>
8.    </span>
9.  </span>
10.
11. > packages/components/switch/src/index.vue: script
12. const props = defineProps({
13.   ...省略代码
14.   checkedText: { type: String, default: '' },
15.   uncheckedText: { type: String, default: '' }
16. })
17.
18. /** 首个文本 */
19. const firstChatAt = (val) => {
20.   const value = val.trim()
21.   return value ? value.charAt(0) : ''
22. }
23.
24. > examples/src/App.vue: template
25. <a-switch checked-text="开启" unchecked-text="关闭"></a-switch>
```

在 defineProps 对象中定义属性 checkedText 和 uncheckedText，分别代表"选中的文字"和"未选中的文字"（第 14、15 行）。在 template 层中绑定 checkedText 和 uncheckedText 时，均调用 firstChatAt 方法获取首个字符（第 4、7 行），这是为了防止用户传入多个字符，导致组件渲染异常（第 19~22 行），渲染效果如图 11-8 所示。

图 11-8 switch 组件的文字渲染

11.5 icon：图标

图标（icon）不仅可以使用户直观地了解 switch 组件的当前状态，而且使 switch 组件看起来更生动有趣，提升用户在与应用程序或网站进行交互时的愉悦感和参与度，如代码清单 library-11-8 所示。

```
代码清单 library-11-8
1.  > packages/components/switch/src/index.vue: template
2.  <span :class="[ns.e('inner')]">
3.    <span :class="[ns.e('inner-checked')]">
4.      <span v-if="checkedText">{{ firstChatAt(checkedText) }}</span>
5.      <a-icon v-if="checkedIcon"><component :is="checkedIcon" /></a-icon>
6.    </span>
7.    <span :class="[ns.e('inner-unchecked')]">
8.      <span v-if="uncheckedText">{{ firstChatAt(uncheckedText) }}</span>
9.      <a-icon v-if="uncheckedIcon"><component :is="uncheckedIcon" /></a-icon>
10.   </span>
11. </span>
12.
13. > packages/components/switch/src/index.vue: template
14. import { AIcon } from "@ui-library/components"
15.
16. > packages/components/switch/src/index.vue: script
17. const props = defineProps({
18.   ...省略代码
19.   checkedIcon: { type: [String, Object], default: '' },
20.   uncheckedIcon: { type: [String, Object], default: '' }
21. })
22.
23. > examples/src/App.vue: template
24. <a-switch :checked-icon="Show" :unchecked-icon="Hide"></a-switch>
25. import { Show, Hide } from "@ui-library/icons"
```

icon 的开发方式和 text 一样，在 defineProps 对象中定义 checkedIcon 和 uncheckedIcon 属性，分别代表"选中的图标"和"未选中的图标"。在渲染图标时，使用我们已经开发完成的 AIcon 组件（第 14 行），用 component 动态组件绑定 is 属性进行渲染，分别绑定属性 checkedIcon 和 uncheckedIcon（第 5、9 行），最后在 examples 演示包中传入图标（第 24、25 行），渲染效果如图 11-9 所示。

图 11-9　switch 组件的图标渲染

11.6 centerIcon：中心圆图标

中心圆图标（centerIcon）可以作为用户与 switch 组件进行交互的目标点，更易于用户识别，用户可以很方便地通过点击中心圆图标来切换开关状态，而不必担心误操作。

通过配置自定义属性将图标渲染在中心圆上，并根据 switch 组件的打开与关闭状态切换中心圆图标，如代码清单 library-11-9 所示。

代码清单 library-11-9
```
1.  > packages/components/switch/src/index.vue: template
2.  <span :class="ns.e('handel')">
3.    <button type="button" :class="[ns.e('button'), ns.is('checked', isChecked)]">
4.      <template v-if="centerIcon">
5.        <a-icon v-if="checkedIcon && isChecked"><component :is="checkedIcon" /></a-icon>
6.        <a-icon v-if="uncheckedIcon && !isChecked"><component :is="uncheckedIcon" /></a-icon>
7.      </template>
8.    </button>
9.    <span :class="[ns.e('inner')]" v-if="!centerIcon">...省略代码</span>
10.
11. > packages/components/switch/src/index.vue: script
12. const props = defineProps({
13.   ...省略代码
14.   centerIcon: Boolean
15. })
```

在 defineProps 对象中自定义属性 centerIcon 为布尔类型，默认值为 false。由于中心圆图标与底层的图标是互斥关系，因此在底层 span 标签中使用 v-if 条件判断变量 centerIcon 是否为 false，如果是，则渲染底层图标，反之则渲染中心圆图标（第 4、9 行）。

目前，在中心圆内部放置了两个图标，需要结合变量 isChecked 的值 true 和 false 来显示不同的图标。如果变量 isChecked 为 true，则渲染 checkedIcon 图标（第 5 行），反之则渲染 uncheckedIcon 图标（第 6 行），渲染效果如图 11-10 所示。

图 11-10 switch 组件的中心圆图标

11.7 disabled：禁用

switch 组件的禁用效果与 checkbox 组件一致，都是在 defineProps 对象中添加属性 disabled，生成 is-disabled 类名，再使用禁用类型的变量渲染组件，如代码清单 library-11-10 所示。

代码清单 library-11-10
```
1.  > packages/components/switch/src/index.vue: template
2.  <div :class="[ns.b()]">
3.    <div :class="[ns.e('input'), ...省略代码, ns.is('disabled', isDisabled)]">
4.      <input type="checkbox" :checked="isChecked" :disabled="isDisabled" />
5.      <span :class="ns.e('handel')">
```

```
6.      <button type="button" :class="[...省略代码, ns.is('disabled', isDisabled)]">
7.        ...省略代码
8.      </button>
9.    </span>
10.   </div>
11. </div>
12.
13. > packages/components/switch/src/index.vue: script
14. const props = defineProps({
15.   ... 省略代码
16.   disabled: Boolean
17. });
18. const isDisabled = computed(() => props.disabled)
19.
20. > packages/theme/src/switch.scss
21. @include e(input) {
22.   ...省略代码
23.   @include s((disabled)) {
24.     #{getVarName('switch', 'bg-color')}: var(#{getVarName('switch', 'disabled-bg-color')});
25.     #{getVarName('switch', 'text-color')}: var(#{getVarName('switch', 'disabled-text-color')});
26.     cursor: not-allowed;
27.   }
28. }
29. @include e(button) {
30.   ...省略代码
31.   @include s((disabled)) {
32.     color: var(#{getVarName('text-color', 'disabled')});
33.   }
34. }}
35. @each $type in $types {
36.   $className: '.a-switch--' + $type;
37.   #{$className} {
38.     @include s((checked)) {...省略代码}
39.     @include s((disabled)) {
40.       #{getVarName('switch', 'bg-color')}: var(#{getVarName('switch', 'disabled-bg-color')});
41.     }
42.   };
43. }
```

在 defineProps 对象中自定义属性 disabled 为布尔类型，默认值为 false（第 16 行）。使用 computed 计算属性返回 props.disabled 的值并赋给变量 isDisabled（第 18 行），将变量 isDisabled 绑定到 switch 组件中（第 3、4、6 行）。在 switch.scss 文件中使用@each 循环生成主题色，同时添加 s((disabled))生成禁用类型的变量（第 39~41 行），为 e(button)和 e(input)添加 s((disabled))，生成中心圆的禁用颜色 color，并覆盖 input 组件的私有变量（第 23~27、31~33 行），渲染效果如图 11-11 所示。

图 11-11　switch 组件的禁用效果

11.8　model：数据绑定

switch 组件仅有打开和关闭两种状态，刚好与布尔类型的 true 与 false 对应。因此，我们只需为 switch 组件添加 click 事件，使 v-model 数据取反切换，如代码清单 library-11-11 所示。

代码清单 library-11-11
```
1.  > packages/components/switch/src/index.vue: template
2.  <div :class="[ns.b()]" @click="handleSwitch">
3.    <div :class="[ns.e('input'), ...省略代码, ns.is('disabled', isDisabled)]">
4.      <input type="checkbox" v-model="modelValue" ...省略代码 />
5.      ...省略代码
6.    </div>
7.  </div>
8.
9.  > packages/components/switch/src/index.vue: script
10. const modelValue = defineModel()              // model
11. const emit = defineEmits(['change']));        // emit
12. const isChecked = computed(() => modelValue.value === true)  // isChecked
13. /** 点击事件 */
14. const handleSwitch = () => {
15.   if(isDisabled.value) { return }
16.   changeEvent()
17. }
18.
19. const changeEvent = () => {
20.   const val = !modelValue.value
21.   modelValue.value = val
22.   emit('change', val)
23. }
```

定义 defineModel() 并赋给变量 modelValue（第 10 行），将变量 modelValue 绑定至 input 标签（第 4 行）。在 switch 组件中添加 click 事件，为其绑定 handleSwitch 方法（第 2 行），在该方法下判断其是否为禁用状态，如果是，则阻止执行（第 15 行），否则继续执行 changeEvent 方法。在 changeEvent 方法中将 modelValue 的值取反后赋给变量 val（第 20 行），然后再次更新 modelValue 的值（第 21 行），最后调用 emit 对象回调父级的 change 事件（第 22 行）。

其中要注意变量 isChecked，在 computed 计算属性中判断 modelValue.value 是否等于 true，这是因为 switch 组件的打开状态以 true 为结果。因此，handleSwitch 方法会使 modelValue 对象的值在 true 与 false 之间切换，当 modelValue 的值为 true 时，表示 switch 组件处于打开状态。

11.9 value：值

switch 组件的打开和关闭状态对应的值分别为 true 和 false。但在实际业务中，后端开发人员不一定会返回 true 和 false，也有可能通过返回 1 和 0 来表示打开和关闭。前端开发人员需要对后端返回的值进行数据类型转换，这种方式对开发人员来说并不友好。因此，组件库应当提供可配置的属性，让开发人员自定义打开和关闭两种状态的值，如代码清单 library-11-12 所示。

代码清单 library-11-12

```
1.  > packages/components/switch/src/index.vue: script
2.  const props = defineProps({
3.    ...省略代码
4.    checkedValue: { type: [Boolean, String, Number], default: true },
5.    unCheckedValue: { type: [Boolean, String, Number], default: false }
6.  })
7.
8.  const isChecked = computed(() => modelValue.value === props.checkedValue)
9.
10. const changeEvent = () => {
11.   const val = isChecked.value ? props.unCheckedValue : props.checkedValue
12.   modelValue.value = val
13. }
14.
15. > examples/src/App.vue
16. <a-switch v-model="check" checked-value="1" unchecked-value="0"></a-switch>
```

在 defineProps 对象中定义属性 checkedValue 和 unCheckedValue 为布尔、字符串、数字类型（第 4、5 行），默认值分别为 true 和 false，与 switch 组件打开和关闭状态的值一致。变量 isChecked 判断 modelValue 的值是否等于 props.checkedValue（第 8 行），它相当于 true。最后在 changeEvent 方法中使用三元运算判断变量 isChecked 的结果，如果是 true，则取 props.unCheckedValue，否则取 props.checkedValue（第 11 行）。获取 checkedValue 和 unCheckedValue 的值，将其再次更新到 modelValue 对象中，即可更新自定义的值（第 12 行）。

在 examples 演示库中使用 checked-value 和 unchecked-value 传入自定义的值（第 16 行）。

11.10 async：异步

在 button 组件、checkbox 组件、radio 组件中均已实现异步（async），都通过判断属性 beforeChange 是否为函数或者 Promise 对象，并加载 loading 图标实现加载交互过程。switch 组件同理，如代码清单 library-11-13 所示。

代码清单 library-11-13

```
1.  > packages/components/switch/src/index.vue: template
2.  <button type="button" ...省略代码, ns.is('disabled', isDisabled || isLoading)]">
3.    <template v-if="isLoading">
```

```
4.      <a-icon><Loading :class="[ns.is('loading-transition', isLoading)]"
/></a-icon>
5.    </template>
6.    <template v-else>
7.      <template v-if="centerIcon">
8.        ...省略代码
9.      </template>
10.   </template>
11. </button>
12.
13. > packages/components/switch/src/index.vue: script
14. import { types } from "@ui-library/utils";
15. import { Loading } from "@ui-library/icons"
16. const props = defineProps({
17.   ...省略代码
18.   beforeChange: [Function, Promise]
19. })
20. const isLoading = ref(false) // loading
21. /** 点击事件 */
22. const handleSwitch = () => {
23.   if(isDisabled.value || isLoading.value) { return }
24.   /** 不是异步 */
25.   const { beforeChange } = props
26.   if(!beforeChange) {
27.     changeEvent()
28.     return
29.   }
30.   /** 判断 Promise 对象 */
31.   const isPromise = types().isPromise(beforeChange())
32.   if(!isPromise) { return flase }
33.   isLoading.value = true
34.   beforeChange(props?.params).then(() => {
35.     changeEvent()
36.     isLoading.value = false
37.   }).catch(() => {
38.     isLoading.value = false
39.   })
40. }
```

根据 UI 组件库设计稿，switch 组件的加载图标渲染在中心圆上，因此在使用 v-if 条件判断变量 isLoading 为 true 时，渲染加载图标（第 3~5 行），否则渲染中心圆的其他图标（第 6~10 行）。

在 defineProps 对象中添加属性 beforeChange 的类别为 Function 和 Promise。在 handleSwitch 方法中判断 props.beforeChange 是否存在，如果不存在，那么可直接调用 changeEvent 方法切换 switch 组件的状态（第 25~29 行）；如果存在，则调用 types() 的 isPromise 方法判断 props.beforeChange 是否是 Promise 对象，并赋给变量 isPromise（第 31 行）。然后判断变量 isPromise 是否为 Promise 对象，如果不是，则阻止程序执行（第 32 行）；如果是，则设置 isLoading

的值为 true，渲染加载图标。在等待回调.then 时重置 isLoading 的值为 false，并调用 changeEvent() 方法切换 switch 组件的状态。如果回调了.catch，那么只需要设置 isLoading 的值为 false。

要重点注意的是，handleSwitch 方法在预先判断"禁用"或"加载"状态时，阻止了所有逻辑，这是为了防止多次触发，也是解决防抖问题的一种方案（第 23 行）。

11.11　transition：过渡动画

过渡动画（transition）是指在 switch 组件的切换过程中，中心圆底层文字和图标的变化效果。动画效果可以帮助用户清楚地识别视觉变化，从而增强对操作的控制感和理解。这种视觉反馈也可以使用户更加专注于状态的变化，提高交互的吸引力和可感知性。

UI 组件库设计稿中设计了"缩放"和"滑动"两种过渡动画效果，我们将使用 CSS3 的 transform 过滤模式实现，如代码清单 library-11-14 所示。

代码清单 library-11-14

```
1.  > packages/components/switch/src/index.vue: template
2.  <span :class="[ns.e('inner'), transitionModule]" v-if="!centerIcon">
3.    <span :class="[...省略代码, ns.is('checked', isChecked)]"> ...省略代码</span>
4.    <span :class="[...省略代码, ns.is('checked', !isChecked)]"> ...省略代码</span>
5.  </span>
6.
7.  > packages/components/switch/src/index.vue: script
8.  const props = defineProps({
9.    ...省略代码
10.   transition: { type: String, default: 'scale' }, // 默认缩放模式
11. })
12. /** 图标动画模式 */
13. const transitionModule = computed(() => `transition-${props.transition}`)
14.
15. > ackages/theme/src/switch.scss
16. @include e(inner) {
17.   ...省略代码
18.   &.transition-scale {
19.     > span {
20.       transform: scale(0);
21.       @include s((checked)) {
22.         transform: scale(1);
23.       }
24.     }
25.   }
26.   &.transition-slide {
27.     > span:first-child {
28.       transform: translateX(-100%);
29.       @include s((checked)) {
30.         transform: translateX(0%);
31.         + span { transform: translateX(100%); }
32.     }
```

```
33.     }
34.   }
35. }
36.
37. > examples/src/App.vue
38. <!-- 默认缩放 -->
39. <a-switch type="warning" checked-text="开"  unchecked-text="关"></a-switch>
40. <!-- 滑动效果 -->
41. <a-switch type="danger" transition="slide" checked-text="开" unchecked-text="关
" ></a-switch>
```

通过类名处理 switch 组件的切换动画，在 defineProps 对象中定义属性 transition 为字符串类型，设置默认值为 scale（第 10 行），使用 computed 计算属性生成类名 transition-scale 并返回给变量 transitionModule（第 13 行），然后绑定到 ns.e('inner')（第 2 行）。接着为中心圆底层的两个 span 标签都生成类名 is-checked（第 3、4 行），但要注意的是，左边为 true，右边为 false。

（1）实现缩放（scale）过渡动画：在 switch.scss 文件中添加类名 transition-scale（第 18~25 行），设置子级的 span 元素的缩放属性 scale 为 0（第 20 行），也就是缩小至 0 的比例。然后使用@include s((checked))生成类名 is-checked，并设置 scale 为 1（第 21~23 行），放大至 1 的比例，从而实现图标或文字的缩小效果。

（2）实现滑动（slide）过渡动画：在 switch.scss 文件中添加类名 transition-slide（第 26~34 行），设置子级的第一个 span 元素的 translateX 为-100%（第 28 行），也就是使元素默认往左侧偏移。然后使用@include s((checked))为第一个 span 元素生成类名 is-checked（第 30 行），并设置 translateX 为 0，也就是还原位置。接着找到同级的下一个 span 元素，设置 translateX 为 100%（第 31 行），使之往右侧偏移。通过设置两个元素同时左右偏移，实现左右滑动的效果。

本章小结

switch 组件的开发重点在于理解组件自身的私有变量的应用，它独立于组件本身的样式处理。通过私有变量实现 switch 组件的尺寸、图标、文字等 UI 效果，并使用 CSS3 的 transition 实现组件的缩放和滑动过渡动画效果。

第 12 章 表单组件

表单（form）组件是网页开发中至关重要的部分，作用是使用户输入数据，并提交数据到服务器进行处理。常见的表单组件包括文本输入框、文本区域输入框、单选按钮、多选框、下拉选择框等。

在使用 form 组件时，要重点考虑可访问性、验证和用户体验。正确的标签、错误信息和验证能够确保用户有效地与表单进行交互。

12.1 渲染 form 组件

本章开发的 form 组件以校验数据为主，以简化版的模式开发，如图 12-1 所示。重点是使读者理解在 UI 组件库中，如何使用 form 组件进行数据校验，并根据校验失败状态抛出自定义错误信息。

图 12-1　form 组件

12.1.1 构建组件

根据主流的第三方 UI 组件库 Ant Design 或 Element plus 的 form 组件规则，其中主要包括 form 组件和 form-item 组件。因此，我们在开发 UI 组件库时也遵循这种做法，如代码清单 library-12-1 所示。

```
代码清单 library-12-1
1.  > packages/components/form/src/form.vue
2.  <form ::class="[ns.b()]">
3.    <slot />
4.  </form>
5.
6.  > packages/components/form/src/formItem.vue
7.  <div :class="[ns.b()]">
8.    <div :class="[ns.e('label')]">
9.      <label>这里是文本</label>
```

```
10.      </div>
11.      <div :class="[ns.e('control')]">
12.        <slot />
13.        <div :class="[ns.e('message')]">失败提示</div>
14.      </div>
15.    </div>
16.
17.  > examples/src/App.vue
18.  <a-form>
19.    <a-form-item>
20.      <a-input />
21.    </a-form-item>
22.  </a-form>
```

在 components 目录下新建 form 目录，并在该目录下新建 form.vue 和 formItem.vue 文件，也就是 form 组件和 formItem 组件。其中，form 组件相当于父级组件，formItem 组件是各个独立的控件，如 input、checkbox、switch 等。

form 组件使用 form 标签作为根元素，使用<slot />插槽渲染（第 2~4 行）。formItem 组件分为左侧文本区域和右侧控件区域（第 8~10、11~14 行），左侧文本使用 label 标签渲染（第 9 行），右侧控件使用<slot />插槽渲染（第 12 行），还需要渲染校验失败的文本提示元素（第 13 行）。

在 examples 演示库中使用 form 组件和 formItem 组件，并结合 input 组件渲染控件，如图 12-2 所示。

图 12-2　form 组件、formItem 组件、input 组件渲染控件

12.1.2　渲染组件

根据 UI 组件库设计稿中的 form 表单形式，CSS 样式主要体现在 formItem 组件的 label 文本区域和失败提示文案，form 组件只是根元素，不需要任何 CSS 样式。因此，下面根据 UI 组

件库设计稿实现 formItem 组件的样式，如代码清单 library-12-2 所示。

代码清单 library-12-2

```scss
1.  > packages/theme/src/common/componentVar.scss
2.  @function formVar($type: '') {
3.    $switch: (
4.      'size': ('default': getVarName('component-size', 'default')),
5.      'font-size': ('default': getVarName('font-size', 'default')),
6.    );
7.    @return $switch;
8.  }
9.
10. > packages/theme/src/formItem.scss
11. @include b(form-item) {
12.   @include set-component-var('form', formVar());
13.   display: flex;
14.   margin-bottom: 24px;
15.   color: var(#{getVarName('text-color', 'primary')});
16.   @include e(label) { width: 100px; }
17.   @include e(control) { ...省略代码 }
18.   @include e(control-inner) {
19.     display: flex;
20.     align-items: center;
21.     width: 100%;
22.     height: var(#{getVarName('form', 'size')});
23.   }
24.   @include e(label-inner) {
25.     display: inline-block;
26.     margin-right: 8px;
27.     line-height: var(#{getVarName('form', 'size')});
28.     height: var(#{getVarName('form', 'size')});
29.     font-size: var(#{getVarName('form', 'font-size')});
30.     width: 100%;
31.     white-space: nowrap;
32.     text-align: right;
33.     @include s((required)) {
34.       &:before {
35.         content: '*';
36.         color: var(#{getVarName('color', 'danger')});
37.         margin-right: 4px;
38.       }
39.     }
40.     @include s((colon)) {
41.       margin-right: 0;
42.       &:after {
43.         content: ':';
44.         margin: 0 8px 0 2px;
45.       }
46.     }
47.   }
48.   @include e(message) { ...省略代码 }
```

```
49.  }
50.
51. > packages/components/form/src/formItem.vue
52. <label :class="[...省略代码, ns.is('required', true), ns.is('colon', colon)]">
53. const props = defineProps({
54.   colon: { type: Boolean, default: true }
55. })
```

　　formItem 组件的可变化属性只有"尺寸"和"文本"，在 componentVar.scss 中定义"size"和"font-size"的私有变量（第 2~7 行）。formItem.scss 文件主要处理 e(label-inner)的 CSS 样式，需要绑定变量的属性是 line-height、height、font-size（第 27~29 行），设置 label 的宽度，默认为 100px（第 16 行），label 文本默认靠右侧对齐（第 32 行）。

　　重点在于 label 文本中的"*"和"："，这两个元素使用伪元素定义。如果存在类名 is-required，则在 label 文本前面生成"*"（第 33~39 行）；如果存在类名 is-colon，则在 label 文本后面生成"："（第 40~46 行）。这样在<label>标签上即可调用 ns.is 方法生成 is-required 和 is-colon 类名（第 52 行），渲染效果如图 12-3 所示。

图 12-3　渲染 formItem 组件

12.1.3　文本区域

　　文本区域是 label 标签的文本渲染区域，用于描述表单字段，帮助用户理解应该在该字段中输入何种信息。清晰明了的 label 文本可以提升用户体验，降低用户在填写表单时出错的可能性。在业务上，为了能够清晰地描述表单字段，会设置可长可短的 label 文本，甚至可以调整 label 文本的对齐方式。因此，在组件中应该提供可配置的属性，供用户配置所需要的效果，如代码清单 library-12-3 所示。

代码清单　library-12-3

```
1. > packages/components/form/src/formItem.vue: template
2. <div :class="[ns.e('label')]" :style="[width]">
3.   <label :class="[...省略代码]" :style="[textAlign]" v-if="label">{{ label }}</label>
4. </div>
5.
6. > packages/components/form/src/formItem.vue: script
7. import { useNamespace, useStyle } from '@ui-library/hook';
8. const uStyle = useStyle();
9. const props = defineProps({
10.   label: { type: String, default: '' },            // 自定义文本
```

```
11.    labelWidth: { type: String, default: '' },    // 文本区域宽度
12.    align: { type: String, default: '' }          // 文本对齐
13. })
14. const width = computed(() => uStyle.width(props.labelWidth))
15. const textAlign = computed(() => uStyle.align(props.align))
```

在 defineProps 对象中定义 label、labelWidth、align 属性。label 用于自定义文本内容，可直接渲染（第 3 行）；labelWidth 用于设置文本区域的宽度，可调用 hook 的 useStyle 对象的 width 方法（第 14 行），然后绑定到 style 对象（第 2 行）；align 用于设置文本的对齐方式，同样调用 useStyle 对象的 align 方法（第 15 行），然后绑定到 label 标签的 style 对象（第 3 行），渲染效果如图 12-4 所示。

图 12-4 设置文本区域

12.2 AsyncValidator：校验库

async-validator 是一个轻量级库，主要功能是对数据进行异步校验。async-validator 提供了一种灵活且强大的方式来定义不同类型数据的校验规则，例如字符串、数字、数组和对象，并支持 Promise 风格的异步校验。

为什么会提到 async-validator？这是因为在现代前端开发中，主流的第三方 UI 组件库 antDesign 和 Element Plus 的表单数据校验模块均采用 async-validator 库。因此，我们在开发 form 表单校验模块时也采用 async-validator 库。

读者可根据下面的代码了解 async-validator 库的基本用法。

```
代码
1.  import AsyncValidator from 'async-validator'
2.
3.  const rules = {
4.    name: { type: 'string', required: true },
5.    age: [
6.       { type: 'number', required: true },
7.       { min: 2, max: 6 }
8.    ]
9.  };
10.
11. const validator = new AsyncValidator(rules);
12. validator.validate({ name: 'muji' }, (errors, fields) => { // 处理结果 });
```

13. `validator.validate({ age: 18 }, (errors, fields) => { // 处理结果 });`

async-validator 库的校验规则定义如变量 rules 所示（第 3~9 行），以 key/value 的方式一一对应，也就是 key 对应校验规则。校验规则可以是对象和数组类型，如属性 name 为对象类型（第 4 行）、属性 age 为数组类型（第 5~8 行）。引入 async-validator 库，在实例化时传入 rules 校验规则（第 11 行），接着调用 validate 方法，传入与 rules 校验规则相同的 key 及需要校验的数据，即可实现校验（第 12、13 行）。

async-validator 库的 validate 方法是一个 Promise 对象，可以根据 .then 和 .catch 处理成功和失败的结果。

提示：读者如需了解更多关于 async-validator 库的知识，可参考链接 12-1。

12.3 rules：数据规则

async-validator 库以 key/value 的形式定义校验规则，在校验数据时，传入与校验规则相同的 key 以及需要校验的 value。那么在 form 表单中该如何传输数据呢？

由于 form 表单是由 form 组件和 formItem 组件组成的，而交互控件在 formItem 组件的插槽中渲染，并且数据也被绑定在控件的 v-model 中，导致控件与 form 和 formItem 两个组件没有关联性。但是我们可以发现，每个控件的 v-model 绑定的 key 并不一样，因此可以把 v-model 的 key 传入 formItem 组件，这样就可以获取 async-validator 库所需的 key。将数据以 JSON 对象的方式传入 form 组件，然后将数据注入 formItem 组件，便可通过 JSON 对象获取指定 key 的数据，满足 async-validator 库的 key/value 的形式。如代码清单 library-12-4 所示。

代码清单 library-12-4

```
1.  > packages/components/form/src/constant.js
2.  export const FORM_KEY = Symbol('formKey')
3.  export const FORM_ITEM_KEY = Symbol('formItemKey')
4.
5.  > packages/components/form/src/composables/use-form-item.js
6.  import { inject } from "vue"
7.  import { FORM_KEY, FORM_ITEM_KEY } from "../constant"
8.  export function useFormItem(){
9.      const formContent = inject(FORM_KEY, undefined)
10.     const formItemContent = inject(FORM_ITEM_KEY, undefined)
11.     return { formContent, formItemContent }
12. }
13.
14. > packages/components/form/src/form.vue
15. import { FORM_KEY } from "./constant"
16. const props = defineProps({
17.     model: { type: Object, default: () => ({}) },
18.     rules: { type: Object, default: () => ({}) }
19. })
20. provide(FORM_KEY, {...toRefs(props)})
```

```
21.
22. > packages/components/form/src/formItem.vue: script
23. import { useFormItem } from "./composables"
24. const { formContent } = useFormItem()
25. const props = defineProps({
26.   ...省略代码
27.   prop: { type: String, default: '' }
28. })
29.
30. console.log('key:', props.prop, '---value:', formContent?.model.value[props.prop])
31.
32. > examples/src/App.vue
33. <a-form :model="formData">
34.   <a-form-item prop="name" label="姓名">...省略代码</a-form-item>
35.   <a-form-item prop="age" label="年龄">...省略代码</a-form-item>
36. </a-form>
37. const formData = ref({ name: "A总", age: 18 })
```

在父级组件传递数据到子级组件时，可以使用 provide 对象指定 key，将数据注入子级组件，子级组件使用 inject 对象接收父级组件的数据。在 src 目录下建立 constant.js，并定义 FORM_KEY 和 FORM_ITEM_KEY 两个变量（第 2、3 行）。以组合式函数的方式获取 form 组件注入的对象（第 6~12 行），useFormItem 组合式函数定义变量 formContent 和 formItemContent，并使用 inject 对象获取数据供子级组件使用。

在 form 组件的 defineProps 对象中定义 model 和 rules 两个属性并赋给变量 props（第 16~19 行），使用 provide 对象注入变量 props（第 20 行）。然后在子级组件 formItem 中引入组合式函数 useFormItem，并解构变量 formContent（第 23、24 行）。在 defineProps 对象中定义属性 prop 接收传入的 key，最后通过解构的变量 formContent?.model 获取 form 组件注入的数据，结合 formItem 组件的 props.prop 获取数据（第 30 行）。在 examples 演示库中根据 form 组件和 formItem 组件定义的规则传入数据（第 33~37 行），打印结果如图 12-5 所示。

```
key: name ---value: A总
key: age ---value: 18
```

图 12-5　获取的数据

12.4　validate：校验函数

校验函数是校验数据的入口，所有需要校验数据的控件都要调用 validate 方法。但并不需要为每个控件添加 validate 方法，而是将其定义在 formItem 组件中，使用 provide 注入的方式在控件中调用，如代码清单 library-12-5 所示。

代码清单 library-12-5

```
1.  > packages/components/form/src/formItem.vue
2.  const validate = (trigger) => {
3.    console.log('触发类型：', trigger)
4.  }
5.  provide(FORM_ITEM_KEY, { validate })
6.
7.  > packages/components/form/index.js
8.  ...省略代码
9.  export * from "./src/composables"
10.
11. > packages/components/input/src/index.vue
12. import { useFormItem } from "@ui-library/components"
13. const { formItemContent } = useFormItem()
14. const props = defineProps({
15.   ...省略代码
16.   validateEvent: { type: Boolean, default: true }
17. });
18. const { isFocus, isBlur, ...省略代码 } = useEvent({
19.   afterBlur(){ props.validateEvent && formItemContent?.validate('blur') }
20. })
21.
22. watch(() => modelValue.value, () => {
23.   props.validateEvent && formItemContent?.validate('change')
24. })
25.
26. > packages/hook/use-event/index.js
27. export const useEvent = (options = {}) => {
28.   ...省略代码
29.   const blurEvent = (e) => {
30.     options?.afterBlur?.() // 失焦后
31.   }
32. }
```

在 formItem 组件中定义 validate 方法，并定义参数 trigger 接收触发类型（第 2~4 行），使用 provide 对象注入 validate 方法（第 5 行）。在 input 组件中引入组合式函数 useFormItem 并解构 formItemContent（第 12、13 行）。在 defineProps 对象中定义属性 validateEvent 为布尔类型，默认值为 true（第 16 行），表示需要校验。

如果 input 组件需要在输入数据的过程中校验数据，那么可以使用 watch 监听 modelValue 的变化，只要 modelValue 的值发生了变化，就调用 formItem 组件中的 validate 方法，并传入触发类型 change（第 22~24 行）。如果 input 组件需要在失焦时校验数据，那么可以在 blurEvent 方法中调用校验方法。但由于 blurEvent 方法是在组合式函数 useEvent 中扩展的，无法直接调用 validate 方法。因此，可以在 useEvent 函数中以 JSON 对象的方式传入自定义方法 afterBlur（第 18~20 行），然后在 blurEvent 方法中执行 afterBlur 方法（第 30 行）。

通过在 formItem 组件中注入 validate 方法，在 input 组件中调用 validate 方法，这样就可以在输入框中输入内容或失焦时打印触发类型结果（第 3 行），如图 12-6 所示。

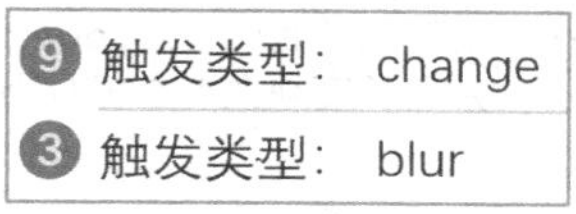

⑨ 触发类型：change

❸ 触发类型：blur

图 12-6　触发类型结果

12.5　trigger：校验规则类型

在 12.4 节中介绍了控件调用 formItem 组件的 validateEvent 方法获取 change 和 blur，在这两种触发方式中，每种控件都是可以自定义的，如 checkbox 组件、radio 组件、select 组件均可定义 change 触发方式。我们在定义校验规则时，可以通过 change 和 blur 确定规则应该用 change 触发还是用 blur 触发，或者是两者均可，如代码清单 library-12-6 所示。

代码清单 library-12-6

```
1.  > packages/components/form/src/formItem.vue: script
2.  const props = defineProps({
3.    ...省略代码
4.    rules: { type: [Object, Array], default: () => ([]) }
5.  })
6.  /** 转换为数组 */
7.  const convertArray = (rules) => {
8.    return rules
9.        ? Array.isArray(rules) ? rules : [rules]
10.       : []
11. }
12. /** 校验规则集合 */
13. const initRules = computed(() => convertArray(props.rules))
14. /** 根据触发类型过滤规则 */
15. const filterRules = (trigger) => {
16.   const rules = initRules.value
17.   return rules.filter((rule) => {
18.     if (!rule.trigger || !trigger) return true
19.     if (Array.isArray(rule.trigger)) {
20.       return rule.trigger.includes(trigger)
21.     } else {
22.       return rule.trigger === trigger
23.     }
24.   })
25. }
26. const validate = (trigger) => {
27.   const rules = filterRules(trigger)
28.   console.log('触发类型：', rules)
29. }
30.
31. > examples/src/App.vue
32. <a-form-item prop="name" label="姓名" label-width="150px" :rules="rules">...省略代码
33. const rules = ref([
```

```
34.       { required: true, message: "请输入姓名", trigger: 'blur' },
35.       { min: 2, max: 8, message: "姓名长度须大于2位且小于8位", trigger: 'change' },
36.    ])
```

在 formItem 组件的 defineProps 对象中定义属性 rules 为对象和数组类型（第 4 行），在 examples 演示库中使用属性 rules 传入校验规则（第 32、33~36 行），校验规则的属性 trigger 分别是 blur 和 change（第 34、35 行），与 formItem 组件的 validate 方法获取的触发类型一致。

在 formItem 组件的 validate 方法中调用 filterRules 方法，并传入参数 trigger 触发类型获取对应的 blur 和 change 类型的校验规则（第 27 行）。由于 defineProps 对象的 rules 可以为对象或数组类型，因此定义了 convertArray 方法，将传入的校验规则转换为数组类型（第 7~11、13 行）。在该方法中使用两次三元运算逻辑，优先判断是否存在校验规则（第 8 行）。如果存在，则再次判断是否为数组类型。如果是，则直接返回；如果不是，则使用"[]"将其包裹为数组类型（第 9 行）。如果不存在校验规则，则返回空数组（第 10 行）。

对于变量 initRules，使用 computed 计算属性调用 convertArray 方法传入 props.rules，通过 convertArray 方法处理的结果均是 Array，如图 12-7 所示。

图 12-7   convertArray 方法转换校验规则为 Array

在 filterRules 方法中，将变量 initRules 的校验规则集合赋给变量 rules（第 16 行），使用 filter 方法过滤规则集合。优先判断规则中是否存在 trigger 属性，如果不存在，也就是没有配置 trigger，则返回 true（第 18 行），表示 change 和 blur 都可以触发；如果 trigger 属性为数组类型，则使用 includes 判断 trigger 属性是否包含与参数 trigger 相同的触发方式，如果包含，则返回 true（第 19 行）。最后判断 trigger 属性的值是否等于参数 trigger 的值，如果等于，则返回 true，否则返回 false（第 22 行）。在过滤完校验规则后，将最终的结果打印到控制台中查看是否正确（第 28 行），如图 12-8 所示。

图 12-8   根据属性 trigger 过滤校验规则

12.6　merge：合并校验规则

由于在 form 组件和 formItem 组件中可以定义属性 rules 传入校验规则，用户在使用组件的过程中也可能向 form 组件和 formItem 组件中传入校验规则，因此需要将两个组件的 rules 属性数据合并（merge），如代码清单 library-12-7 所示。

代码清单 library-12-7

```
> packages/components/form/src/formItem.vue
const initRules = computed(() => {
  const propRules = convertArray(props.rules)   // formItem 组件的校验规则
  const formRules = fromContent?.rules?.value   // form 组件的校验规则
  if(formRules && props.prop) {
    const _rules = formRules[props.prop]
    _rules && propRules.push(...convertArray(_rules))
  }
  return propRules
})

> examples/src/App.vue
<a-form :model="formData" :rules="formRules">
  <a-form-item prop="name" label="姓名" :rules="rules">...省略代码</a-form-item>
  <a-form-item prop="age" label="年龄">...省略代码</a-form-item>
</a-form>
const formRules = ref({
  name: [{ min: 2, max: 8, message: "这是 form 组件定义的校验规则", trigger: 'change' }]
})
const rules = ref([
  { required: true, message: "请输入姓名", trigger: 'blur' },
  { min: 2, max: 8, message: "姓名长度须大于 2 位且小于 8 位", trigger: 'change' },
])
```

在 examples 演示库中定义 JSON 对象格式的校验规则（第 17~19 行），其中，name 与 prop 属性的值对应（第 14 行）。在 formItem 组件的 initRules 方法中定义变量 propRules，获取 props.rules 的校验规则（第 3 行）；定义变量 formRules，获取 form 组件 rules 属性的校验规则（第 4 行）。接着当判断变量 formRules 和 props.prop 都存在时，通过 props.prop 获取变量 formRules 指定 key 的校验规则（第 6 行），将其追加到变量 propRules 中（第 7 行），最后返回合并后的校验规则（第 9 行）。

通过合并校验规则，再把结果传入 filterRules 方法，即可得到新的校验规则的集合，如图 12-9 所示。

```
触发类型:  change ---- 校验规则:
▼ (2) [Proxy(Object), Proxy(Object)] i
  ▶ 0: Proxy(Object) {min: 2, max: 8, message: '姓名长度须大于2位且小于8位', trigger: 'change'}
  ▶ 1: Proxy(Object) {min: 2, max: 8, message: '这是合并的校验规则', trigger: 'change'}
    length: 2
  ▶ [[Prototype]]: Array(0)
触发类型:  blur ---- 校验规则:
▼ [Proxy(Object)] i
  ▶ 0: Proxy(Object) {required: true, message: '请输入姓名', trigger: 'blur'}
    length: 1
  ▶ [[Prototype]]: Array(0)
```

图 12-9 合并后的校验规则

12.7 validate：数据校验

数据校验建立在对校验规则和 async-validator 库的处理之上。12.2 节中介绍了 async-validator 库的使用，12.6 节中介绍了校验规则的合并处理，下面来实现数据校验结果的提示逻辑，如代码清单 library-12-8 所示。

代码清单 library-12-8

```
1.  > packages/components/form/src/formItem.vue: template
2.  <transition :name="`${ns.namespace}-form-message`">
3.    <div v-if="validateMessage" :class="[ns.e('message')]">{{ validateMessage }}</div>
4.  </transition>
5.
6.  > packages/components/form/src/formItem.vue: script
7.  const validateMessage = ref('')              // 校验结果
8.  const validate = (trigger) => {
9.    const rules = filterRules(trigger)         // 校验规则
10.   const propName = props.prop                // 获取 key
11.   const formModel = fromContent?.model?.value  // 获取数据集合
12.   // 实例化库
13.   const validator = new AsyncValidator({
14.     [propName]: rules
15.   })
16.   // 开始校验
17.   return validator.validate(
18.     { [propName]: formModel[propName] },
19.     { firstFields: true }
20.   ).then(() => {
21.     onalidateSuccess()
22.   }).catch(({ errors, fields }) => {
23.     validateFailed(errors)
24.   })
25. }
26.
```

```
27.  const onalidateSuccess = () => {
28.    validateMessage.value = ''
29.  }
30.
31.  const validateFailed = (errors) => {
32.    validateMessage.value = errors?.[0].message
33.  }
```

validate 方法是 12.4 节定义的校验函数入口（第 8 行），只要控件触发 blur 事件或 change 事件，就会回调 validate 方法。变量 rules 已经存储了处理后的校验规则，下面根据 12.2 节中的 async-validator 库的格式来处理校验逻辑。

在 validate 方法中先获取 props.prop，也就是将由 formItem 组件传入的 prop 属性的名称赋给变量 propName（第 10 行），再获取由 form 组件传入的数据集合（第 11 行），然后实例化 async-validator 校验库，并将以 JSON 对象格式传入的变量 propName 和 rules 校验规则赋给变量 validator（第 13、14 行）。接着调用 async-validator 库实例化对象的 validate 方法，并传入 propName 和要校验的数据 formModel[propName]（第 17、18 行），同时设置 firstFields 为 true（第 19 行），表示在指定字段的第一个规则验证错误时执行回调，不再处理相同字段的验证规则。

最后回调 .then 和 .catch（第 20、22 行），分别触发 onalidateSuccess 和 validateFailed 方法。onalidateSuccess 方法只用于清空变量 validateMessage 的值（第 28 行），而 validateFailed 方法用于为变量 validateMessage 赋值（第 32 行），也就是获取校验规则 message 属性的值，将其更新后即可显示校验失败的错误提示（第 3 行），如图 12-10 所示。

图 12-10　校验失败的错误提示

提示：form 组件校验失败的文本显示和隐藏动画使用了 Vue.js 3 内置的组件 transition，文本的动画效果的 CSS 样式在 transition.scss 文件中。本节不展开讲解 transition 组件的应用，详见 13.3 节。

12.8　submit：提交校验

提交校验是指在点击保存表单或提交表单的按钮时，对表单中的所有数据再一次进行校验。这么做的原因是确保用户已经填写了有效数据，防止出现误写、漏写等情况。

12.8.1 存储 formItem 组件数据

在点击保存表单或提交表单的按钮时校验数据，其实是在 form 组件中触发每个 formItem 组件的 validate 方法，因此需要把 formItem 组件的 validate 方法存储到 form 组件中，如代码清单 library-12-9 所示。

```
代码清单 library-12-9
1.  > packages/components/form/src/form.vue
2.  //所有字段
3.  const modelFields = []
4.  //获取所有字段函数
5.  const pushField = (context) => modelFields.push(context)
6.  //注入对象
7.  provide(FORM_KEY, {...toRefs(props), pushField})
8.  //暴露对象
9.  defineExpose({ validate })
10.
11. > packages/components/form/src/formItem.vue: script
12. const context = reactive({
13.   ...toRefs(props),
14.   validate
15. })
16. provide(FORM_ITEM_KEY, context)
17. onMounted(() => {
18.   props.prop && formContent.pushField(context)
19. })
```

在 form 组件中存储 formItem 组件的数据，需要在 form 组件中定义方法，由 provide 对象注入（第 5、7 行）。formItem 组件挂载完成后，如果存在 props.prop，则调用 form 组件注入的 pushField 方法（第 18 行），把 formItem 组件的 props 对象和 validate 方法作为参数传给 pushField 方法，如图 12-11 所示。这样 form 组件便可以使用 push 方法将 formItem 组件的数据存储到变量 modelFields 中。

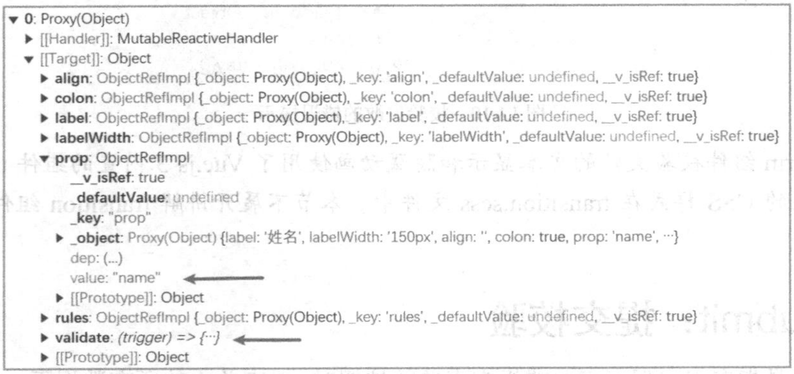

图 12-11 formItem 组件的 props 对象和 validate 方法

12.8.2 调用 form 组件校验

在点击保存表单或提交表单的按钮时校验数据，需要 form 组件暴露一个方法供外部使用，在该方法中循环变量 modelFields，触发每个对象的 validate 方法，也就是 formItem 组件的 validate 方法，如代码清单 library-12-10 所示。

代码清单 library-12-10

```
> packages/components/form/src/form.vue
const validate = async (fields) => {
  let verificationError = [];
  for(const field of modelFields) {
    try {
      await field.validate()
    } catch (fields) {
      verificationError = [...verificationError, ...fields]
    }
  }
  if(!verificationError.length) return true
  return Promise.reject(verificationError)
}
provide(FORM_KEY, {...toRefs(props), pushField})
defineExpose({ validate })

> examples/src/App.vue: template
<a-form ref="formRef" :model="formData" :rules="formRules">
...省略代码
<a-form-item><a-button @click="submit">提交</a-button></a-form-item>
</a-form>

> examples/src/App.vue: script
const formRules = ref({
  name: [{ required: true, message: "请输入姓名", trigger: 'blur' }],
  age: [{ required: true, message: "请输入姓名", trigger: 'blur' }]
})

const submit = () => {
  formRef.value.validate().then(res => {
    console.log('成功')
  }).catch((err) => {
    console.log(111, err)
  })
}
```

在 form 组件中定义 validate 方法，并使用 defineExpose 对象将其暴露出去（第 2、15 行）。在 validate 方法中声明变量 verificationError，用于存储校验错误的数据，可供外部使用（第 3 行）。接着使用 for...of 循环变量 modelFields（第 4 行），由于 formItem 组件的 validate 方法返回的是 Promise 对象，因此需要结合 try...catch 处理逻辑（第 5、6 行）。在调用 validate 方法时，如果返回的结果是 reject，则表示错误状态，便会进入 catch，并将错误信息存储至变量

verificationError（第7~9行）。

  for...of 循环结束后，判断变量 verificationError 的长度是否存在，如果不存在，则表示全部校验通过，直接返回 true（第 11 行）。如果变量 verificationError 中有数据存在，则表示有错误信息，就返回 Promise.reject，并把错误信息带出去（第 12 行）。在 examples 演示库中调用 form 组件的 validate 方法，回调.then 表示校验通过（第 30、31 行），回调.catch 表示有错误，可以打印出校验错误信息（第 32、33 行），如图 12-12 所示。

```
错误信息:   ▼ (2) [{…}, {…}]  ⓘ
            ▶ 0: {message: '请输入姓名', fieldValue: '', field: 'name'}
            ▶ 1: {message: '请输入姓名', fieldValue: '', field: 'age'}
```

图 12-12　校验错误信息

12.8.3　指定字段校验

  指定字段校验是指在 form 表单中存在多个字段且只针对某几个特定的字段进行校验。被指定的字段可能需要满足定制化的验证规则，为特定需求提供个性化的校验功能，以适应特定业务场景的需求，如代码清单 library-12-11 所示。

代码清单 library-12-11

```
1.  > packages/components/form/src/form.vue
2.  const validate = async (fields) => {
3.    // 存储需要校验的字段
4.    const validateFields = filterFields(fields) || modelFields
5.    let verificationError = [];
6.    for(const field of validateFields) {
7.      try {
8.        await field.validate()
9.      } catch (fields) {
10.       verificationError = [...verificationError, ...fields]
11.     }
12.   }
13.   if(!verificationError.length) return true
14.   return Promise.reject(verificationError)
15. }
16. // 过滤需要校验的字段
17. const filterFields = (fields) => {
18.   if(!fields) { return null }
19.   const fieldArr = modelFields.filter((v) => fields.includes(v.prop))
20.   return !!fieldArr.length ? fieldArr : null
21. }
22.
23. > examples/src/App.vue
24. ...省略代码
25. const submit = () => {
26.   formRef.value.validate(['name']).then(...省略代码)
27. }
```

变量 modelFields 中已经存储了所有的校验规则且为数组类型，每个对象中都有属性 prop，如图 12-10 所示。将与属性 prop 相对应的对象过滤出来作为新的校验对象，即可实现对指定字段的校验。

在 examples 演示库中调用 form 组件的 validate 方法时，以数组的格式传入 name（第 26 行），在 form 组件的 validate 方法中定义参数 fields（第 2 行）。接着调用 filterFields 方法，在该方法中判断参数 fields 是否存在，如果不存在，则返回 null（第 18 行）。如果 fields 中存在数据，则使用 filter 循环变量 modelFields，在循环过程中判断属性 prop 的值是否包含在参数 fields 中（第 19 行），如果是，则返回 true，否则返回 false，并将过滤的结果赋给变量 fieldArr。最后使用三元运算判断 fieldArr.length 是否存在，如果存在，则返回 fieldArr，否则返回 null（第 20 行）。

变量 validateFields 通过"或"运算判断 filterFields 方法是否存在返回值，如果存在，则使用 filterFields 方法的返回值，否则仍然使用变量 modelFields（第 4 行），将 for...of 循环的变量改为 validateFields 即可（第 6 行）。字段 name 的校验错误信息如图 12-13 所示。

图 12-13　字段 name 的校验错误信息

12.9　reset：重置

重置（reset）功能允许用户将表单中所有已输入的字段恢复为初始状态，即清除用户输入的内容，使表单回到最初的空白或初始值状态。重置功能通常与提交功能相对应，帮助用户更轻松地管理表单内容，提供更好的用户体验，如代码清单 library-12-12 所示。

代码清单 library-12-12
```
1.  > packages/components/form/src/form.vue
2.  const reset = () => {
3.    const validateFields = modelFields
4.    for(const field of validateFields) {
5.      field?.resetField()
6.    }
7.  }
8.  defineExpose({ validate, reset })
9.
10. > packages/components/form/src/formItem.vue
```

```
11. const props = defineProps({...省略代码})
12. // 初始值
13. let initValue = null
14. // 校验结果
15. const validateMessage = ref('')
16. // 重置字段
17. const resetField = () => {
18.   props.prop && (formContent.model.value[props.prop] = initValue)
19. }
20.
21. onMounted(() => {
22.   if(props.prop) {
23.     formContent.pushField(context)
24.     initValue = formContent?.model?.value?.[props.prop]
25.   }
26. })
```

重置功能与提交功能的逻辑类似，都是在 form 组件中调用 formItem 组件的方法重写数据。在 form 组件中定义 reset 方法并暴露给外部使用（第 2、8 行），在 reset 方法中定义变量 validateFields，用于存储 modelFields（第 3 行）。然后与 validate 方法同理，使用 for...of 循环调用 formItem 组件的 resetField 方法（第 4~6 行）。

重置可以理解为对表单数据的还原，无论是空的值还是有默认数据。因此，当在生命周期函数 onMounted 中判断 props.prop 存在时，获取当前字段的数据存储至变量 initValue（第 24 行）。当重置方法 resetField 被触发时，依然判断 rops.prop 是否存在，如果存在，则将变量 initValue 存储的数据重新赋给对应的 formContent.model.value（第 17~19 行）。

12.10 required：必填标识

必填标识（required）的作用是为用户提供一种提示，可用于改善用户体验，引导用户迅速填写必要字段，减少因疏忽而导致表单无法提交的情况，如代码清单 library-12-13 所示。

代码清单 library-12-13
```
1. > packages/components/form/src/formItem.vue: template
2. <div :class="[ns.b(), ns.is('required', isRequired)]">...省略代码</div>
3.
4. > packages/components/form/src/formItem.vue: script
5. const isRequired = computed(() => initRules.value.some((rule) => rule.required))
```

是否需要显示必填标识的条件来自变量 initRules 的校验规则，如果在校验规则集合中存在 required 属性，则表示显示必填标识。因此，可用 some 方法进行判断（第 5 行）。

12.11 size：尺寸

form 组件和 formItem 组件是父子级的关系，与 checkbox 组件和 checkboxGroup 组件是相同的。form 组件和 formItem 组件都以插槽的模式渲染控件，因此在 form 组件中添加 size 属性时，需要以 provide 的方式注入，然后在控件中引入 useFormItem 组合式函数并解构 size 属性，如代码清单 library-12-14 所示。

代码清单 library-12-14

```
1.  > packages/components/form/src/formItem.vue: template
2.  <div :class="[ns.b(), ns.is('required', isRequired), ns.m('size', 
controlSize)]">
3.  
4.  > packages/components/form/src/formItem.vue: script
5.  const controlSize = computed(() => formContent?.size?.value)
6.  
7.  > packages/components/input/src/index.vue: script
8.  import { useFormItem } from "@ui-library/components "
9.  const { formContent, formItemContent } = useFormItem()
10. const controlSize = computed(() => props.size || formContent?.size?.value)
11. 
12. > packages/components/button/src/index.vue
13. import { useFormItem } from "@ui-library/components"
14. const { formContent } = useFormItem()
15. const controlSize = computed(() => isGroupSize || formContent?.size?.value)
```

引入 form 组件的组合式函数 useFormItem，并解构 formContent（第 8、9、13、14 行）。定义变量 controlSize，使用 computed 计算属性结合"或"运算获取 size 属性的值，并将其绑定至组件的 ns.m('size', controlSize)（第 5、10、15 行）。

提示：截止到本章，我们开发的组件有 input 组件、checkbox 组件、radio 组件、button 组件等，在为它们添加 size 属性时，都需要引入组合式函数 useFormItem 并解构 formContent，再为上述组件绑定 formContent 的 size 属性，因在语法上均一致，故本书不重复讲解。

本章小结

本章以介绍 form 组件的校验逻辑为主，通过校验库 Validator 实现各种数据的校验逻辑。通过开发 form 组件，可以更深入地理解校验规则的处理逻辑，以及触发校验规则的方式、合并校验规则、提交校验、重置、必填项等行为逻辑。

第 13 章
消息提示组件

消息提示（message）组件是应用程序中常见的交互元素，通常用于提醒用户有关的重要信息、警告或通知。通过弹出消息框或通知栏等形式，向用户提供直观的可视化反馈，以便用户及时理解和响应。清晰、明确的消息提示有助于提升用户体验，使用户更直观地与应用程序互动。

13.1　createVNode 函数

createVNode 函数是 Vue.js 3 中一个重要的底层 API，用于创建虚拟节点（VNode）。它允许开发人员以编程的方式生成 Vue.js 3 组件或 HTML 的虚拟元素，从而实现更灵活的动态渲染机制。

为什么要了解 createVNode 函数呢？回看前几章中开发的组件，如 button 组件、checkbox 组件、switch 组件等，这些类型的组件在实际业务中会产生数据交互、UI 交互、人机交互等。因此，在项目中需要引入 template 模块层，使用全闭合或半闭合标签的模式渲染组件。返观 message 组件，它的作用是展示通知和反馈信息，比如信息提示、警告、通知等，设计定位为用户界面的直接反馈工具，从而增强用户体验。由此可知，message 组件只用于消息提示，并没有其他特别的功能，那么在业务上将 message 组件引入 template 模块层，然后使用 v-if 或 v-show 显示或隐藏组件的方式实现消息提示，不仅会显得过于臃肿，还会增加用户的维护成本。因此，我们将使用 Vue.js 3 中的 createVNode 函数生成 message 组件，用户在使用过程中只需要调用 message 组件的方法，即可实现消息提示。

13.1.1　基本语法

根据 Vue.js 3 官方介绍，createVNode 函数有 3 个参数，分别是虚拟节点类型、虚拟节点相关的属性和事件监听器、子节点。如代码清单 library-13-1 所示。

代码清单　library-13-1

```
1. > examples/src/message.vue
2. <template><div>这是一个 Message 组件</div></template>
3.
4. > examples/src/App.vue: template
5. <a-button @click="test">消息提示</a-button>
6.
7. > examples/src/App.vue: script
8. import { createVNode, render } from 'vue'
9. import MessageNode from "./message.vue"
```

```
10.
11.  const test = () => {
12.    // 创建 div 元素
13.    const container = document.createElement('div');
14.    // 获取 body
15.    const body = document.body;
16.    // 创建虚拟节点
17.    const vnode = createVNode(
18.      MessageNode,
19.      {}
20.    )
21.    // 渲染组件到 container 上
22.    render(vnode, container);
23.    // 添加 container 到 body 中
24.    body.appendChild(container.firstElementChild);
25.  }
```

在渲染虚拟节点前，需要引入 createVNode 和 render 两个函数（第 8 行），再引入自定义组件 MessageNode（第 9 行）。定义 test 方法，并在该方法内创建 div 元素赋给变量 container，获取 body 元素（第 13、15 行）。接着调用 createVNode 函数（第 17 行），其中第 1 个参数用于渲染 MessageNode 组件（第 18 行），第 2 个参数则是 JSON 对象（第 19 行）。然后使用 render 函数将虚拟节点渲染到创建的元素 container 上（第 22 行）。最后将元素 container 的第一个子元素 firstElementChild 添加到 body 元素中，这样就完成了虚拟节点的渲染，如图 13-1 所示。

图 13-1　使用 createVNode 和 render 函数渲染虚拟节点

13.1.2　属性及事件

createVNode 函数的属性和事件由第 2 个参数传入。属性与常规组件一样，在父级组件传入，使用 props 对象接收子级组件，事件使用 "on 前缀加事件名称" 的驼峰式写法，如代码清单 library-13-2 所示。

代码清单 library-13-2
```
1.  > examples/src/message.vue: template
2.  <template><div>{{ content }}</div></template>
3.
4.  > examples/src/message.vue: script
5.  const props = defineProps({
6.    content: { type: String, default: '' }
7.  })
8.
9.  > examples/src/App.vue: script
```

```
10.  const test = () => {
11.    ...省略代码
12.    const vnode = createVNode(
13.      MessageNode,
14.      {
15.        content: "第 2 个参数传入了属性",
16.        onClick: () => {
17.          alert('测试 click 事件')
18.        }
19.      }
20.    )
21.  }
```

MessageNode 组件的 defineProps 对象定义属性 content 为字符串类型（第 6 行），由 createVNode 函数的第 2 个参数传入 content 属性（第 15 行），使之在 MessageNode 组件中渲染。click 事件由 onClick 绑定函数触发（第 16~18 行），如图 13-2 所示。

```
▶ <div id="app" data-v-app> ⋯ </div>
  <script type="module" src="/src/main.js?t=1715502345915"></script>
▼ <div>
    <div>第2个参数传入了属性</div>
  </div>
```

图 13-2　createVNode 函数的属性及事件

13.2　渲染 message 组件

在 UI 组件库设计稿中，message 组件有 4 种状态，分别是"信息""成功""警告"和"错误"。每种状态的图标或背景都有一种颜色，并且分为"无图标"和"关闭"两种类型。message 组件如图 13-3 所示，其中标注了 message 组件的尺寸。我们使用 createVNode 函数和 render 函数，结合 UI 组件库设计稿标注的尺寸来渲染 message 组件。

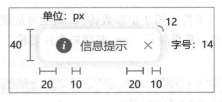

图 13-3　message 组件

13.2.1　构建组件

message 组件采用虚拟节点的模式进行渲染，本节将使用 createVNode 函数和 render 函数来构建 message 组件，如代码清单 library-13-3 所示。

代码清单 library-13-3

```js
1.  > packages/components/message/src/method.js
2.  import { createVNode, render } from 'vue'
3.  import MessageNode from "./message.vue"
4.  const createMessage = (options = {}) => {
5.    const container = document.createElement('div');
6.    const appendTo = document.body;
7.    const vnode = createVNode( MessageNode, {} )
8.    render(vnode, container); // 渲染组件到 container 上
9.    appendTo.appendChild(container.firstElementChild); // 添加 container 到 body
10. }
11. const message = (params = {}) => { createMessage() }
12. export default message
13.
14. > packages/components/message/index.js
15. import Method from "./src/method.js"
16. import { functionInstall } from "@ui-library/utils"
17. export const AMessage = functionInstall(Method, '$message') // 提供按需加载的方式
18. export default AMessage // 导出组件
19.
20. > packages/utils/install.js
21. export const componentInstall = (com) => {...省略代码 }
22. // 函数注册
23. export const functionInstall = (fn, name) => {
24.   fn.install = (app) => {
25.     fn._context = app._context // 挂载 Vue.js 3 实例
26.     app.config.globalProperties[name] = fn // 挂载全局
27.   }
28.   return fn
29. }
30.
31. > examples/src/App.vue
32. 方式1：全局方法
33. import { getCurrentInstance } from "vue"
34. const { appContext } = getCurrentInstance()
35. const test = () => {
36.   appContext.config.globalProperties.$message()
37.   console.log(appContext.config.globalProperties.$message._context) // 访问 Vue.js 3 实例
38. }
39.
40. 方式2：按需加载
41. import { AMessage } from '@ui-library/components';
42. const test = () => {
43.   AMessage()
44.   console.log('AMessage()', AMessage._context) // 访问 Vue.js 3 实例
45. }
```

message/src 目录中分别有 message.vue 和 message.js 文件，message.js 文件的 createVNode 函数的逻辑基本上与 13.1.1 节中的代码逻辑一致，只是在 message 方法中调用了 createMessage

方法（第 11 行），然后暴露 message 方法（第 12 行），message.vue 文件便是要渲染的组件。message/index.js 文件不再引入.vue 后缀类型的文件，而是引入 method.js 文件并命名为 Method（第 15 行），然后调用 functionInstall 传入 Method，并传入自定义名称$message，$message 用于 Vue.js 3 实例的全局注册（第 17 行）。

在 functionInstall 方法中定义 fn 和 name 参数，fn 接收的是 Method，name 接收的是$message。与 componentInstall 方法一致，将 install 作为属性赋给 fn（第 24 行），然后将 app._context 赋给 fn._context（第 25 行），接着将参数 name 的值挂载到 Vue.js 3 实例的全局方法的名称$message 上，最后返回 fn。components/index.js 和 packages/components.js 两个文件的引入方式与其他组件一致。

在 examples 演示库中使用"全局方法"和"按需加载"两种方式渲染 message 组件，并访问.context 实例（第 31~45 行），如图 13-4 所示。

图 13-4　message 组件结构及 Vue.js 3 实例

13.2.2　渲染组件

根据 UI 组件库设计稿中 message 组件的效果，message 组件分为"左侧图标"、"关闭图标"和"文本"3 个元素，并且是并排显示的。因此，建立 3 个 div 元素并引入 a-icon 组件渲染图标，如代码清单 library-13-4 所示。

```
代码清单　library-13-4
1.  > packages/components/message/src/message.vue
2.  <div :class="[ns.b()]" >
3.     <div :class="[ns.e('icon')]"><a-icon><Info /></a-icon></div>
4.     <div :class="[ns.e('content')]">这是一条信息</div>
5.     <div :class="[ns.e('close')]"><a-icon><Close /></a-icon></div>
6.  </div>
7.
8.  > packages/theme/src/message.scss
9.  @include b(message) {
10.    @include set-component-var('message', messageVar());
11.    ...省略代码;
12.    position: fixed;                    // 居于可视区定位
```

```
13.       left: 50%;                    // 左侧偏移 50%
14.       top: 16px;                    // 顶部偏移 16px
15.       z-index: 10001;               // 层级设置
16.       white-space: nowrap;          // 强制不换行
17.       transform: translateX(-50%);  // x轴-50%
18.       transition: .3s;
19.       @include e(wrap) {...省略代码}
20.       @include e(close) {
21.         ...省略代码
22.         opacity: .5;                // 50%透明度
23.       }
24.       @include e(icon) {
25.         ...省略代码
26.         @include sn-e(content) { margin-left: 10px; }
27.       }
28.       @include e(content) {...省略代码}
29.     }
```

  message 组件使用 1 个根元素 div 包裹 3 个子级 div 元素（第 2~6 行），根元素 div 使用 ns.b() 生成类名 a-message，子级 div 元素生成类名 a-message__icon、a-message__content、a-message__close。子级的第 1 个和第 3 个 div 元素分别使用 a-icon 组件渲染 Info 图标和 Close 图标。message 组件的渲染效果如图 13-5 所示。

图 13-5　message 组件的渲染效果

  本书中不会逐一讲解 message 组件的 CSS 样式，只罗列出重点的 CSS 样式说明，具体如下。

◎ **定位**：message 组件脱离文档流并居于浏览器可视区域，需使用 fixed 定位（第 12 行）。

◎ **垂直居中**：要使脱离文档的 fixed 定位元素实现垂直居中的效果，可设置 left 属性为 50%，使元素居于父级定位元素的一半向右侧偏移（第 13 行），并结合 transform 属性设置 translateX(-50%)，使元素居于自身的一半向 x 轴左侧偏移（第 17 行）。

◎ **文字不行换**：使用 white-space 属性设置值为 nowrap，强制文字不换行（第 16 行）。

◎ **偏移**：top 和 z-index 属性只用于产生偏移效果，在实际开发中，top 属性和 z-index 属性要通过计算得出结果，再使用行间样式写入 message 组件（第 14、15 行）。

◎ **图标与文本间距**：UI 组件库设计稿中标注了图标与文本的间距是 10px，并且图标元素与文本元素是兄弟节点关系，可以使用兄弟节点选择器"+"，选择紧跟在当前元素之后的元素。也就是找到跟随在图标元素之后的文本元素，再设置 10px 的 margin-left 左边距（第 24~27 行）。如果不存在图标元素，自然也无法找到文本元素，这也符合 UI 组件库设计稿 "无图标" 类型的效果。

◎ **关闭图标**：关闭图标相比于其他图标来说没那么明显，因此设置 opacity 属性为 0.5，即透明度为 50%（第 22 行）。

13.3 transition：过渡动画

过渡动画是指在用户界面中，message 组件由浏览器可视区顶部向下滑动的动画效果，平滑和自然的过渡动画效果既可以提升用户体验，使消息在界面上的呈现更生动、更吸引人，也可以使用户操作更流畅和连贯。

13.3.1 transition 组件

在 Vue.js 3 中，官方提供了内置组件 transition 和 transition-group 实现过渡动画。transition 会在一个元素或组件进入和离开 DOM 时应用动画。transition-group 会在一个 v-for 列表中的元素或组件被插入、移动或移除时应用动画。据此，我们选择 transition 组件即可，如代码清单 library-13-5 所示。

代码清单 library-13-5
```
1.  > packages/components/message/src/message.vue: template
2.  <transition :name="ns.b()"> // 生成 name 属性名称 a-message
3.      <div :class="[ns.b()]">...省略代码</div>
4.  </transition>
```

13.3.2 动画实现

过渡动画是通过 CSS 实现的，因此需要 class 类名。transition 组件的类名使用 name 属性生成，为 name 属性绑定 ns.b()方法便会生成 a-message。通过 name 属性分别为 "进入" 和 "离开" 自动生成 3 个类名：a-message-enter-from、a-message-enter-to、a-message-enter-active 和 a-message-leave-from、a-message-leave-to、a-message-leave-active。需要注意的是，类名 a-message 之后的部分是 transition 组件自动生成的，使用类名时需书写完整，如代码清单 library-13-6 所示。

代码清单 library-13-6
```
1.  > packages/components/message/src/message.vue: template
2.  <transition :name="ns.b()"> // a-message
3.      <div :class="[ns.b()]" :style="[customStyle]" v-show="visible">...省略代码</div>
4.  </transition>
5.
6.  > packages/components/message/src/message.vue: script
7.  const customStyle = computed(() => {
8.      return {
```

```
9.            top: '16px',
10.           zIndex: 10000
11.       }
12.   })
13.   const visible = ref(false)
14.   onMounted(() => {
15.       visible.value = true
16.   })
17.
18.   > packages/theme/src/transition.scss
19.   ...省略代码
20.   div.#{$namespace}-message-enter-from,
21.   div.#{$namespace}-message-leave-to {
22.       opacity: 0;
23.       transform: translate(-50%, -100%);
24.   }
```

在渲染 message 组件时提到了 top 属性和 z-index 属性使用计算后的值写入行间样式，因此先使用 computed 对象将 top 和 z-index 返回给变量 customStyle（第 7~12 行），然后将变量 customStyle 绑定到 message 组件中（第 3 行）。

transition 组件的"进入"和"离开"动作可由 v-if、v-show、<component>和改变特殊的 key 属性中的任意一种触发。因此，定义变量 visible 为 false（第 13 行），使用 v-show 绑定至 message 组件（第 3 行），然后在生命周期函数 onMounted 加载完成时设置变量 visible 的值为 true（第 14~16 行），最后加载 transition.scss 文件的 CSS 样式便能实现过渡动画效果（第 20~24 行）。

13.3.3 动画过程

transition 组件的动画效果是由 name 属性生成的类名和 message 组件类名相互结合产生的。将 message.scss 文件 transition 属性值改为 10s，使动画持续 10 秒，然后点击按钮生成 message 提示，在浏览器控制台中即可看到生成的 DOM 元素，如图 13-6 所示。

> ▶<div class="a-message a-message-enter-active a-message-enter-to" style="top: 16px; z-index: 10000;"> ⋯ </div> flex

图 13-6 message 组件的 DOM 元素

在图 13-6 中，类名是在 13.3.2 节中提到的"进入"阶段生成的，但我们会发现并没有 a-message-enter-from。这是因为 a-message-enter-from 是进入动画的起始状态，在元素插入之前被添加，并且在元素插入完成后的下一帧就被直接移除了，导致我们无法看到类名 a-message-enter-from。在类名 a-message-enter-from 中添加样式 opacity 的透明度为 0，translate 的 y 轴为 −100%，这两个值就是动画的"开始值"。当 a-message-enter-from 被移除时，opacity 和 translate 两个属性就要走向"结束值"，而结束值就是类名 a-message 的 opacity 和 translate，分别是透明度为 1 和 y 轴为 0。这就是 transition 组件"进入"动画的过程，如图 13-7 所示。

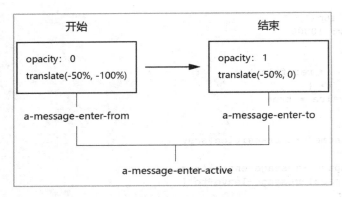

图 13-7　transition 组件"进入"动画的过程

transition 组件"离开"动画的过程和"进入"动画相似，只是使用的类名不同。为 message 组件的生命周期函数 onMounted 添加定时器，设置变量 visible 为 false，如代码清单 library-13-7 所示。

```
> packages/components/message/src/message.vue: script
onMounted(() => {
    visible.value = true
    setTimeout(() => {
        visible.value = false
    }, 12000)
})
```

代码清单 library-13-7

需要注意的是，setTimeout 设置了 12 秒的时延，这是因为 message 组件的过渡动画的持续时间是 10 秒，因此要过 10 秒后方可执行关闭动画。当执行到关闭动画时，message 组件的 DOM 元素会被添加上"离开"的类名，如图 13-8 所示。

▶<div class="a-message a-message-leave-active a-message-leave-to" style="top: 16px; z-index: 10000;"> ⋯ </div> flex

图 13-8　transition 组件"离开"动画

a-message-leave-from 是离开动画的起始状态，会在离开过渡动画被触发时被添加，并且在下一帧后被移除，因此也无法看到该类名。由于没有定义类名 a-message-leave-from，因此离开动画的 opacity 和 translate 两个属性的"开始值"来自类名 a-message。当离开动画执行时，会添加上类名 a-message-leave-to，该类名会覆盖类名 a-message 的值，再一次变回透明度为 0 和 y 轴为–100%的状态，执行关闭动画。transition 组件"离开"动画的过程如图 13-9 所示。

图 13-9　transition 组件"离开"动画的过程

13.3.4　钩子函数

transition 组件的作用是为元素添加进入和离开的过渡动画效果。transition 组件中提供了在过渡动画不同阶段的钩子函数，代码如下。

```
1.  <Transition
2.    @before-enter="onBeforeEnter"
3.    @enter="onEnter"
4.    @after-enter="onAfterEnter"
5.    @enter-cancelled="onEnterCancelled"
6.    @before-leave="onBeforeLeave"
7.    @leave="onLeave"
8.    @after-leave="onAfterLeave"
9.    @leave-cancelled="onLeaveCancelled"
10. >
11.   <!-- ... -->
12. </Transition>
13.
14. // 在元素被插入 DOM 之前被调用，用于设置元素的 "enter-from" 状态
15. function onBeforeEnter(el) {}
16.
17. // 在元素被插入 DOM 之后的下一帧被调用，用于开始进入动画
18. function onEnter(el, done) {
19.   done() // 调用回调函数 done 表示过渡结束。如果与 CSS 结合使用，则这个回调是可选参数
20. }
21. // 当进入过渡完成时调用
22. function onAfterEnter(el) {}
23.
24. // 当进入过渡且在过渡完成之前被取消时调用
```

```
25. function onEnterCancelled(el) {}
26.
27. // 在 leave 钩子之前调用
28. function onBeforeLeave(el) {}
29.
30. // 在离开过渡开始时调用，用于开始离开动画
31. function onLeave(el, done) {
32.   done()  // 调用回调函数 done 表示过渡结束。如果与 CSS 结合使用，则这个回调是可选参数
33. }
34.
35. // 在离开过渡完成，且元素已从 DOM 中移除时调用
36. function onAfterLeave(el) {}
37.
38. // 仅在 v-show 过渡中可用
39. function onLeaveCancelled(el) {}(
```

提示：如需了解更多关于 transition 组件的应用，可参考链接 13-1。

13.4 attribute：初始化属性

初始化属性用于定义 message 组件的默认效果，通过默认的属性配置，用户只需要调用 message 组件提供的方法即可渲染组件。如果用户需要定制 message 组件的效果，只需要根据提供的属性对默认属性的值进行覆盖，如代码清单 library-13-8 所示。

代码清单 library-13-8

```
1.  > packages/components/message/src/method.js
2.  const messageDefaults = { type: 'info', content: '' }
3.
4.  /** 初始化配置 */
5.  const initOptions = (params) => {
6.    // message 类型: string, vnode
7.    const options = !params || types().isString(params) ? { message: params } : params
8.    // 配置
9.    const config = { ...messageDefaults, ...options }
10.   return config
11. }
12.
13. const createMessage = (options = {}) => {
14.   const vnode = createVNode(
15.     MessageNode,
16.     { ...options }
17.   )
18. }
19.
20. const message = (params = {}) => {
21.   const normalized = initOptions(params)
22.   createMessage(normalized)
23. }
24.
```

```
25. > examples/src/App.vue
26. AMessage({content: '这是一条提示消息'})
```

在 method.js 文件中定义变量 messageDefaults，初始化属性 type 和 content（第 2 行）。然后声明函数 initOptions，使用扩展运算符合并变量 messageDefaults 和变量 options 的数据后再返回出去（第 9、10 行）。在 message 方法中调用函数 initOptions 并传入参数，initOptions 函数会自动处理结果并返回给变量 normalized（第 21 行），接着将得到的结果传入 createMessage 函数（第 22 行）。在 createMessage 函数中同样使用扩展运算符传入 createVNode 函数的第 2 个参数（第 16 行）。

在 examples 演示库中调用 AMessage 方法时可配置属性 content，传入自定义文本。当数据进入 initOptions 函数时（第 5 行），在函数中使用三元运算判断参数 params 是否存在，如果不存在 params 或者 params 为字符串类型时，使用 JSON 对象组合数据并赋给变量 options（第 7 行），否则保持参数 params 不变，最后扩展合并。message 组件的初始化属性定义效果如图 13-10 所示。

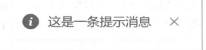

图 13-10　初始化属性定义

13.5　z-index：层级顺序

层级顺序也可以理解为堆叠顺序，如同纸张一样，z-index 的值越大，位置越靠上。z-index 非常重要，是面向整个 UI 组件库的全局属性。除 message 组件使用外，后续介绍的 modal 组件、drawer 组件中均会应用 z-index 属性。因此，在每次打开居于浏览器可视区窗口定位的组件时，都需要对 z-index 的值累加一次，使它的值比上一次的值大，如代码清单 library-13-9 所示。

代码清单 library-13-9
```
1.  > packages/hook/use-zindex/index.js
2.  import { ref, computed } from "vue"
3.  export const initZindex = 3000        // 全局默认值
4.  export const zIndex = ref(0)          // 累加值
5.  export const useZindex = () => {      // hook
6.    const currentZindex = computed(() => initZindex + zIndex.value)  // 当前的
7.    const nextZindex = () => zIndex.value++                          // 累加的
8.    return { currentZindex, nextZindex }
9.  }
10.
11. > packages/components/message/src/message.vue
12. ...省略代码;
13. import { useZindex } from "@ui-library/hook";
14. const { nextZindex, currentZindex } = useZindex()
```

```
15.    const customStyle = computed(() => {
16.      return {
17.        top: '16px',
18.        zIndex: currentZindex.value
19.      }
20.    })
21.
22.    onMounted(() => {
23.      nextZindex()
24.    })
```

在 hook 目录下新建 use-zindex 目录和 index.js 文件,作为全局的 z-index 属性逻辑文件。在 index.js 文件中定义两个变量 initZindex 和 zIndex(第 3、4 行),再定义 useZindex 组合式函数(第 5 行)。其中,nextZindex 方法用于对变量 zIndex 累加 1(第 7 行),变量 currentZindex 使用 computed 计算属性使变量 initZindex 与 zIndex 累加后的值相加,得到最新的结果(第 6 行),然后返回 currentZindex 和 nextZindex(第 8 行)。

在 message 组件中引入 useZindex 组合式函数,并解构 currentZindex 和 nextZindex(第 13、14 行)。重点在于生命周期函数 onMounted,需要先执行 nextZindex 方法使变量 zIndex 累加 1,变量 customStyle 的 zIndex 属性才能获取变量 currentZindex 的值。z-index 属性的效果如图 13-11 所示。

```
<div class="a-message" style="top: 16px; z-index: 3003;"> ⋯ </div>
```

图 13-11 z-index 属性的效果

13.6 top:顶部偏移

在 message 组件中采用 fixed 定位模式,通过设置属性 top 的值,可以调整组件与浏览器可视区窗口顶部的偏移距离,从而产生由上至下的滑动过渡动画效果。13.3 节中提到,过渡动画需要存在"开始值"和"结束值",因此,我们有必要了解当存在多个 message 组件时,每个 message 组件居于什么位置,也就是确定开始和结束位置的值。

message 组件的偏移位置如图 13-12 所示。图中展示了 3 个 message 组件,已知第 1 个 message 组件开始位置的值 top 为 0,结束位置的值是 offset 属性的值 16px,居于浏览器可视区顶部由上至下滑动;第 2 个 message 组件的虚线框表示开始位置,也就是从第 1 个 message 组件的底部开始,结束位置是第 1 个 message 组件底部的值加上 offset 属性的值;第 3 个 message 组件的虚线框表示开始位置,也就是从第 2 个 message 组件的底部开始,结束位置是第 2 个 message 组件底部的值加上 offset 属性的值。

图 13-12　message 组件的偏移位置

由上述分析可以推断，在对 message 组件的 top 属性值进行定位时，需要获取上一个 message 组件的底部位置。因此，我们首先要存储 message 组件，再获取指定的 message 组件，从而获取位置数据。

13.6.1　存储 message 组件

message 组件的存储过程相对简单，在每次执行到 createMessage 函数并渲染到页面后，便可存储 message 组件，如代码清单 library-13-10 所示。

代码清单 library-13-10

```
1.  > packages/components/message/src/message.vue
2.  const props = defineProps({
3.      id: { type: String, default: "" }
4.  });
5.
6.  > packages/components/message/src/method.js
7.  import { messageInstances } from "./instance"
8.  /** 唯一 id */
9.  let onlyId = 0;
10.
11. const createMessage = (options = {}) => {
12.     // id 唯一值
13.     const id = `message_${onlyId++}`
14.     const vnode = createVNode(MessageNode, { ...options, id })
15.     ...省略代码
16.     // 获取组件对象
17.     const vm = vnode.component
18.     const instance = {
19.         id,
20.         vnode,
21.         vm,
22.         handler: {
23.             close: () => {
24.                 vm.setupState.visible = false
25.             }
26.         },
27.         props: vm.props,
```

```
28.   }
29.   return instance
30. }
31.
32. const message = (params = {}) => {
33.   const normalized = initOptions(params)
34.   const message = createMessage(normalized)
35.   messageInstances.push(message)
36. }
```

message 组件在页面中会出现很多次，为了区分每个组件，需要添加自定义属性 id 作为唯一标识。声明变量 onlyId 为 0（第 9 行），在每次调用 createMessage 函数时使变量 onlyId 累加 1，并赋给变量 id（第 13 行），将其作为 createVNode 函数的第 2 个参数传入变量 id（第 14 行）。为 message 组件的 defineProps 对象定义相同名称的属性 id，接收 createVNode 函数第 2 个参数传入的 id（第 3 行）。

在 createMessage 函数完成对 message 组件虚拟节点的渲染后，将虚拟节点的组件 vnode.component 赋给变量 vm（第 17 行）。message 组件的偏移位置如图 13-13 所示，在 vnode.component 中要重点使用的参数是 props 和 setupState。

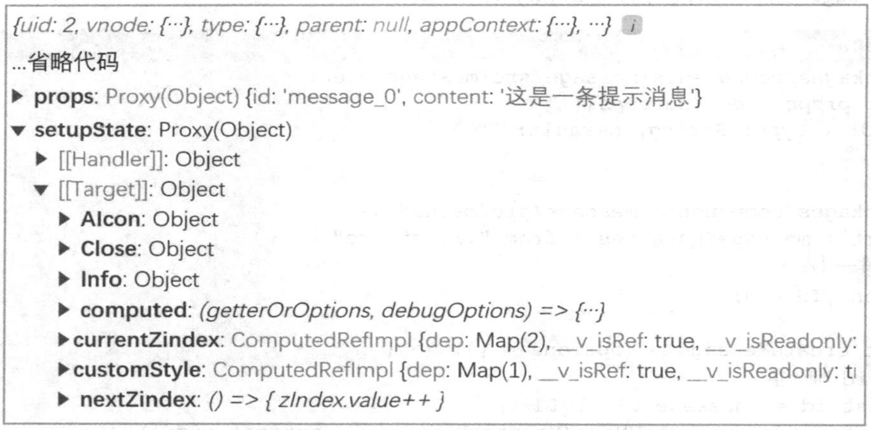

```
{uid: 2, vnode: {…}, type: {…}, parent: null, appContext: {…}, …} ℹ
…省略代码
▶ props: Proxy(Object) {id: 'message_0', content: '这是一条提示消息'}
▼ setupState: Proxy(Object)
   ▶ [[Handler]]: Object
   ▼ [[Target]]: Object
      ▶ AIcon: Object
      ▶ Close: Object
      ▶ Info: Object
      ▶ computed: (getterOrOptions, debugOptions) => {…}
      ▶ currentZindex: ComputedRefImpl {dep: Map(2), __v_isRef: true, __v_isReadonly:
      ▶ customStyle: ComputedRefImpl {dep: Map(1), __v_isRef: true, __v_isReadonly: t
      ▶ nextZindex: () => { zIndex.value++ }
```

图 13-13　message 组件的偏移位置

接着定义变量 instance 存储 id、vnode、vm、handler、props 共 5 个对象（第 18~28 行），我们也可以根据实际开发情况存储所需的对象，并不是所有对象都要被用到，但 id 是必需项。最后返回变量 instance（第 29 行），使用 message 方法获取返回的 message 实例，并用 push 传给变量 messageInstances（第 7、35 行）。

13.6.2　计算 top 偏移

计算 top 偏移，其实就是计算当前 message 组件与上一个 message 组件的数据。在 13.6.1 节中完成了 message 组件的存储以及 message 组件唯一标识属性 id 的更新，已经满足了 top 值的计算条件。因此，下面根据 13.6.1 节的逻辑思路来计算 top 值，如代码清单 library-13-11 所示。

代码清单 library-13-11
```
> packages/components/message/src/instance.js
// 获取实例对象（当前实例和上一个实例）
export const getInstance = (id) => {
    const index = messageInstances.findIndex(item => item.id === id)
    // 当前实例
    const current = messageInstances[index]
    // 上一个实例
    let prev = index > 0 ? messageInstances[index - 1] : null
    return { current, prev }
}
// 获取上一个实例对象的偏移
export const getPrevBottomOffset = (id) => {
    const { prev } = getInstance(id)
    // 如果不存在，则为 0
    if (!prev) { return 0 }
    return prev.vm.setupState?.bottomPosition
}
> packages/components/message/src/message.vue
import { useResizeObserver } from '@vueuse/core'
import { getPrevBottomOffset } from "./instance"
...省略代码
const visible = ref(false)
const messageRef = ref(null)
const height = ref(0)
// 上一个 message
const prevButtonPosition = computed(() => getPrevBottomOffset(props.id))
// 偏移更新
const offsetTop = computed(() => props.offset + prevButtonPosition.value)
// 从底部到可视区顶部的距离
const bottomPosition = computed(() => height.value + offsetTop.value)
// 样式
const customStyle = computed(() => {
    return { top: offsetTop.value + 'px', ...省略代码 }
})

onMounted(() => ...省略代码)
useResizeObserver(messageRef, (entries) => {
    const entry = entries[0]
    height.value = entry.contentRect.height
})
```

在 instance.js 文件中定义 getInstance、getPrevBottomOffset 两个方法（第 2~17 行）。在 message.vue 文件中引入 getPrevBottomOffset 方法（第 20 行），首先获取上一个 message 组件，定义变量 prevButtonPosition，使用 computed 计算属性调用 getPrevBottomOffset 方法并传入参数 props.id（第 26 行），在 getPrevBottomOffset 方法中又调用了 getInstance 方法，并传入 id（第

13 行)。

　　在 getInstance 方法中，使用 findIndex 方法过滤变量 messageInstances 存储的 message 组件，并与传入的参数 id 相匹配（第 4 行）。如果匹配成功，那么变量 index 的值就是当前 message 组件的索引值，获取上一个 message 组件，也就是将索引值减 1；如果匹配不成功，就赋值为 null（第 8 行）。最后返回当前 message 组件和上一个 message 组件，也就是变量 current 和 prev（第 9 行）。

　　回到 getPrevBottomOffset 方法中，需要解构出变量 prev。如果 prev 不存在，则返回为 0（第 15 行）；如果 prev 存在，则获取上一个 message 组件底部位置的值 setupState?.bottomPosition（第 16 行）。当 message.vue 文件的变量 prevButtonPosition 获取到上一个 message 组件底部的偏移值后，便可开始计算当前 message 组件的开始位置。

　　在 message.vue 文件中定义变量 offsetTop，使用 computed 计算属性将 defineProps 对象 offset 属性的值与变量 prevButtonPosition 的值相加（第 28 行）。由于 offset 属性的默认值是 16，prevButtonPosition 的值为 0，所以当前 message 组件（也就是第 1 个 message 组件）的开始位置为 16px。最后将变量 offsetTop 的值绑定到变量 customStyle 的 top 中（第 33 行）。

　　在获取每个 message 组件的底部偏移数据时，用到了 @vueuse/core 插件。在 UI 组件库根目录执行指令 pnpm install @vueuse/core@10.9.0 -w，安装 @vueuse/core 插件。在 message.vue 文件中引入 useResizeObserver 对象，跟踪 DOM 元素的变化尺寸（第 19 行）。为 useResizeObserver 对象的第 1 个参数传入需要跟踪的对象，也就是 message.vue 的根元素 div（第 37、23 行）；第 2 个参数是回调函数，会返回指定对象的尺寸，只需获取高度赋给变量 height（第 38、39 行）。在变量 bottomPosition 中使用 computed 计算属性获取变量 height 和变量 offsetTop 两个属性相加的值（第 30 行），变量 bottomPosition 就是当前 message 组件的底部偏移值。

　　提示：@vueuse/core 提供了很多可重用的 Vue Composables，涵盖广泛的功能，如状态管理、副作用管理、DOM 操作、计时器、窗口大小等。这些功能性 Composables 可以很好地与 Vue.js 3 组件结合使用，帮助用户更好地组织和管理代码逻辑，详见链接 13-2。

13.7　autoClose：自动关闭

　　message 组件的自动关闭动作是一种反馈机制，是指在特定条件下从用户界面中自动移除弹窗，如代码清单 library-13-12 所示。这种交互动作是用户体验设计中不可或缺的一部分，能够为用户提供清晰的反馈和控制权。

代码清单 library-13-12

```
1.  > packages/components/message/src/method.js
2.  const createMessage = (options = {}) => {;
3.    // 提供自定义关闭动作回调
4.    const userOnClose = options?.onClose
5.    const vnode = createVNode(MessageNode, {
6.        ...省略代码
```

```
7.      onClose(){
8.        userOnClose?.()
9.        closeMessage(instance)
10.     },
11.   }
12.  )
13. }
14.
15. // 关闭弹窗
16. const closeMessage = (instance) => {
17.   const idx = messageInstances.indexOf(instance)
18.   if (idx === -1) return
19.   messageInstances.splice(idx, 1)
20.   instance?.handler?.close()
21. }
22.
23. > packages/components/message/src/message.vue: template
24. <transition :name="ns.b()" @leave="onClose">...省略代码</transition>
25.
26. > packages/components/message/src/message.vue: script
27. import { useResizeObserver, useTimeoutFn } from '@vueuse/core'
28. const props = defineProps({
29.   ...省略代码
30.   duration: { type: Number, default: 3000 },
31.   onClose: Function,
32. });
33. const startTime = async () => {
34.   if(props.duration === 0) { return false }
35.   ;({ stop } = useTimeoutFn(() => {
36.     close()
37.   }, props.duration))
38. }
39. const close = () => visible.value = false
40. onMounted(() => startTime())
```

　　message 组件的关闭过程是使变量 visible 的值变为 false，然后从变量 messageInstances 中删除当前存储的 message 组件。关闭过程的主要逻辑是 method.js 的 closeMessage 方法，在该方法中使用 indexOf 方法判断当前是否存在 message 组件，如果不存在，则不需要处理（第 18 行）；如果存在，则调用 splice 方法从数组中删除 message 组件（第 19 行），然后执行 close 方法（第 20 行）。

　　closeMessage 方法由 createVNode 函数的第 2 个参数 onClose 执行（第 7~10 行）。onClose 还会执行 userOnClose()，userOnClose() 由用户在关闭 message 组件后进行回调，如果用户在使用 message 组件时定义了属性 onClose 为函数，那么当 message 组件关闭时，onClose 会被执行，否则不执行（第 4 行）。

　　自动关闭过程使用 @vueuse/core 提供的 useTimeoutFn 对象定时器（第 27 行），定义方法 startTime 解构 stop（第 33、35 行），并设置 props.duration 为 3000 毫秒，也就是在 3 秒后自动

执行 close，使 visible 的值为 false（第 37 行）。要注意的是，当 if 条件判断 props.duration 为 0 时，会阻止自动关闭（第 34 行），这是为用户提供的参数，可自定义是否选择自动关闭状态。

最后在生命周期函数 onMounted 中执行 startTime 方法（第 40 行），在倒计时结束后使变量 visible 为 false。当 message 组件关闭时，就会触发 transition 组件的钩子 @leave（第 24 行）。钩子 @leave 调用 createVNode 函数传入的 onClose，在 onClose 方法中执行 closeMessage 方法删除存储的 message 对象。

13.8 handleClose：手动关闭

手动关闭是指通过点击 message 组件的关闭图标实现关闭过程。需要注意的是，当鼠标悬停在 message 组件上时，需停止自动关闭动作；当鼠标离开时，再次开启自动关闭动作，如代码清单 library-13-13 所示。

代码清单 library-13-13
```
1.  > packages/components/message/src/message.vue: template
2.  <transition :name="ns.b()" @leave="onClose">
3.    <div ref="messageRef" @mouseenter="onMouseEnter" @mouseleave="startTime">
4.      ...省略代码
5.      <div @click.stop="close" v-if="showClose"><a-icon><Close /></a-icon></div>
6.    </div>
7.  </transition>
8.
9.  > packages/components/message/src/message.vue: script
10. const props = defineProps({
11.   ...省略代码
12.   showClose: Boolean
13. });
14. const startTime = async () => {
15.   if(props.duration === 0) { return false }
16.   ;({ stop } = useTimeoutFn(() => {
17.     close()
18.   }, props.duration))
19. }
20. const onMouseEnter = () => stop?.()
21. const close = () => visible.value = false
```

为 message 组件的根元素添加"鼠标进入"事件 @mouseenter 和"鼠标离开"事件 @mouseleave（第 3 行）。当鼠标悬停在提示窗口时，触发 onMouseEnter 方法，在该方法中调用 useTimeoutFn 解构的 stop() 函数清除定时器，停止自动关闭动作（第 20 行）。当鼠标离开时，再次触发 startTime 方法。

然后为关闭图标添加 @click 事件触发 close 方法，使变量 visible 为 false，关闭 message 组件（第 21 行）。最后为关闭图标添加 v-if 指令绑定属性 showClose，可由用户配置是否显示关闭图标（第 5 行）。

13.9　allClose：全部关闭

全部关闭动作是指用户一次性关闭所有 message 组件，方便快速地清除所有消息提示，而无须逐个关闭 message 组件。将 message 组件全部关闭，就是触发每个 message 组件的 handler.close()方法，如代码清单 library-13-14 所示。

代码清单 library-13-14

```
1.  > packages/components/message/src/method.js
2.  ...省略代码
3.  export function closeAll() {
4.    for (const instance of messageInstances) {
5.      instance.handler.close()
6.    }
7.  }
8.  const message = (params = {}) => { ...省略代码 }
9.  message.closeAll = closeAll
```

在每次生成 message 组件时，都会为变量 messageInstances 追加 message 组件实例。然后在 closeAll 方法中使用 for...of 循环变量 messageInstances，触发每个 message 组件的 handler.close() 方法（第 4~6 行）。最后将 closeAll 方法挂载到 message 对象中（第 9 行）。

13.10　theme：主题

根据 UI 组件库设计稿，message 组件的主题有"图标颜色"和"背景颜色"，两种主题类型均通过属性 type 的值而产生变化。图标主题将采用"映射组件"的方式实现，如代码清单 library-13-15 所示。

代码清单 library-13-15

```
1.  > packages/utils/componentsType.js
2.  import { Info, Warning, Success, Danger, Error, Close } from "@ui-library/icons"
3.  export const typeIcon = {
4.    info: Info,
5.    warning: Warning,
6.    success: Success,
7.    danger: Danger,
8.    error: Error,
9.    close: Close,
10. }
11.
12. > packages/components/message/src/message.vue: template
13. <div ref="messageRef" :class="[ns.b(), ns.m(type)]" ...省略代码>
14.   <div :class="[ns.e('icon')]"><a-icon><component :is="icon" /></a-icon></div>
15.   <div :class="[ns.e('content')]">{{ content }} - {{ currentZindex }}</div>
16.   <div :class="[ns.e('close')]" @click.stop="close" v-if="showClose">
17.     <a-icon><component :is="typeIcon['close']" /></a-icon>
18.   </div>
```

```
19.   </div>
20.
21. > packages/components/message/src/message.vue: script
22. import { typeIcon } from "@ui-library/utils"
23. const props = defineProps({
24.     type: { type: String, default: '' }
25. });
26. const icon = computed(() => typeIcon[props.type])
27.
28. > packages/theme/src/message.scss
29. @include b(message) {...省略代码}
30. @each $type in $types {
31.   $className: '.a-message--' + $type;
32.   #{$className} {
33.       @include set-component-var('message', messageVar($type), 'type');
34.   };
35. }
```

在 utils 目录下建立 componentsType.js 文件，引入上述图标，并定义变量 typeIcon 暴露不同类型的图标（第 2~10 行）。

在 message 组件中引入 typeIcon，为 defineProps 对象定义属性 type 接收外部传入的数据，使用 computed 计算属性根据 props.type 映射 typeIcon 指定的图标组件，赋给变量 icon（第 26 行）。通过动态组件 component 处理 message 组件图标，为 is 属性绑定变量 icon（第 14 行），使用 typeIcon['close']关闭图标（第 17 行），最后生成样式（第 30~35 行）。不同主题的 message 组件的渲染效果如图 13-14 所示。

图 13-14　不同主题的 message 组件

13.11　background：背景颜色

message 组件的背景颜色主要在视觉对比、情感色调、品牌一致性、无障碍考虑、视觉层次结构和情境相关性等方面影响着消息的传达效果和视觉吸引力。用户可以通过背景颜色区分不同消息的优先级，同时背景颜色可以确保信息易于辨识和理解。

在处理 message 组件的背景颜色时，只需添加布尔类型的属性 background。若设置属性 background 的值为 true，则启用带有背景颜色的 message 组件，如代码清单 library-13-16 所示。

代码清单 library-13-16
```
1.  > packages/utils/themeType.js
2.  export const themeType = { info: 'primary' }
3.
4.  > packages/components/message/src/message.vue: template
5.  <div :class="[ns.b(), ns.m(theme), ns.is('background', background)]">...</div>
6.
7.  > packages/components/message/src/message.vue: script
8.  import { typeIcon, themeType } from "@ui-library/utils"
9.  /** props */
10. const props = defineProps({
11.     ...省略代码
12.     background: Boolean,
13. });
14.
15. const theme = computed(() => themeType?.[props.type] || props.type)
16.
17. > packages/theme/src/message.scss
18. @include b(message) {
19.     ...省略代码
20.     @include s((background)) {
21.         box-shadow: none;
22.         border: none;
23.         background-color: var(#{getVarName('message', 'bg-color')});
24.         color: #fff;
25.         #{getVarName('message', 'icon-color')}: var(#{getVarName('color', 'white')});
26.     }
27. }
```

message 组件默认 type 属性是 info，而 UI 组件库的主题配置中并没有 info 这个类型，但通过 UI 组件库可知，info 显示的主题是 primary，因此可以在 utils 目录下建立 themeType.js 文件，将 info 映射为 primary（第 2 行）。

在 message 组件中引入 themeType，再定义变量 theme，使用 computed 计算属性根据 props.type 的值映射 themeType 指定的类型（第 15 行），然后将 ns.m(type)更换为 ns.m(theme)（第 5 行），并绑定 ns.is('background', background)，生成类名 is-background。

在 message.scss 文件中使用@include s((background))重写背景颜色、字体颜色、图标颜色等属性（第 20~26 行）。不同背景颜色的 message 组件的渲染效果如图 13-15 所示。

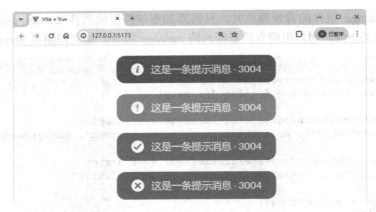

图 13-15　不同背景颜色的 message 组件

13.12　主题方法

主题方法是指在 message 对象中挂载与主题类型相同的方法，通过调用主题方法来调用指定类型的 message 组件，无须再通过传入的 type 属性识别主题，如代码清单 library-13-17 所示。

代码清单 library-13-17
```
> packages/components/message/src/method.js
// 主题类型
const messageTheme = ['info', 'success', 'warning', 'error']
const message = (params = {}) => {...省略代码}
// 主题方法
messageTheme.forEach(theme => {
  message[theme] = (options = {}) => {
    const config = initOptions(options)
    return message({...config, type: theme})
  }
})

> examples/src/App.vue
AMessage.info({...省略代码})
AMessage.success({...省略代码})
AMessage.warning({...省略代码})
AMessage.danger({...省略代码})
```

在 method.js 文件中定义变量 messageTheme，其中包含 info、success、warning、error 共 4 种主题类型（第 3 行）。然后使用 forEach 方法循环变量 messageTheme 为 message 对象添加对应主题的方法（第 6、7 行），调用 initOptions 方法传入配置项，将结果赋给变量 config（第 8 行）。最后扩展变量 config，并组合上 type 属性即可，渲染效果与图 13-15 一致。

本章小结

　　message 组件采用 createVNode 虚拟节点的模式渲染组件，读者可以通过虚拟节点的方式理解 Vue.js 3 的 createVNode 函数和 render 函数的应用，以及 transition 过渡动画组件由初始到结束的整体演变过程。

第 14 章
模态框组件

模态框是一种使用 HTML、CSS 和 JavaScript 构建的对话框组件。模态框被渲染在浏览器可视区窗口中，通常用于单独显示内容，并且可以在不离开浏览器可视区窗口的情况下实现一些互动。

模态框是满屏组件，背景是黑色且透明的底色，整体内容窗口居于浏览器可视区窗口的中上部分，如图 14-1 所示。

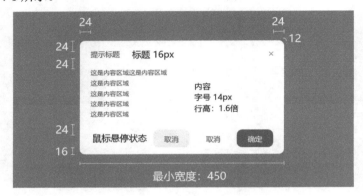

图 14-1　模态框

图 14-1 是模态框的基础效果，其中详细标注了每个元素的尺寸。模态框共有 3 个部分，分别是头部、内容区、脚部。在 UI 组件库设计稿中可以看到，模态框可以没有脚部，使内容在满屏内滚动。

UI 组件库设计所展示的效果都是静态的，很多交互效果无法直接体现，如自定义宽度、自定义脚部、隐藏/显示按钮、修改按钮文本、点击遮罩层是否关闭弹窗、是否显示关闭按钮等。这些交互效果将在后续章节中逐一进行讲解。

14.1　mask：遮罩层

遮罩层（mask）是在网页开发中的一种辅助性组件，它能够在可视区的内容上覆盖一个满屏半透明的层，通常用于显示弹出消息、模态窗口，或者调暗背景内容以突出显示特定组件。

从图 14-1 中可以看到，遮罩层和模态框如同两张叠放的纸张，遮罩层在底层，模态框在顶层。

14.1.1 构建组件

mask 组件是一层半透明的背景，交互动作不多，可采用虚拟节点生成。模态框中有较多信息和交互功能，可采用插槽的方式传入 mask 组件中渲染，两者结合便可实现 mask 组件的遮罩效果，如代码清单 library-14-1 所示。

代码清单 library-14-1

```
1.  > packages/components/mask/src/mask.js
2.  import { createVNode, defineComponent, h, renderSlot } from 'vue'
3.  import { useNamespace } from '@ui-library/hook';
4.  export default defineComponent({
5.    name: "a-mask",
6.    props: {
7.      mask: { type: Boolean, default: true }
8.    },
9.    setup(props, { slots }){
10.     const ns = useNamespace("mask");
11.     return () => {
12.       return createVNode(
13.         'div',
14.         { class: [ns.b()] },
15.         [
16.           h( 'span', { class: props.mask ? ns.e('wrap') : '' }),
17.           renderSlot(slots, 'default')
18.         ]
19.       )
20.     }
21.   }
22. })
23.
24. > packages/components/mask/index.js
25. import Mask from "./src/mask.js"
26. // 提供按需加载的方式
27. export const AMask = Mask
28.
29. > examples/src/App.vue
30. <a-mask></a-mask>
31. import { AMask } from "@ui-library/components"
```

mask 组件的结构和 message 组件一致，同样是载入 js 文件（第 1 行）。在 mask.js 文件中使用常规语法写入 name、props、setup 对象（第 5、6、9 行），在 setup 对象中直接返回 createVNode 函数的虚拟节点（第 12 行）。

createVNode 函数的第 1 个参数是 div，也就是渲染一个虚拟的 DOM 元素，而不是一个组件（第 13 行）；在第 2 个参数中添加 class 生成类名（第 14 行）；第 3 个参数是数组类型的，用于渲染子元素，也就是在第 1 个参数生成的 DOM 元素中插入子元素。第 3 个参数的第 1 个对象使用 Vue.js 3 的 h 函数渲染虚拟节点 span，并根据 props.mask 对象判断是否生成类名.a-mask_wrap（第 2、16 行）；第 3 个参数的第 2 个对象使用 Vue.js 3 的 renderSlot 函数渲染默认插槽，也就是<slot />（第2、17行）。

最后在mask/index.js文件中提供按需加载的方式。需要注意的是，mask组件只在UI组件库内部供其他组件生成遮罩层，不需要提供给用户使用。在examples演示库中引入AMask组件，测试渲染mask组件的结构，如图14-2所示。

```
<div id="app" data-v-app>
▼ <div class="a-mask">
    <span class="a-mask__wrap"></span>
    "这里是插槽内容"
  </div>
</div>
```

图14-2 mask组件的结构

14.1.2 渲染组件

mask组件在可视区的内容上覆盖一个满屏的半透明层，并且以fixed固定定位的模式铺满整个屏幕，和模态框以叠加的方式形成上下层关系。因此，可以使用mask组件的子元素span标签来渲染底层的半透明层，如代码清单library-14-2所示。

代码清单 library-14-2
```
1.  > packages/components/mask/src/mask.js
2.  ...省略代码
3.  import { useZindex } from "@ui-library/hook"
4.  export default defineComponent({
5.    ...省略代码
6.    setup(props, { slots }){
7.      const ns = useNamespace("mask");
8.      const { nextZindex, currentZindex } = useZindex()
9.      nextZindex()
10.     return () => {
11.       return createVNode('div',
12.         {
13.           class: [ns.b()],
14.           style: [{zIndex: currentZindex.value}]
15.         },
16.         ...省略代码
17.       )
18.     }
19.   }
20. })
21.
22. > packages/theme/src/mask.scss
23. @include b(mask) {
24.   position: fixed;
25.   left: 0;
26.   top: 0;
27.   width: 100vw;
```

```
28.     height: 100vh;
29.     z-index: 100;
30.     overflow-y: auto;
31.     @include e(wrap) {
32.       position: fixed;
33.       left: 0;
34.       top: 0;
35.       width: 100%;
36.       height: 100%;
37.       z-index: -1;
38.       background-color: rgba(0, 0, 0, .4);
39.     }
40.  }
```

要使元素位于浏览器可视区窗口中，需要使用 position:fixed 固定 mask 组件的位置，并设置 left 和 top 为 0，表示居于左上角（第 24~26 行）。宽度和高度的单位使用 vw 和 vh，也就是以居于可视区窗口的比例为单位计算（第 27、28 行），如 1vw = 1/100 视口宽度、1vh = 1/100 视口高度，以此类推。为 mask 组件添加 y 轴滚动属性 overflow-y，使 modal 组件的内容高度大于 mask 组件的高度，可上下滚动（第 30 行）。

子元素 span 也使用 fixed 固定位置，从 left 和 top 为 0 开始计算（第 32~34 行），宽度和高度需自动适应父级的宽度和高度，因此可设为 100%（第 35、36 行）。属性 z-index 的值为-1，定位为父元素，位于底层，不会影响到顶层的元素（第 37 行），透明度为 40%的黑色背景（第 38 行），渲染效果如图 14-3 所示。

最后引入 hook 的 useZindex 组合式函数，并解构 nextZindex 和 currentZindex（第 3、8 行），为生成虚拟元素的 div 添加 style（第 14 行），为 z-index 属性绑定 currentZindex，设置 div 的层级，执行 nextZindex()方法使全局变量 zIndex 不断累加（第 9 行）。

图 14-3　modal 组件结合 mask 组件的渲染效果

14.2 modal：对话框

将对话框（modal）组件与 mask 组件结合使用，只需在 modal 组件引入 mask 组件，如代码清单 library-14-3 所示。

代码清单 library-14-3
```
1. > packages/components/modal/src/index.vue: template
2. <template>
3.   <a-mask>这里是modal组件</a-mask>
4. </template>
5. > packages/components/modal/src/index.vue: script
6. import { AMask } from "@ui-library/components"
7.
8. > examples/src/App.vue
9. <a-modal></a-modal>
```

为 modal 组件引入 mask 组件（第 6 行），并写入测试内容（第 3 行）。在 examples 演示包中即可使用 modal 组件，如图 14-4 所示。

```
<div id="app" data-v-app>
  ▼ <div class="a-mask" style="z-index: 3001;">
      <span class="a-mask__wrap"></span>
      "这里是modal组件"
    </div>
</div>
```

图 14-4 引入 mask 组件后的 modal 组件

可以看到 mask 组件被成功渲染，并且 modal 组件传入的插槽元素也被正常渲染。图 14-4 中渲染的 span 标签和插槽元素与代码清单 library-14-1 中第 16、17 行的顺序是一致的。如果需要渲染更多的子元素，可将其写入 createVNode 函数的第 3 个参数。

14.2.1 构建结构

模态框分为头部、内容区和脚部，因此下面优先处理 modal 组件这 3 个部分的结构，如代码清单 library-14-4 所示。

代码清单 library-14-4
```
1. > packages/components/modal/src/index.vue
2. <a-mask>
3.   <div :class="[ns.b()]">
4.     <div :class="[ns.e('wrap')]">
5.       <div :class="[ns.e('header')]">头部</div>
6.       <div :class="[ns.e('body')]">内容区</div>
7.       <div :class="[ns.e('footer')]">脚部</div>
8.     </div>
```

```
9.     </div>
10. </a-mask>
```

为 modal 组件引入 mask 组件，将 modal 组件的整体结构作为插槽内容传入 mask 组件。要注意 modal 组件结构的最外层 ns.b()，该层主要用于产生"满屏内容滚动"的效果，设置高度为 100%，将 modal 组件固定在浏览器可视区。子级 ns.e('wrap') 是 modal 组件的主体，也就是整体的白色圆角矩形，然后将其分为头部、内容区和脚部。modal 组件的结构如图 14-5 所示。

```html
<div class="a-mask" style="z-index: 3001;">
  <span class="a-mask__wrap"></span>
  <div class="a-modal">
    <div class="a-modal__wrap">
      <div class="a-modal__header">头部</div>
      <div class="a-modal__body">内容区</div>
      <div class="a-modal__footer">脚部</div>
    </div>
  </div>
</div>
```

图 14-5　modal 组件的结构

14.2.2　渲染组件

在 UI 组件库设计稿中，modal 组件位于可视区垂直居中的位置，背景为白色，各个元素的尺寸也有相应标注。下面据此来渲染 modal 组件，如代码清单 library-14-5 所示。

代码清单 library-14-5
```
1. > packages/components/modal/src/index.vue: template
2. <div :class="[ns.e('wrap')]">
3.   <div :class="[ns.e('header')]">
4.     <h4 :class="[ns.e('title')]">{{ title }}</h4>
5.     <a-icon :class="[ns.e('close')]"><Close /></a-icon>
6.   </div>
7.   <div :class="[ns.e('body')]"><slot /></div>
8.   <div :class="[ns.e('footer')]">
9.     <a-button text size="large">取消</a-button>
10.    <a-button type="primary" size="large">确定</a-button>
11.  </div>
12. </div>
13.
14. > packages/components/modal/src/index.vue: script
15. const props = defineProps({
16.   title: { type: String, default: '' }
17. })
18.
19. > packages/theme/src/modal.scss
20. @include b(modal) {
21.   padding: 15vh 0;
```

```scss
22.     margin: 0 auto;
23.     width: 450px;
24.     @include e(body) {
25.       ...省略代码
26.       padding: 0 24px 24px;
27.       word-break: break-word;   // 强制换行
28.     }
29.     @include e(footer) {
30.       padding: 0 24px 16px;
31.       text-align: right;
32.     }
33.     @include e(title) {
34.       ...省略代码
35.       white-space: nowrap;      // 不换行
36.     }
37.   }
```

modal 组件的头部包含 HTML 元素 h4 和组件 a-icon，分别对应标题和关闭图标，并分为左右两侧（第 4、5 行）。通过绑定属性 title 渲染 h4 标题（第 4 行）；在内容区使用<slot />渲染外部传入的内容（第 7 行）；在脚部引入 a-button 组件并设置 size 为 large（第 9、10 行）。

在 modal.scss 文件中设置 modal 组件根元素的上下内边距均为 15vh（第 21 行），使 modal 组件的顶部和底部与浏览器可视区的边界产生一定的距离，而非紧贴可视区的边界。在标题区域添加属性 white-space: nowrap，使文本不换行（第 35 行）。在内容区添加属性 word-break: break-word，使文本换行（第 27 行），并为内容区的右、下、左设置 24px 的内边距，顶部为 0（第 26 行）。这是因为头部已经设置内边距，所以内容区顶部不需要设置内边距。脚部的内边距与内容区的做法一致（第 30 行），设置属性 text-align: right，使内容默认靠右侧对齐（第 31 行）。modal 组件的渲染效果如图 14-6 所示。

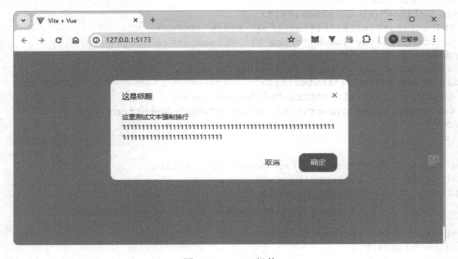

图 14-6　modal 组件

14.3　teleport：传送

传送（teleport）组件是 Vue.js 3 内置的新特性组件，允许开发人员将 teleport 组件包裹的子级组件"传送"给指定的 DOM 元素，用于确保被传送的元素不受父元素的样式或层叠上下文的错误影响，如代码清单 library-14-6 所示。

代码清单 library-14-6

```
1.  > packages/components/modal/src/index.vue
2.  <teleport to="body">   // 或指定 ID 的 DOM 元素 #idName
3.    <a-mask vertical>
4.      <div :class="[ns.b()]">...省略代码</div>
5.    </a-mask>
6.  </teleport>
```

modal 组件使用 teleport 组件包裹所有元素，并添加属性 to，设置 to 的值为 body（第 2 行），teleport 组件便会使 modal 组件作为 body 元素的子级进行渲染，如图 14-7 所示。

图 14-7　使用 teleport 组件的 modal 组件

14.4　transition：过渡动画

transition 组件的过渡动画可由 v-if、v-show、<component>和改变特殊的 key 属性中的任意一个触发。在执行过渡动画的过程中，会自动添加"进入"和"离开"的类名。在 modal 组件中使用过渡动画效果，就是为 mask 组件添加 v-show 指令，并结合 CSS 属性 transiton，使 mask 组件和 modal 组件整体产生过渡动画。因此，需要为 mask 组件添加过渡动画的时间，如代码清单 library-14-7 所示。

代码清单 library-14-7

```
1.  > packages/components/modal/src/index.vue: template
2.  <teleport to="body">
3.    <transition :name="ns.b()">
4.      <a-mask v-show="visible"><div :class="[ns.b()]">...省略代码</div></a-mask>
5.    </transition>
6.  </teleport>
7.
8.  > packages/components/modal/src/index.vue: script
9.  const visible = defineModel()
10. const cancel = () => {
```

```
11.     visible.value = false
12.   }
13.
14. > packages/theme/src/mask.scss
15. @include b(mask) { ...省略代码; transition: .2s; }
16.
17. > packages/theme/src/modal.scss
18. .a-modal-enter-from, .a-modal-leave-to { opacity: 0; }
19. .a-modal-enter-active .a-modal { animation: dialogFadeIn .2s; }
20. .a-modal-leave-active .a-modal { animation: dialogFadeOut .2s; }
21. @keyframes dialogFadeIn {  // 进入动画
22.   0% { transform: translateY(-30px); opacity: 0; }
23.   100% { transform: translateY(0); opacity: 1; }
24. }
25.
26. @keyframes dialogFadeOut {  // 离开动画
27.   0% { transform: translateY(0); opacity: 1; }
28.   100% { transform: translateY(-30px); opacity: 0; }
29. }
30.
31. > examples/src/App.vue
32. <a-modal v-model="show" title="这是标题">这里测试文本强制换行</a-modal>
33. <a-button @click="show = true">打开</a-button>
34. import { ref } from "vue"
35. const show = ref(false)
```

为 modal 组件添加 transition 过渡动画组件，并为属性 name 绑定 ns.b() 生成类名 a-modal（第 3 行）。为 mask 组件添加 v-show 指令，绑定变量 visible（第 4、9 行）。然后在 examples 演示库中引用 a-modal 组件时，使用 v-model 的变量 show（第 32、35 行）。最后在按钮上添加 @click 事件，使变量 show 为 true，便可显示 a-modal 组件（第 33 行）。

modal 组件使用的 v-show 指令被绑定在 mask 组件中，因此产生过渡动画的时间需写在 mask.scss 文件中，这也是遮罩层公共组件的统一时间（第 15 行）。为 modal 组件添加过渡动画的效果如图 14-8 所示。

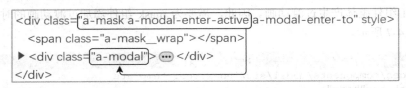

图 14-8　为 modal 组件添加过渡动画

transition 组件在"进入"和"离开"两个状态中会生成类名 a-modal-enter-from 和 a-modal-leave-to，可以设置透明度为 0，即消失状态（第 18 行）。由于 transition 组件生成的类名在 mask 组件中，因此需要通过 mask 组件找到 modal 组件，为 modal 设置由上至下的小距离的偏移过渡动画效果。

类名 a-modal-enter-active 是进入动作，通过该类名找到 a-modal，为其添加 CSS 3 动画

animation，并指定动画名称为 dialogFadeIn，动画时长为 200 毫秒（第 19 行）；类名 a-modal-leave-active 是离开动作，通过该类名找到 a-modal，为其添加 CSS 3 动画 animation，并指定动画名称为 dialogFadeOut，动画时长为 200 毫秒（第 20 行）。

自定义动画 dialogFadeIn 和 dialogFadeOut 的 0%和 100%，分别代表"开始"和"结束"，修改 CSS 属性 translateY 的 y 轴偏移和 opacity 透明度（第 21~24、26~29 行）。

14.5 footer：脚部

modal 组件的脚部（footer）内容，通常包括各种操作按钮或者其他交互元素。用户通过点击按钮来执行相应的操作，比如确认、取消、关闭等。这需要用户能够灵活地实现脚部的自定义、按钮文本配置。

14.5.1 按钮属性

按钮属性可以帮助用户更方便地进行操作，提升用户体验。按钮属性包括设置按钮和确认按钮的文本内容、显示/隐藏及其他属性，通过这些属性可以定制 modal 组件脚部按钮的外观和行为，如代码清单 library-14-8 所示。

代码清单 library-14-8

```
1.  > packages/components/modal/src/modal.js
2.  export const modalProps = {
3.      title: { type: String, default: '' },                    // 标题
4.      footer: { type: Boolean, default: true },                // 是否渲染脚部
5.      cancelButtonText: { type: String, default: '取消' },     // 取消按钮文本
6.      cancelButtonShow: { type: Boolean, default: true },      // 取消按钮是否显示
7.      confirmButtonText: { type: String, default: '确定' },    // 确定按钮文本
8.      confirmButtonShow: { type: Boolean, default: true }      // 确定按钮是否显示
9.  }
10.
11. > packages\components\modal\src\index.vue: template
12. <div :class="[ns.e('footer')]" v-if="footer">
13.   <a-button text size="large" @click="cancel" v-if="cancelButtonShow">
14.     {{ cancelButtonText }}
15.   </a-button>
16.   <a-button type="primary" size="large" v-if="confirmButtonShow">
17.     {{ confirmButtonText }}
18.   </a-button>
19. </div>
20.
21. > packages\components\modal\src\index.vue: script
22. import { modalProps } from "./modal"
23. const props = defineProps(modalProps)
24.
25. > examples/src/App.vue
26. <a-modal v-model="visible" title="这是标题" :cancel-button-show="false">
```

modal 组件 defineProps 对象的自定义属性较多，可将属性抽离至 modal.js 文件（第 2~9 行），再引入 modal 组件中使用（第 22、23 行），后续开发的新的自定义属性可以添加至 modal.js 文件中。

首先在 modal.js 文件中新增自定义 footer、cancelButtonText、cancelButtonShow、confirmButtonText、confirmButtonShow 等属性，分别为：是否渲染脚部、取消按钮文本、是否显示取消按钮、确定按钮文本、是否显示确定按钮。然后为上述属性分别绑定取消按钮和确定按钮（第 12~18 行）。

最后在 examples 演示库中引用 a-modal 组件，配置 modal.js 文件定义的属性（第 26 行）。

14.5.2　脚部插槽

脚部插槽是自定义 modal 组件的脚部元素。通过在 modal 组件中定义名称为 "footer" 的插槽，可以在 modal 组件的脚部渲染自定义内容，增强 modal 组件的灵活性，使其适应不同场景下的需求，如代码清单 library-14-9 所示。

代码清单 library-14-9
```
1.  > packages/components/modal/src/modal.js
2.  <div :class="[ns.e('footer')]" v-if="footer">
3.    <template v-if="$slots.footer">
4.      <slot name="footer" />
5.    </template>
6.    <template v-else>
7.      <a-button text size="large" v-if="cancelButtonShow">...省略代码</a-button>
8.      <a-button type="primary" size="large" v-if="confirmButtonShow">...省略代码</a-button>
9.    </template>
10. </div>
11.
12. > examples/src/App.vue
13. <a-modal v-model="visible" title="这是标题">
14.   这里是 modal 内容区渲染的文本
15.   <template #footer>
16.     <a-button type="success" size="large">自定义取消按钮</a-button>
17.     <a-button type="primary" size="large">自定义确定按钮</a-button>
18.   </template>
19. </a-modal>
```

在 ns.e('footer') 区域中新增两个 <template> 模板标签（第 3、6 行），在第一个 <template> 中使用 v-if 条件判断是否存在名称为 footer 的插槽（第 3 行），如果存在，则渲染 <slot /> 插槽（第 4 行），否则渲染 modal 组件的脚部（第 6 行）。

在 examples 演示库中使用 #footer 自定义脚部插槽元素（第 15~18 行），modal 组件的脚部自定义插槽的渲染效果如图 14-9 所示。

图 14-9　modal 组件的脚部自定义插槽

14.5.3　按钮事件

modal 组件脚部按钮事件的作用主要是实现用户与 modal 组件之间的交互，用户可以通过事件对 modal 组件进行确认、取消或关闭操作，从而完成相关任务或退出对话框。如代码清单 library-14-10 所示。

代码清单 library-14-10
```
1.  > packages/components/modal/src/composables/use-modal-event.js
2.  import { getCurrentInstance } from "vue"
3.  export function useModalEvent({ props, visible }){
4.      const { emit } = getCurrentInstance()
5.      // 取消按钮事件
6.      const useBeforeCancel = () => {
7.          useClose()
8.          emit('cancel')
9.      }
10.     // 确定按钮事件
11.     const useBeforeChange = () => { emit('ok') }
12.     // 关闭窗口
13.     const useClose = () => visible.value = false
14.     return { useBeforeCancel, seBeforeChange, useClose }
15. }
16.
17. > packages/components/modal/src/composables/use-modal.js
18. import { useModalEvent } from "./use-modal-event"
19. export function useModal({ props, visible }){
20.     const { useBeforeCancel, useBeforeChange, useClose } = useModalEvent({ props, visible })
21.     return { useBeforeCancel, useBeforeChange, useClose }
22. }
23.
24. > packages/components/modal/src/index.vue: template
25. <a-icon :class="[ns.e('close')]" @click="useClose"><Close /></a-icon>
26. <a-button text size="large" v-if="cancelButtonShow" @click="useBeforeClose">
27. <a-button type="primary" size="large" v-if="confirmButtonShow" @click="useBeforeChange">
28.
29. > packages/components/modal/src/index.vue: script
30. import { useModal } from "./composables/use-modal"
```

```
31.    const props = defineProps(modalProps)
32.    const emit = defineEmits(['ok', 'cancel'])
33.    const visible = defineModel()
34.    const { useBeforeCancel, useBeforeChange, useClose } = useModal({props, visible})
```

脚部按钮事件的开发方式和 checkbox 组件相同，在 composables 目录下定义 use-modal.js 文件作为主入口，并定义 use-modal-event.js 文件作为处理事件的模块。

在 use-modal-event.js 文件中定义组合式函数 useModalEvent，并接收参数 props 和 visible（第 3 行），在函数内定义 useBeforeCancel、useBeforeChange 和 useClose 方法，分别为"取消按钮事件"、"确定按钮事件"和"关闭窗口"（第 6、11、13 行）。

在 use-modal.js 文件中引入组合式函数 useModalEvent，解构该函数的 3 个方法并返回（第 18、20、21 行）。在 modal 组件中引入组合式函数 useModal，传入参数 props 和 visible，并解构 useBeforeCancel、useBeforeChange、useClose 方法（第 30、34 行）。接着为右上角的关闭图标添加@click 事件，绑定 useClose 方法（第 25 行）；为取消事件按钮和确定事件按钮分别绑定 useBeforeClose 和 useBeforeChange 方法（第 26、27 行）。

当 useBeforeCancel 方法被触发时，会调用 useClose 方法使 defineModel 双向绑定数据 visible 的值为 false，从而关闭窗口（第 7、13 行），同时使用 emit 对象传递 cancel 事件（第 8 行），供父级组件使用。当 useBeforeChange 方法被触发时，仅使用 emit 对象传递 ok 事件（第 11 行），供父级组件使用。

在完成了按钮事件的绑定后，便可在父级组件中使用@cancel 和@ok 绑定对应的方法，实现相关的业务逻辑。

14.6　loading：加载

为 modal 组件的确定按钮添加加载效果，可以提升用户体验，增强系统透明度，为用户提供更好的操作反馈，以及防止连续点击或其他的错误操作。

14.6.1　confirmLoading 属性

confirmLoading 是供用户手动触发加载效果的自定义属性，用户可根据自己的业务场景手动启用或关闭加载效果，如代码清单 library-14-11 所示。

代码清单 library-14-11
```
1.  > packages/components/modal/src/composables/use-modal-state.js
2.  import { computed } from "vue"
3.  export function useModalState({ props }){
4.    const isLoading = computed({
5.      get(){
6.        return props.confirmLoading
7.      }
8.    })
9.    return { isLoading }
```

```
10. }
11.
12. > packages/components/modal/src/composables/use-modal.js
13. import { useModalState } from "./use-modal-state"
14. export function useModal({ props, visible }){
15.     const { isLoading } = useModalState({ props })
16.     return { ...省略代码, isLoading }
17. }
18.
19. > packages/components/modal/src/modal.js
20. export const modalProps = { ...省略代码, confirmLoading: Boolean }
21.
22. packages/components/modal/src/index.vue
23. <a-button ...省略代码 :loading="isLoading">{{ confirmButtonText }}</a-button>
24. const { ...省略代码, isLoading } = useModal({props, visible})
25.
26. > examples/src/App.vue: template
27. <a-modal v-model="visible" :confirm-loading="loading" @ok="handleOk" ...省略代码>
28.
29. > examples/src/App.vue: script
30. const loading = ref(false)
31. const handleOk = () => {
32.   loading.value = true
33.   setTimeout(() => {
34.     loading.value = false
35.   }, 2000)
36. }
```

在defineProps对象中添加自定义属性confirmLoading，设置为布尔类型（第20行）。新建use-modal-state.js文件作为状态模块，在模块中定义组合式函数useModalState（第3行），并使用computed计算属性的get方法返回props.confirmLoading的状态给变量isLoading（第5~7、4行），然后返回isLoading（第9行）。

在use-modal.js文件中引入use-modal-state模块并解构isLoading（第13、15行）。在modal组件中，同样解构isLoading并绑定至确定按钮的loading属性（第23、24行）。在examples演示库中引用a-modal组件，为组件添加属性confirm-loading，绑定变量loading，并为modal组件回调的@ok事件绑定自定义方法handleOk（第27行）。当点击确定按钮时，便会调用handleOk方法修改变量loading的值，使确定按钮产生加载效果，如图14-10所示。

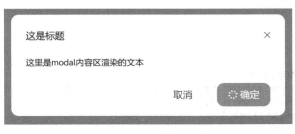

图14-10　确定按钮的加载效果

14.6.2 beforeChange 属性

beforeChange 是一个自定义属性，通常用于执行在模态框中进行操作之前的其他动作，使确定按钮自动处于加载状态。当满足所执行的动作后，确定按钮的加载状态会自动关闭，从而减少用户手动设置 confirmLoading 属性的动作，如代码清单 library-14-12 所示。

代码清单 library-14-12

```
1.  > packages/components/modal/src/modal.js
2.  export const modalProps = {...省略代码, beforeChange: Function }
3.
4.  > packages/components/modal/src/composables/use-modal-state.js
5.  export function useModalState({ props, visible }){
6.    const loading = ref(false)
7.    const isLoading = computed({
8.      get(){
9.        return props.confirmLoading || loading.value
10.     },
11.     set(val) {
12.       loading.value = val
13.     }
14.   })
15. }
16.
17. > packages/components/modal/src/composables/use-modal-event.js
18. import { getCurrentInstance } from "vue"
19. import { types } from "@ui-library/utils"
20. export function useModalEvent({ props, visible, isLoading }){
21.   const { emit } = getCurrentInstance()
22.   const useBeforeChange = () => {
23.     const isFunction = types().isFunction(props.beforeChange)
24.     if(isFunction) {
25.       isLoading.value = true
26.       props.beforeChange().then(() => {
27.         isLoading.value = false
28.         emit('ok')
29.       }).catch(() => {
30.         isLoading.value = false
31.       })
32.       return
33.     }
34.     emit('ok')
35.   }
36. }
37.
38. > examples/src/App.vue
39. <a-modal v-model="visible" :before-change="handleBeforeChange"></a-modal>
40. const handleBeforeChange = () => {
41.   return new Promise((resolve, reject) => {
42.     setTimeout(() => {
43.       reject()
```

```
44.      }, 2000)
45.    })
46.  }
```

beforeChange 属性的交互动作与 button 组件的属性 beforeChange 一致。在 modal 组件中，也需要在 defineProps 对象中添加属性 beforeChange，设置为函数类型（第 2 行）。在 use-modal-state.js 文件中自定义变量 loading（第 6 行），在 isLoading 对象中添加 set 对象，更新变量 loading 的值（第 11~13 行）。在 get 对象中使用"或"运算监听变量 loading，只要 loading 的值发生变化，那么 get 对象便会获取新的值返回给 isLoading（第 9 行）。

在 useBeforeChange 方法中优先判断 props.beforeChange 是否是一个函数（第 23 行）。如果是，则先更新 isLoading 为 true，启用加载状态（第 25 行）；再执行 props.beforeChange 函数，并在回调.then 时重新更新 isLoading 为 false，取消加载状态（第 26、27 行），并执行 emit('ok')（第 28 行）。如果回调的是.catch，那么同样更新 isLoading 为 false，取消加载状态，但不需要执行 emit('ok')（第 30 行）。

在 examples 演示库中使用 modal 组件时，传入属性 before-change 并绑定 handleBeforeChange 函数（第 39 行），在该函数中返回 Promise 对象（第 40~46 行）。

14.7　event：事件回调

modal 组件事件回调可以为用户带来更友好的用户体验和交互控制。通过在打开或关闭动作时回调来执行特定的操作，例如，在打开时加载数据或设置弹窗内容，而在关闭时可以进行相应的善后处理，如清空数据或重置页面状态等。这些事件回调使得 modal 组件更加灵活，帮助开发人员更好地控制弹窗的行为和内容，从而提升用户体验，如代码清单 library-14-13 所示。

代码清单　library-14-13
```
1.  > packages/components/modal/src/index.vue
2.  <template>
3.    <teleport to="body">
4.      <transition
5.        :name="ns.b()"
6.        @enter="useEnter"
7.        @after-enter="useAfterEnter"
8.        @before-leave="useBeforeLeave"
9.        @after-leave="useAfterLeave"
10.     >...省略代码</transition>
11.   </teleport>
12. </template>
13.
14. > packages/components/modal/src/composables/use-modal.js
15. export function useModal({ props, visible }){
16.   ...省略代码
17.   const useEnter = () => emit('open')      // 开始进入动画
18.   const useAfterEnter = () => emit('opened')  // 动画过渡完成后
19.   const useBeforeLeave = () => emit('close')  // 离开动画之前
```

```
20.        const useAfterLeave = () => emit('closed')   // 离开动画完成后
21.        return { ...省略代码, useEnter, useAfterEnter, useBeforeLeave, useAfterLeave }
22.    }
```

modal 组件的打开和关闭回调事件使用 Vue.js 3 内置的 transition 组件钩子函数实现。13.3 节已经介绍了 transition 组件有 8 个钩子函数，modal 组件回调会用到其中 4 个，分别是@enter、@after-enter、@before-leave 和@after-leave（第 6~9 行），依次对应开始进入动画、动画过渡完成后、离开动画之前和离开动画完成后，对应绑定的函数依次是 useEnter、useAfterEnter、useBeforeLeave 和 useAfterLeave（第 17~20 行）。

14.8　maskClose：遮罩关闭

用户在点击模态框的透明遮罩层时，默认情况下是关闭 modal 组件的，这种交互方式更简便和流畅，让用户能够更自然地关闭模态框。但并非所有的业务都需要通过点击遮罩层关闭 modal 组件，因此需要提供配置属性，提供用户点击遮罩层但不关闭 modal 组件的交互，如代码清单 library-14-14 所示。

代码清单 library-14-14

```
1.  > packages/components/mask/src/mask.js
2.  export default defineComponent({
3.    name: "a-mask",
4.    props: {
5.      ...省略代码
6.      maskClose: { type: Boolean, default: true }
7.    },
8.    setup(props, {emit, slots }){
9.      const ns = useNamespace("mask");
10.     // 关闭事件
11.     const onClose = () => {
12.       if(!props.maskClose) { return }
13.       emit('close')
14.     }
15.     return () => {...省略代码}
16.   }
17. })
18.
19. > packages/components/modal/src/index.vue
20.   <transition ...省略代码>
21.     <a-mask v-show="visible" v-bind="$attrs" @close="useClose">...省略代码</a-mask>
22.   </transition>
23.
24. > examples/src/App.vue
25.   <a-modal ...省略代码 :mask-close="false">modal 内容区渲染的文本</a-modal>
```

遮罩层的关闭事件通过绑定的 mask 组件实现，因此在 mask 组件的 props 对象中自定义属性 maskClose 为布尔类型，默认值为 true（第 6 行）。当在 onClose 方法中判断 props.maskClose

为 false 时，阻止执行（第 12 行），也就是不执行 emit('close')，因此不会发生关闭弹窗交互。

在 modal 组件中使用 v-bind 绑定属性$attrs，向父级组件传递不包含在 definePorps 对象中的属性，使属性直接传递到子级组件（第 21 行）。在 examples 演示库中，为 a-modal 组件设置属性 mask-close 为 false，即可阻止点击遮罩层关闭弹窗（第 25 行）。

14.9 unmount：销毁

modal 组件的销毁机制通过销毁<slot />插槽渲染的元素对象来实现，销毁机制可以释放资源并减少内存占用，优化页面性能，提升用户体验。关闭并销毁 modal 组件的<slot />插槽元素可以清理和重置组件内部状态，确保下次打开页面时是全新的状态，减少潜在的错误和不一致的情况，如代码清单 library-14-15 所示。

代码清单 library-14-15
```
1.  > packages/components/modal/src/index.vue
2.  <teleport to="body">
3.    <transition ...省略代码 @before-enter="useBeforeEnter">
4.      <a-mask v-show="visible" v-bind="$attrs" @close="useClose">
5.        <div :class="[ns.b()]">
6.          <div :class="[ns.e('wrap')]">
7.            <div :class="[ns.e('header')]"> ...省略代码</div>
8.            <div :class="[ns.e('body')]">
9.              <div :class="[ns.e('render-content')]" v-if="rendered"><slot /></div>
10.           </div>
11.           <div :class="[ns.e('footer')]" v-if="footer">...省略代码</div>
12.         </div>
13.       </div>
14.     </a-mask>
15.   </transition>
16. </teleport>
17.
18. > packages/components/modal/src/composables/use-modal.js
19. export function useModal({ props, visible }){
20.   ...省略代码
21.   const rendered = ref(false)
22.   // 进入动画之前
23.   const useBeforeEnter = () => rendered.value = true
24.   // 离开动画完成后
25.   const useAfterLeave = () => {
26.     props.unmountOnClose && (rendered.value = false)
27.     emit('closed')
28.   }
29. }
30.
31. > packages/components/modal/src/modal.js
32. export const modalProps = {
33.   ...省略代码
```

```
34.        unmountOnClose: Boolean
35.  }
```

Vue.js 3 可使用 v-if 指令销毁一个元素,如果 v-if 的条件为 false,那么 Vue.js 3 的底层逻辑会移除页面中的 DOM 元素,达到销毁的目的。因此,在 use-modal.js 文件中定义变量 rendered 为 false(第 21 行),在 modal 组件的 ns.e('body')元素中添加一个子级 div,并绑定 ns.e('render-content')生成类名(第 9 行)。

由于在打开 modal 组件时必定会渲染<slot />插槽元素,因此需要在 transition 组件中添加 @before-enter 钩子函数绑定方法 useBeforeEnter(第 3 行)。这样每次打开 modal 组件时都会触发 useBeforeEnter 方法,使变量 rendered 为 true,即可渲染<slot />插槽元素(第 23 行)。

由于在关闭 modal 组件时必定会触发 useAfterLeave 方法,因此可以在该方法中判断 defineProps 对象自定义属性 unmountOnClose 的状态(第 34 行)。如果 unmountOnClose 为 true,则表示需要销毁,设置变量 rendered 为 false 便能销毁元素,如图 14-11 所示。

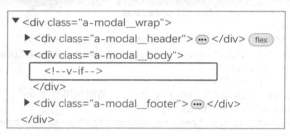

图 14-11　关闭 modal 组件时销毁<slot />插槽元素

14.10　width:宽度

modal 组件的宽度属性 width 可以根据具体的需求进行灵活调整,合理配置属性 width 也有助于保持界面的美观性和一致性,使 modal 组件在各种情景下都有良好的展示效果,如代码清单 library-14-16 所示。

代码清单 library-14-16

```
1.  > packages/components/modal/src/modal.js
2.  export const modalProps = {
3.     ...省略代码
4.     width: { type: String, default: '' }
5.  }
6.
7.  > packages/components/modal/src/composables/use-modal.js
8.  import { useStyle } from "@ui-library/hook"
9.  export function useModal({ props, visible }){
10.    const uStyle = useStyle()
11.    ...省略代码
12.    // 宽度
13.    const width = computed(() => {
```

```
14.            return uStyle.width(props.width)
15.       })
16.       return { ...省略代码, width }
17. }
18.
19. > packages/components/modal/src/index.vue
20. <a-mask v-show="visible" v-bind="$attrs" @close="useClose">
21.   <div :class="[ns.b()]" :style="[width]"> ...省略代码</div>
22. </a-mask>
```

更新 DOM 的 style 属性，可在 use-modal.js 文件引入 hook 钩子函数包的 useStyle 组合函数（第 8 行），再调用 uStyle.width 方法传入 defineProps 对象的自定义属性 width（第 4、14 行），使用 computed 计算属性将获取的结果赋给变量 width（第 13 行），最后将变量 width 绑定到 modal 组件的 ns.b()节点的 style 属性上（第 21 行）。modal 组件的宽度设置效果如图 14-12 所示。

图 14-12　modal 组件的宽度设置

14.11　fixedScreen：固定屏

modal 组件的固定屏设计可以确保用户在滚动查看内容时始终能够看到头部和脚部信息，不会因为滚动而看不到关键导航或操作按钮。固定头部和脚部的内容可以帮助用户聚焦于当前内容，保证重要的引导和操作功能始终都在可见范围之内，提高用户操作的便捷性和效率。固定 modal 组件的头部和脚部内容能够优化用户交互体验，提高界面设计的可用性和友好度。实现逻辑如代码清单 library-14-17 所示。

代码清单 library-14-17
```
1. > packages/components/modal/src/modal.js
2. export const modalProps = {
3.    ...省略代码
4.    fixedScreen: Boolean
5. }
6.
7. > packages/components/modal/src/index.vue
8. <a-mask v-show="visible" v-bind="$attrs" @close="useClose">
9.   <div :class="[ns.b(), ns.is('fixed-screen', fixedScreen)]" :style="[width]">
```

```
10.       ...省略代码
11.     </div>
12.  </a-mask>
13.
14. > packages/theme/src/modal.scss
15. @include b(modal) {
16.    ...省略代码
17.    @include s((fixed-screen)) {  // 设置高度为100%
18.      height: 100%;
19.      padding: 10vh 0;
20.      @include e(wrap, false) {  // 弹性盒模型
21.        display: flex;
22.        flex-direction: column;  // 使子级的头部、内容区、脚部，三者呈纵向布局
23.        max-height: 100%;
24.      }
25.      @include e(body, false) {  // 将body设置为flex: 1，跟随窗口大小调节
26.        flex: 1;
27.        height: 0;
28.        display: flex;
29.      }
30.      @include e(render-content, false) {  // 为内容渲染层设置overflow-y，纵向滚动
31.        overflow-y: auto;
32.        width: 100%;
33.      }
34.    }
35. }
```

在 modal.js 文件中自定义属性 fixedScreen 为布尔类型，默认值为 false（第 4 行），并使用 nb.s 方法将其绑定到 modal 组件中（第 9 行）。如果使用 modal 组件并设置了属性 fixedScreen，则会自动生成类型 is-fixed-screen。然后通过 CSS 调整 modal 组件的布局（第 17~34 行），如图 14-13 所示，modal 组件的固定屏滚动效果如图 14-14 所示。

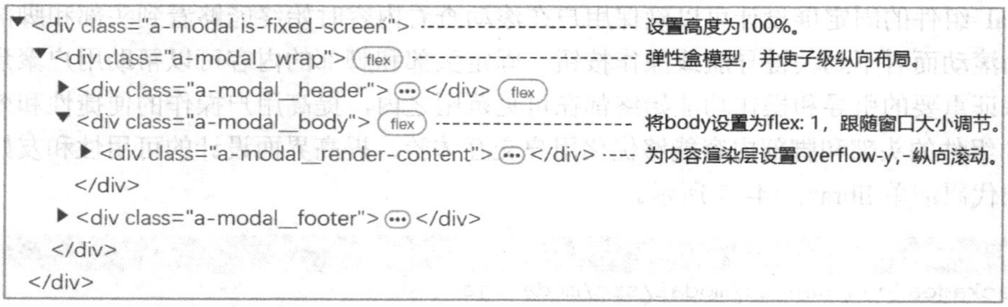

图 14-13 CSS 调整 modal 组件的布局

图 14-14　modal 组件的固定屏滚动效果

本章小结

本章主要介绍以 slot 插槽模式渲染模态框的主体内容，包括脚部插槽的渲染，还介绍了 beforeChange 异步函数的处理方式，以及打开和关闭模态框的事件回调。

第 15 章 对话框组件

messageBox 组件是一种常见的对话框组件，用于在应用程序中显示重要的消息警告或确认信息。它通常包括一个消息文本和一组按钮，例如"确定"和"取消"按钮，以便用户可以执行相应的操作。

15.1 构建组件

messageBox 组件通常用于显示简单的消息提示，例如信息提示、警告提示或确认提示，一般只包含简短的文本信息和提示标题，适用于简单的操作反馈场景。messageBox 组件的 UI 设计效果如图 15-1 所示。

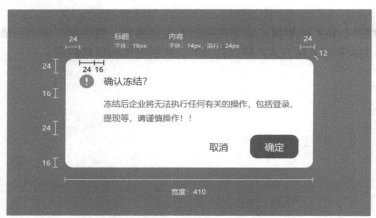

图 15-1 messageBox 组件的 UI 设计效果

根据 UI 组件库设计稿，messageBox 组件的 UI 效果和 modal 组件在整体上是相似的，只是 messageBox 组件不需要头部，且内容文本信息是固定模式的。

messageBox 组件的开发模式与 message 组件相同，因此可以将 modal 组件和 message 组件合并为 messageBox 组件，如图 15-2 所示。

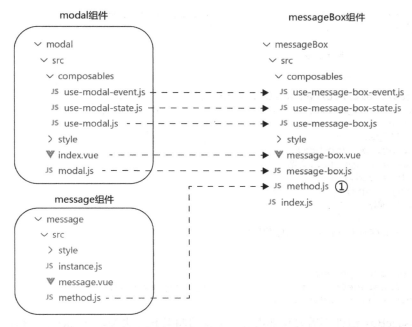

图 15-2　将 modal 组件和 message 组件合并为 messageBox 组件

从图 15-2 中可以看到，modal 组件的整体结构没有发生任何改变，只是将文件名称改为 messageBox 组件对应的文件名称，并将 message 组件的 method.js 文件完整复制到 messageBox 组件中，如序号①所示。

method.js 文件使用 createVNode 函数生成虚拟节点，createVNode 函数的第 1 个参数载入的组件便是 message-box.vue 文件。createVNode 函数在将 messageBox 组件挂载至页面时，会触发 messageBox 组件的生命周期函数 onMounted，在生命周期函数中将变量 visible 设置为 true。该方法与 message 组件一致，如代码清单 library-15-1 所示。

代码清单 library-15-1
```
1.  > packages/components/messageBox/src/message-box.vue: template
2.  <template>
3.    <transition :name="ns.b()">
4.      <a-mask v-show="visible" v-bind="$attrs" @close="useClose">
5.        <div :class="[ns.b()]">
6.          <div :class="[ns.e('wrap')]">
7.            <div :class="[ns.e('body')]">这里是固定内容区域</div>
8.            <div :class="[ns.e('footer')]" v-if="footer">...省略代码</div>
9.          </div>
10.       </div>
11.     </a-mask>
12.   </transition>
13. </template>
```

```
14. > packages/components/messageBox/src/message-box.vue: script
15. ...省略代码
16. onMounted(() => {
17.     visible.value = true
18. })
19.
20. > packages/theme/src/messageBox.scss
21. @include b(message-box) {
22.   margin: 0 auto;
23.   width: 410px;
24.   box-sizing: border-box;
25.   @include e(wrap) {...省略代码}
26.   @include e(body) {
27.     padding: 24px;
28.     ...省略代码;
29.   }
30.   @include e(footer) {...省略代码}
31. }
32.
33. > examples/src/App.vue
34. import { AMessageBox } from "@ui-library/components";
35. const open = () => AMessageBox()
```

由于 messageBox 组件不需要头部内容，因此可将其直接去掉。在 messageBox.scss 文件中复制 modal.scss 文件，然后删除 s((fixed-screen))、e(header)、e(title)和 e(close)，仅保留 e(wrap)、e(body)和 e(footer)，最后将 e(body)的属性 padding 的 4 个内边距都设置为 24px（第 27 行）。

在 examples 演示库中引入 AMessageBox 组件，便可通过点击事件调用 messageBox 组件的渲染，该方式与 message 组件一致。messageBox 组件的渲染效果如图 15-3 所示。

图 15-3　messageBox 组件的渲染效果

15.2　title：标题

messageBox 组件的标题用于简要描述消息内容或类型，可以让用户迅速了解消息的性质，从而有助于用户更快地做出反应。messageBox 组件的标题由"图标"和"文本"组成。其中，图标和 message 组件的图标一致，可分为警告、确认、错误或者提示；文本则为字符串类型。

由于 messageBox 组件的图标与 message 组件的图标完全一致。因此，下面遵循 message 组件的模式进行开发，如代码清单 library-15-2 所示。

代码清单 library-15-2

```
1.  > packages/components/messageBox/src/message-box.vue: template
2.  <div :class="[ns.e('wrap')]">
3.    <div :class="[ns.e('body')]">
4.      <div :class="[ns.e('header')]">
5.        <a-icon :class="[ns.m(theme)]"><component :is="icon" /></a-icon>
6.        <div :class="[ns.e('title')]">{{ title }}</div>
7.      </div>
8.      <div :class="[ns.e('content')]">这里是内容描述</div>
9.    </div>
10.   <div :class="[ns.e('footer')]" v-if="footer">...省略代码</div>
11. </div>
12.
13. > packages/components/messageBox/src/message-box.vue: script
14. ...省略代码
15. import { typeIcon, themeType } from "@ui-library/utils"
16. // icon
17. const icon = computed(() => typeIcon[props.icon])
18. const theme = computed(() => themeType?.[props.icon] || props.icon)
19.
20. > packages\components\messageBox\src\message-box.js
21. export const messageBoxProps = {
22.   ...省略代码
23.   icon: { type: String, default: 'warning' },
24.   title: { type: String, default: '' }
25. }
26.
27. > packages/theme/src/messageBox.scss
28. @include b(message-box) {
29.   ...省略代码
30.   @include e(header) {
31.     display: flex;
32.     align-items: center;
33.     .a-icon {
34.       font-size: 22px;
35.       line-height: 1;
36.       margin-right: 12px;
37.       color: var(#{getVarName('messageBox', 'icon-color')});
38.     }
39.   }
40.   @include e(title) {
41.     font-size: 16px;
42.     font-weight: 600;
43.     line-height: 1.5;
44.   }
45. }
46.
47. @each $type in $types {
48.   $className: '.a-message-box--' + $type;
49.   #{$className} {
```

```
50.      @include set-component-var('messageBox', messageBoxVar($type), 'type');
51.    };
52.  }
```

图标和文本位于 messageBox 组件的头部，可将两者放在一个父元素下作为子级显示。在上述代码清单中，定义 div 元素，并绑定 ns.e('header')生成类名 a-message-box__header（第 4~7 行）。图标的渲染使用 a-icon 组件包裹 Vue.js 3 的动态组件 component，并使用属性 is 绑定变量 icon（第 5 行），变量 icon 使用 computed 计算属性通过 props.icon 映射变量 typeIcon 的图标（第 17 行）。

使用 div 元素绑定 ns.e('title')生成类名 a-message-box__title，可以直接使用 props.title 属性渲染文本部分（第 6、24 行）。最后为标题的图标和文本添加 CSS 样式，messageBox 组件的标题的渲染效果如图 15-4 所示。

图 15-4   messageBox 组件的标题的渲染效果

15.3   content：内容描述

messageBox 组件的内容用于传达具体的信息、警告或提示，帮助用户清楚地了解当前发生的问题或需要采取的动作，让用户更清晰地理解消息框的意图，并有效地进行交互操作，提升用户体验。

内容既可以是常规的普通字符串文本，也可以是使用 Vue.js 3 的 h 函数生成的带有 HTML 格式或事件的标签内容文本，如代码清单 library-15-3 所示。

代码清单 library-15-3
```
1.  > packages/components/messageBox/src/message-box.vue: template
2.  <div :class="[ns.e('wrap')]">
3.    <div :class="[ns.e('body')]">
4.      <div :class="[ns.e('header')]">...省略代码</div>
5.      <div :class="[ns.e('content')]">
6.        <template v-if="isStringContent">{{ content }}</template>
7.        <component v-if="isVNodeContent" :is="content" />
8.      </div>
9.    </div>
10.   <div :class="[ns.e('footer')]" v-if="footer">...省略代码</div>
11. </div>
12.
13. Packages/components/messageBox/src/message-box.js
```

```
14. export const messageBoxProps = {
15.     ...省略代码
16.     content: { type: [String, Object], default: '' }
17. }
18.
19. > packages/components/messageBox/src/message-box.vue: script
20. import { onMounted, computed, isVNode } from "vue"
21. import { typeIcon, themeType, types } from "@ui-library/utils"
22. ...省略代码
23. // isString
24. const isStringContent = computed(() => types().isString(props.content))
25. const isVNodeContent = computed(() => isVNode(props.content))
26.
27. > packages/theme/src/messageBox.scss
28. @include b(message-box) {
29.     ...省略代码
30.     @include e(content) {
31.       margin: 12px 0 0 34px;
32.       line-height: 24px;
33.       font-size: 14px;
34.       color: var(#{getVarName('neutral-color', 'light-3')});
35.     }
36. }
37.
38. > examples/src/App.vue
39. import { AMessageBox } from "@ui-library/components";
40. const open = () => {
41.     AMessageBox({
42.       title: "这是标题",
43.       content: h(
44.             'div',
45.             { style: 'color:red;' },
46.             '冻结后企业将无法执行任何有关的操作,包括登录、提现等,请谨慎操作!!'
47.       ),
48.       content: '冻结后企业将无法执行任何有关的操作,包括登录、提现等,请谨慎操作!!'
49.     })
50. }
```

在 message-box.js 文件中添加属性 content,设置其为字符串和对象两种类型(第 16 行)。在 messageBox 组件中引入 Vue.js 3 的 isVNode 对象(第 20 行),用于判断 props.content 是否是 h 函数节点,并将结果赋给变量 isVNodeContent(第 25 行)。然后在 ns.e('content')中添加动态组件 component,使属性 is 直接绑定 content,并使用 v-if 判断变量 isVNodeContent 是否为 true,如果为 true,则渲染(第 7 行)。

对于常规的字符串,使用 types()的 isString()方法判断其是否为字符串类型,并将结果赋给变量 isStringContent(第 24 行),然后在 ns.e('content')中添加<template>占位模板渲染内容,并使用 v-if 判断变量 isStringContent 是否为 true,如果为 true,则渲染(第 6 行)。

在 examples 演示库中使用 messageBox 组件定义属性 content,即可使用 h 函数或字符串的模式(第 43~48 行)。messageBox 组件的内容渲染效果如图 15-5 所示。

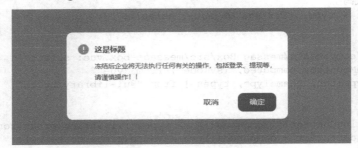

图 15-5　messageBox 组件的内容渲染效果

本章小结

messageBox 组件是 message 组件和 modal 组件的结合体，渲染模式和 message 组件一致，采用 createVNode 虚拟节点渲染。与 modal 组件相比，messageBox 组件只是去除了头部信息，并且将内容改为属性配置的模式。

第 16 章 抽屉组件

抽屉（drawer）组件是一种常见的界面元素，通常用于显示或隐藏自定义的扩展内容。它通常以抽屉滑动的形式进出浏览器可视区，具有空间利用优势，帮助用户快速访问隐藏的内容，同时不会占用主要的屏幕空间，在用户体验上也比较流畅。drawer 组件的默认打开方式如图 16-1 所示。

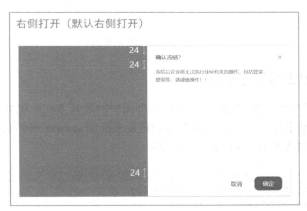

图 16-1　drawer 组件的默认打开方式

16.1　构建组件

根据 UI 组件库设计稿，drawer 组件和 modal 组件的功能基本一致，标题、关闭按钮、取消按钮、确定按钮，以及内容区都通过插槽的模式渲染，并且需要 mask 组件的透明遮罩层。

通过对比 drawer 组件和 modal 组件，可以分析出以下几个不同点。

（1）交互

modal 组件由可视区中间打开。

drawer 组件可由可视区上、右、下、左 4 个方向以滑动的形式打开。

（2）尺寸设置

modal 组件仅可以设置宽度。

drawer 组件可以设置宽度和高度。

（3）定位模式

modal 组件使用属性 margin，使组件居于可视区中间。

drawer 组件使用属性 position，使组件居于可视区上、右、下、左 4 个位置。

由此，我们可以将 modal 组件复制一份并改为 drawer 组件，然后根据上述的不同点进行调整。drawer 组件的结构如图 16-2 所示，对应分支为本书配套代码中的 library-16-1。

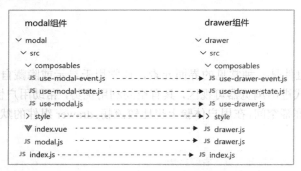

图 16-2　drawer 组件的结构

16.2　placement：方向

placement 属性通常用于指定 drawer 组件在页面中的展开方向，比如指定其从左往右、从右往左、从上往下或从下往上展开，根据布局的需要来控制 drawer 组件的展开方向，以适应不同的界面设计需求，如代码清单 library-16-2 所示。

代码清单　library-16-2
```
1.  > packages/components/drawer/src/drawer.js
2.  export const drawerProps = {
3.      ...省略代码
4.      placement: { type: String, default: 'right' },
5.  }
6.
7.  > packages/components/drawer/src/index.vue
8.  <div :class="[ns.b(), ns.m('placement', placement)]" :style="[width]">
9.    <div :class="[ns.e('wrap')]">...省略代码</div>
10. </div>
```

在 drawer.js 文件中添加属性 placement 为字符串类型，设置默认值为 right，使 drawer 组件默认由可视区的右侧向左侧滑动展开。然后使用 ns.m 方法绑定属性 placement，生成类名 a-drawer--placement_right，如图 16-3 所示。

```
<div class="a-drawer a-drawer--placement_right" style="width: 410px;">
  ▼ <div class="a-drawer__wrap"> flex
    ▶ <div class="a-drawer__header">...</div> flex
    ▶ <div class="a-drawer__body">...</div>
    ▶ <div class="a-drawer__footer">...</div>
    </div>
  </div>
```

图 16-3　为 drawer 组件生成类名

16.2.1 absolute：绝对定位

absolute 是 position 属性中的一个值，absolute 可以使元素脱离文档流，使元素根据父元素的定位而定位。为了使 drawer 组件实现上、右、下、左 4 个方向的滑动展开效果，需要设置 drawer 组件的根元素为"绝对定位"，如代码清单 library-16-3 所示。

代码清单 library-16-3

```scss
> packages/theme/src/drawer.scss
@include b(drawer) {
  position: absolute;
  box-sizing: border-box;
  @include e(wrap) {
    background-color: #fff;
    height: 100%;
    display: flex;
    flex-direction: column;
  }
  @include e(header) {...省略代码}
  @include e(body) {
    ...省略代码;
    flex: 1;
    height: 0;
    overflow-y: auto;
  }
}
```

在 drawer.scss 文件中为 b(drawer)元素设置属性 position，并设置值为 absolute，使元素脱离文档流，根据 mask 组件进行定位（第 3 行）。在 e(wrap)中，设置属性 display 的值为 flex、属性 flex-direction 的值为 column（第 8、9 行），使元素变为弹性盒模型，并使"头部"、"内容区"和"脚部" 3 个子元素呈纵向排列。

由于在内容区 e(body)中渲染外部传入的插槽元素，且无法确定插槽元素的内容是否会超出内容区的区域。因此，需要在内容区中设置纵向的滚动属性 overflow-y 为 auto，并设置属性 flex 的值为 1，使内容区的高度为自动适应。drawer 组件的绝对定位如图 16-4 所示。

图 16-4　drawer 组件的绝对定位

16.2.2　position：位置

为 drawer 组件设置了绝对定位后，要使绝对定位的元素产生位置变化，需要结合 top、right、bottom 和 left 属性调整元素的位置，如代码清单 library-16-4 所示。

代码清单 library-16-4
```
1.  > packages/theme/src/drawer.scss
2.  @include b(drawer) {
3.    @include e(wrap) {...省略代码}
4.    @include m(placement, 'right') { top: 0; bottom: 0; right: 0; }
5.    @include m(placement, 'left') { top: 0; bottom: 0; left: 0; }
6.    @include m(placement, 'top') { left: 0; top: 0; right: 0; }
7.    @include m(placement, 'bottom') { left: 0; bottom: 0; right: 0; }
8.    @include e(header) {...省略代码}
9.  }
10. 
11. > examples\src\App.vue
12. <a-drawer v-model="open" title="Drawer组件" placement="right"></a-drawer>
13. <a-drawer v-model="open" title="Drawer组件" placement="left"></a-drawer>
14. <a-drawer v-model="open" title="Drawer组件" placement="top"></a-drawer>
15. <a-drawer v-model="open" title="Drawer组件" placement="bottom"></a-drawer>
```

在 drawer.scss 文件中添加 right、left、top、bottom 位置的类名样式，对应生成的类名为 a-drawer-placement_right、a-drawer-placement_left、a-drawer-placement_top、a-drawer-placement_bottom（第 4~7 行）。右侧和左侧属性为纵向布局，可设置 top 和 bottom 为 0，对右侧设置 right 为 0，对左侧设置 left 为 0（第 4、5 行）。顶部和底部属性为横向布局，可设置 left 和 right 为 0，对顶部设置 top 为 0，对底部设置 bottom 为 0（第 6、7 行）。

在 examples 演示库中调用 drawer 组件时，使用属性 placement 传入 4 个不同方向的值，使 drawer 组件出现在不同的位置，如图 16-5 所示。

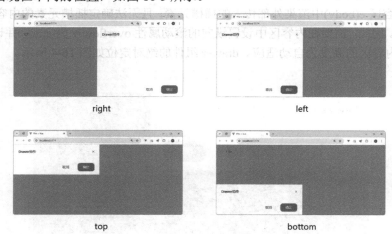

图 16-5　drawer 组件的位置

16.3　size：尺寸

drawer 组件的 size 属性用于指定抽屉展开的宽度或高度。通过设置 size 属性，可以自由调整 drawer 组件的大小，以确保符合设计需求。在默认情况下，可根据 UI 组件库设计稿中的样式，设置 drawer 组件的固定宽度或高度为 410px，也可根据业务需求手动配置属性 size，使之达到自定义宽度或高度，如代码清单 library-16-5 所示。

代码清单 library-16-5

```
1.  > packages/components/drawer/src/composables/use-drawer.js
2.  export function useDrawer({ props, visible }){
3.      ...省略代码
4.      // 离开动画完成后
5.      const useAfterLeave = () => { ...省略代码 }
6.      // 尺寸
7.      const size = computed(() => {
8.          if(['left', 'right'].includes(props.placement)) {
9.              return uStyle.width(props.size)
10.         }
11.         if(['top', 'bottom'].includes(props.placement)) {
12.             return uStyle.height(props.size)
13.         }
14.     })
15.     return { size, ...省略代码 }
16. }
17.
18. > packages/components/drawer/src/drawer.js
19. export const drawerProps = {
20.     ...省略代码
21.     size: { type: String, default: '410px' }
22. }
23.
24. > packages\components\drawer\src\index.vue: template
25. <a-mask v-show="visible" v-bind="$attrs" @close="useClose">
26.   <div :class="[ns.b(), ns.m('placement', placement)]" :style="[size]">
27.     ...省略代码
28.   </div>
29. </a-mask>
30.
31. > packages\components\drawer\src\index.vue: script
32. const { ...省略代码, size } = useDrawer({props, visible})
```

根据 placement 属性的配置，可使 drawer 组件沿上、右、下、左 4 个方向滑动展开，并通过 CSS 样式的 top、right、bottom、left 属性设置显示位置。由此可以分析出，左、右方向展开的 drawer 组件需设置宽度，上、下方向展开的 drawer 组件需设置高度。在为 drawer 组件配置宽度或高度时，只需判断属性 placement 的值即可。

在 use-drawer.js 文件中定义变量 size，使用 computed 计算属性判断 placement 的值。当

placement 的值为 left 或 right 时，调用 uStyle.width(props.size)返回宽度样式（第 8~10 行）；当 placement 的值为 top 或 bottom 时，调用 uStyle.height(props.size)返回高度样式（第 11~13 行）。

最后在 drawer.js 文件中添加属性 size 并将其设置为字符串类型,默认值为 410px（第 21 行）。在 drawer 组件中解构 use-drawer.js 文件的变量 size（第 32 行），并将其绑定至 drawer 组件的根元素中（第 26 行）。如果用户没有配置 size 属性，那么 drawer 组件的高度或宽度默认为 410px；如果用户配置了 size 属性，则根据用户配置的 size 值设置高度或宽度。

16.4 transition：过渡动画

drawer 组件的展开和收起动画效果都是抽屉从屏幕边缘滑入或滑出，用户可以清晰地看到 drawer 组件是如何打开或关闭的。滑动动画效果既有助于提高用户对应用程序界面的感知和理解，又能使界面转换更加流畅，从而提升用户体验。

drawer 组件的展开和收起效果的开发逻辑和 modal 组件一致，都要定义过渡动画的"开始值"和"结束值"，如代码清单 library-16-6 所示。

代码清单 library-16-6

```
1.  > packages\components\drawer\src\index.vue
2.  <teleport to="body">
3.    <transition :name="ns.b()" ...省略代码>   // name 属性生成类名 a-drawer
4.      ...省略代码
5.    </transition>
6.  </teleport>
7.
8.  > packages/theme/src/drawer.scss
9.  @use "./common/mixins.scss" as *;
10. @use "./common/componentVar.scss" as *;
11. @include b(drawer) {
12.   position: absolute;
13.   box-sizing: border-box;
14.   transition: .4s;
15.   transform: translate(0, 0);
16.   ...省略代码
17. }
18.
19. .a-drawer-enter-from, .a-drawer-leave-to {
20.   opacity: 0;
21.   .a-drawer--placement_left { transform: translateX(-100%); }
22.   .a-drawer--placement_right { transform: translateX(100%); }
23.   .a-drawer--placement_top { transform: translateY(-100%); }
24.   .a-drawer--placement_bottom { transform: translateY(100%); }
25. }
26. .a-drawer-enter-active,
27. .a-drawer-leave-active {
28.   overflow: hidden;
29.   transition: .4s;
```

```
30.    .a-drawer { opacity: 1; }
31. }
```

drawer 组件的展开和收起的滑动动画效果将采用 CSS3 提供的 translateX 和 translateY 属性来实现，这两个属性可用于元素的 x 轴和 y 轴偏移，可以是正数或负数。drawer 组件上、右、下、左方向的展开均在浏览器的可视区内，也是最终所在的位置，可以将其理解为"结束值"，也就是 x 轴和 y 轴的值都为 0（第 15 行）。

13.3 节中提到，Vue.js 3 提供的<transition>组件可以使用 name 属性结合"进入"动画，生成 a-drawer-enter-from 类名，并在元素插入完成后的下一帧就直接移除类名 a-drawer-enter-from。因此，可以使用类名 a-drawer-enter-from 作为 drawer 组件动画效果的"开始值"。

1. 左侧

左侧是指抽屉由左至右滑动进入浏览器可视区。在生成类名 a-drawer-enter-from 时，找到后代类名 a-drawer-placement_left，设置 x 轴为–100%（第 21 行），使 drawer 组件默认向左侧偏移自身宽度的位置，也就是定义"开始值"。当类名 a-drawer-enter-from 被移除时，x 轴的值再次回到 0，也就是"结束值"。有了"开始值"和"结束值"，再为 drawer 组件添加 transition 动画过渡属性的时间为 400 毫秒（第 14 行），就会产生滑动动画效果。

2. 右侧

右侧与左侧同理，先找到后代类名 a-drawer--placement_right，设置 x 轴为 100%（第 22 行），使 drawer 组件默认向右侧偏移自身宽度的位置。当类名 a-drawer-enter-from 被移除时，x 轴的值再次回到 0。

3. 顶部

先找到后代类名 a-drawer--placement_top，设置 y 轴为–100%（第 23 行），使 drawer 组件默认向上偏移自身高度的位置。当类名 a-drawer-enter-from 被移除时，y 轴的值再次回到 0。

4. 底部

底部和顶部同理，先找到后代类名 a-drawer--placement_bottom，设置 y 轴为 100%（第 24 行），使 drawer 组件默认向下偏移自身高度的位置。当类名 a-drawer-enter-from 被移除时，y 轴的值再次回到 0。

本章小结

本章介绍的 drawer 组件与 modal 组件属于相同类型的组件。modal 组件在浏览器可视区中间打开，可以通过设置属性 placement 使 drawer 组件从上、右、下、左 4 个方向展开，并且可使用属性 size 设置 drawer 组件展开的大小。同样采用 slot 插槽的模式渲染 drawer 组件的主体内容。

第 17 章 构建UI组件库文档

UI 组件库文档通常是指用于构建用户界面的各种组件的技术文档，其作用是帮助开发人员了解并有效地使用组件，以便构建一致、可靠的用户界面。

UI 组件库文档通常包括以下内容。
- 属性和方法：描述每个组件可接收的属性和可调用的方法，以便开发人员在其代码中正确地使用组件。
- 事件：包括组件所触发的事件，以及如何监听和处理事件的说明。
- 用法示例：通过示例代码演示如何在实际应用程序中使用各个组件。
- 最佳实践指南：提供关于如何更好地利用组件库的建议，以便开发人员构建一致且易于维护的用户界面。

通过 UI 组件库文档，开发人员不仅可以更快速地构建符合设计规范且一致的界面，而且能够更好地了解组件的功能和限制。

17.1 VitePress

VitePress 是一款为 Vue.js 3 设计的静态网站生成器。它基于 Vite 构建工具，并使用 Markdown 文件编写内容。VitePress 提供了内置的 Markdown 渲染和代码高亮功能，编写文档和撰写博客都非常简便。VitePress 还支持自定义主题和插件，支持用户根据自己的需求进行扩展。此外，VitePress 的开发模式支持热模块替换（Hot Module Replacement，HMR）、可视化编辑器以及预渲染功能，这些功能使得它在快速构建现代静态网站上具有较高的效率和灵活性。因此，我们将使用 VitePress 构建 UI 组件库文档。

提示：了解更多有关 VitePress 的知识点，可访问其官方网站，见链接 17.1。

17.1.1 初始化文档

在 3.2 节中已经介绍了在 ui-library 根目录下建立 docs 目录作为 UI 组件库文档。由于 UI 组件库文档需要使用 node_modules 组件依赖，因此要初始化 package.json 文件。进入 docs 目录，在终端执行 npm init -y 指令，生成 package.json 文件，如图 17-1 所示。

使用 VitePress 创建 UI 组件库文档，需要在终端执行指令 npm add -D vitepress@1.0.0-rc.25，安装 VitePress，如图 17-2 所示。VitePress 安装完成后，再执行 npx vitepress init 指令，进入 VitePress 项目构建向导。

```
PS D:\ui-library\docs> npm init -y
Wrote to D:\ui-library\docs\package.json:

{
  "name": "docs",
  "version": "1.0.0",
  "main": "index.js",
  "scripts": {
    "test": "echo \"Error: no test specified\" && exit 1",
    "docs:dev": "vitepress dev",
    "docs:build": "vitepress build",
    "docs:preview": "vitepress preview"
  },
  "keywords": [],
  "author": "",
  "license": "ISC",
  "devDependencies": {
    "vitepress": "^1.0.0-rc.25"
  },
  "description": ""
}
```

图 17-1　初始化 package.json 文件

```
PS D:\ui-library\docs> npm add -D vitepress@1.0.0-rc.25

added 71 packages, and audited 72 packages in 7s

10 packages are looking for funding
  run `npm fund` for details

1 high severity vulnerability

To address all issues, run:
  npm audit fix

Run `npm audit` for details.
PS D:\ui-library\docs> npx vitepress init
```

图 17-2　安装 VitePress

具体的配置步骤如下。

（1）当前目录的初始化配置

对于"VitePress 应该在哪里初始化配置"的问题，可直接输入"./"，表示在当前目录下安装，如图 17-3 所示。

```
Run `npm audit` for details.
PS D:\ui-library\docs> npx vitepress init


┌  Welcome to VitePress!
│
◆  Where should VitePress initialize the config?
│  ./
```

图 17-3　当前目录的初始化配置

(2)项目标题和描述

Site title(设置项目标题)、Site description(设置项目描述)为选填内容,在完成项目构建后,也可以通过.vitepress/config.mts文件修改,如图17-4所示。

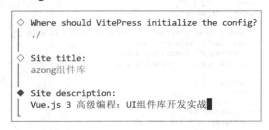

图17-4 设置项目标题和描述

(3)主题

主题有以下3种选项。

◎ Default Theme (Out of the box, good-looking docs):默认主题 + 开箱即用。
◎ Default Theme + Customization:默认主题 + 自定义。
◎ Custom Theme:自定义主题。

这里选择第一项即可,如图17-5所示。

图17-5 选择主题

(4)是否使用TypeScript

对于"配置文件和主题文件是否使用TypeScript"的问题,选择Yes或No都可以,如图17-6所示。

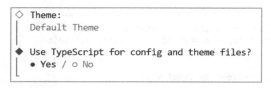

图17-6 是否使用TypeScript

(5)添加到package.json

对于"是否将VitePress npm 脚本添加到package.json"的问题,选择Yes。这样package.json文件的script对象中会自动生成docs:dev、docs:build和docs:preview指令。全部安装完成后,

执行 npm run docs:dev 运行项目,如图 17-7 所示。对应分支为本书配套代码中的 library-17-1。

```
◇ Use TypeScript for config and theme files?
│ Yes
│
◇ Add VitePress npm scripts to package.json?
│ Yes
│
└ Done! Now run npm run docs:dev and start writing.
```

图 17-7 运行项目

17.1.2 配置导航栏

VitePress 导航栏的配置属性是 nav,配置的数据为数组类型的对象,对应显示的位置是页面右上角。在 docs 目录下新建 guide 目录,并新建 index.md 文件,使用 Markdown 语法在 index.md 文件中写入任意内容,如代码清单 library-17-2 所示。

代码清单 library-17-2
```
1.   > docs/guide/index.md
2.   # 这是一级标题
3.   ## 这是二级标题
4.   
5.   > docs/.vitepress/config.mts
6.   import { defineConfig } from 'vitepress'
7.   export default defineConfig({
8.     ...省略代码
9.     themeConfig: {
10.      logo: '/images/logo.png',
11.      nav: [
12.        { text: '指南', link: '/guide/' },
13.        { text: '组件', link: '/markdown-examples' }
14.      ],
15.    }
16.  })
```

.vitepress/config.mts 是站点配置文件,属性 nav 用于配置站点右上角的导航栏(第 11~14 行)。属性 text 是导航栏的文本;属性 link 是点击文本时跳转的链接,也就是文件的路径,跳转链接始终以"/"开头。导航"指南"是文件夹,link 的值指向文件夹/guide/,并默认读取以 index 开头的文件(第 12 行)。导航"组件"是文件,link 的值指向文件/markdown-examples(第 13 行)。两者的区别在于,前者是文件夹,末尾带"/";后者是文件,末尾不带"/"。点击导航"指南"便会跳转页面,并载入/guide/index.md 文件,如图 17-8 所示。

提示:如需了解更多关于导航栏的配置信息,可查看 VitePress 官网,见链接 17-2。

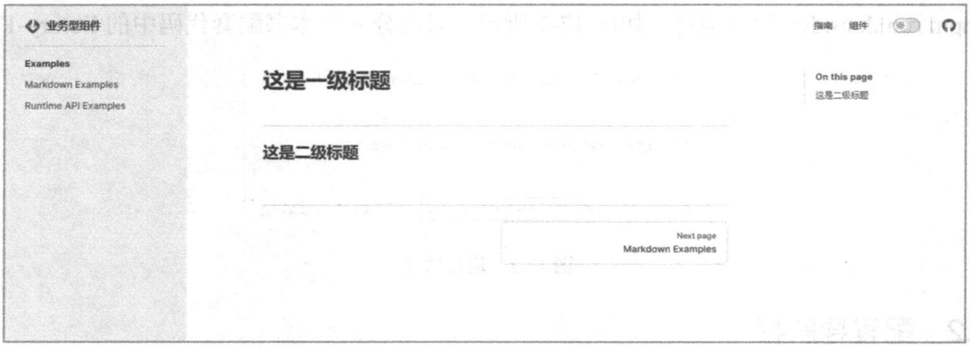

图 17-8　载入/guide/index.md 文件

17.1.3　配置侧边栏

侧边栏是在站点左侧位置显示的树形菜单，最多可以有 6 级子菜单，超出 6 级会被忽略不显示。配置侧边栏的属性是 themeConfig.sidebar。目前在 config.mts 文件中已经配置了"指南"和"组件"两个导航，但无论点击哪个导航菜单，侧边栏显示的菜单都一致。这是因为属性 sidebar 配置的侧边栏是全局的，所有的导航栏都使用同样的配置，如果要为不同的导航配置不同的侧边栏菜单，需要使侧边栏和导航栏一一对应，如代码清单 library-17-3 所示。

代码清单 library-17-3
```
> docs/.vitepress/config.mts
export default defineConfig({
    ...省略代码
    themeConfig: {
      logo: '/images/logo.png',
      nav: [
        { text: '指南', link: '/guide/' },
        { text: '组件', link: '/components/' }
      ],
      sidebar: {
        '/guide/': [
          {
            text: '基础',
            items: [
              { text: '设计', link: '/guide/design' },
              { text: '安装', link: '/guide/install' },
              { text: '快速开始', link: '/guide/start' }
            ]
          },
          {
            text: '进阶',
            items: [
              { text: '国际化', link: '/guide/lang' },
              { text: '主题', link: '/guide/theme' },
```

```
25.            { text: '暗黑模式', link: '/guide/diablo' }
26.        ]
27.      }
28.    ],
29.    '/components/': [
30.      {
31.        text: '基础',
32.        items: [
33.          { text: 'Button 按钮', link: '/components/button' },
34.          { text: 'Checkbox 多选框', link: '/components/checkbox' },
35.          { text: 'Input 输入框', link: '/components/input' },
36.          { text: 'Select 下拉选项', link: '/components/select' },
37.          { text: 'Icon 图标', link: '/components/icon' }
38.        ]
39.      }
40.    ]
41.   }
42.  }
43. })
```

首先将属性 themeConfig.sidebar 改为 JSON 类型格式，添加"/guide/"和"/components/"两个 key（第 11、29 行），它们需要与导航栏的 link 属性的值一致（第 7、8 行）。然后为/guide/和/components/添加侧边栏菜单数据（第 10~42 行），配置侧边栏菜单的渲染效果如图 17-9 所示。

图 17-9　配置侧边栏菜单

17.2　解析 Markdown 文件

　　Markdown 文件是一种使用普通文本格式编写的文件，通常包含文本内容和用于指定文本样式的标记。Markdown 文件通常以".md"为文件扩展名，可以通过各种编辑器和工具进行编辑和解析。

　　Markdown 文件具有易读易写的特性，使用一些简单的符号和约定来指定文本的样式和结构。例如，使用"*"表示斜体或粗体文本，使用"#"表示标题，使用"\"表示代码块等。由于其简洁性和易读性，Markdown 常用于撰写文档、博客文章和技术文档。

　　常见的 Markdown 文件包括 README.md（用于项目文档）、blog posts、技术文档等。解析

Markdown 文件通常涉及将其转换为 HTML 或其他格式，以便在 Web 页面上显示或打印。

17.2.1 主题入口

在 17.1.1 节中构建 UI 组件库文档时，选择"默认主题+开箱即用"的主题模式，这是 VitePress 提供的内置主题，UI 组件库设计稿中的文档如图 17-10 所示。这种模式无法满足 UI 组件库文档的展示需求，因此需要采用自定义主题的模式实现 UI 组件库文档的定制化。

图 17-10 UI 组件库设计稿中的文档

VitePress 自定义主题的入口是 .vitepress\theme 目录的 index 文件。VitePress 检测 .vitepress\theme 目录存在主题入口文件时，会使用自定义主题而不是默认主题，我们便可以扩展默认主题，在其基础上实现更高级的自定义。由于在初始化文档时已经自动生成了 theme 目录主题入口以及相应的代码逻辑，因此只需要删除 index.ts 文件的 Layout 部分的代码，保留代码清单 library-17-4 即可。

代码清单 library-17-4
```
1.  > docs/.vitepress/theme/index.ts
2.  import type { Theme } from 'vitepress'
3.  import DefaultTheme from 'vitepress/theme'
4.  import './style.css'
```

```
5.  export default {
6.    extends: DefaultTheme,
7.    enhanceApp({ app, router, siteData }) {
8.      // ...
9.    }
10. } satisfies Theme
```

17.2.2　注册全局组件

UI 组件库是一个比较庞大的项目，每种组件都存在不同的状态和交互逻辑。在展示 UI 组件库文档时，不但要渲染组件效果，还要解析组件的代码。因此在开发 UI 组件库文档时，需要使用"配置+动态组件"的方式读取组件路径，并使用 Markdown-it 解析组件。使用 Markdown-it 解析组件的规则是统一的，可以通过注册全局组件的方法实现，如代码清单 library-17-5 所示。

代码清单 library-17-5
```
1.  > .vitepress/components/demo/index.vue
2.  <template>注册全局组件</template>
3.
4.  > docs/.vitepress/theme/index.ts
5.  import type { Theme } from 'vitepress'
6.  import DefaultTheme from 'vitepress/theme'
7.  import Deom from "../components/demo/index.vue"
8.  import './style.css'
9.  export default {
10.   extends: DefaultTheme,
11.   enhanceApp({ app, router, siteData }) {
12.     app.component('Demo', Deom)
13.   }
14. } satisfies Theme
```

在 .vitepress 目录下新建 components/demo 目录，并建立 index.vue 文件，然后引入 theme/index.ts 主题入口文件（第 7 行）。在 enhanceApp 对象中使用 app.component 注入全局组件，并定义组件名称为 Demo（第 12 行）。

17.2.3　markdown-it-container

使用 VitePress 撰写技术文档，是因为 VitePress 提供了内置的 Markdown 解析器和一些预设的主题，并提供对 Vue.js 3 组件的完整支持。虽然 VitePress 内置了 Markdown 扩展功能，但由于 UI 组件库文档需要在最大程度上实现定制模式及自由布局，即便采用 VitePress 内置的 Markdown 解析器，依然无法满足需求，因此要借助 markdown-it-container 插件实现对 Markdown 文件的解析，如代码清单 library-17-6 所示。

代码清单 library-17-6
```
1.  > docs/.vitepress/config.mts
2.  import { defineConfig } from 'vitepress'
3.  import MdContainer from "markdown-it-container"
```

```
4.   export default defineConfig({
5.     themeConfig: {...省略代码},
6.     markdown: {
7.       config: (md) => {
8.         md.use(MdContainer, 'demo', {
9.           render(tokens: string, idx: string){
10.            if (tokens[idx].nesting === 1) {
11.              return '<Demo>';
12.            } else {
13.              return '</Demo>';
14.            }
15.          }
16.        })
17.      },
18.    },
19.  })
20.
21.  > docs/package.json
22.  {
23.    ...省略代码
24.    "dependencies": {
25.      "markdown-it-container": "^4.0.0"
26.    }
27.  }
28.
29.  > docs/components/button.md
30.  # Button 按钮
31.  ## 基础按钮
32.  :::demo
```

使用 markdown-it-container 插件前需要先安装依赖包，指令为 npm install markdown-it-container@4.0.0 -S。然后引入 config.mts 文件，并定义名称为 MdContainer（第 3 行）。VitePress 内置的 Markdown 扩展中的"高级配置"可以实现 markdown 选项，支持自定义更多功能（第 6~18 行）。

属性 markdown 选项中的 config 是一个函数（第 7 行），参数 md 是 markdown-it 对象。markdown-it 对象的 use 方法用于安装插件，markdown-it 会将插件集成到其解析过程中，以便在解析 Markdown 时应用插件中定义的逻辑。

因此在上述代码清单中，use 方法的第一个参数是 markdown-it-container，也就是 MdContainer；第二个参数是解析 Markdown 文件指定的字符":::demo"（第 8 行）；第三个参数是解析 Markdown 文件指定区域渲染内容的规则（第 9~15 行）。可以注意到，第三个参数的 render 根据条件返回了<Demo>，也就是渲染 theme/index.ts 注册的全局组件 Demo。当点击侧边栏的"Button 按钮"组件时，会在 components/button.md 文件中的":::demo"区域渲染<Demo>组件。markdown-it-container 解析 Markdown 文件的效果如图 17-11 所示。

图 17-11　markdown-it-container 解析 Markdown 文件

提示：markdown-it-container 是 markdown-it 的一个插件，支持用户在 Markdown 中创建自定义容器，这些容器用于封装具有特定类的内容块。

17.2.4　tokens 容器

tokens 是在 Markdown 解析过程中生成的对象，用于表示 Markdown 文件中的各种元素，如标题、段落、列表、链接等。在 markdown-it-container 插件中，特定的标记会被解析成 tokens，以便在后续的处理中进行进一步操作。

在 17.2.3 节中，md.use 方法的第二个参数是 "demo"，也就是解析 :::demo 范围的自定义容器，解析成功后会生成 tokens 集合，然后通过 tokens 集合的索引获取自定义容器的数据，如代码清单 library-17-7 所示。

代码清单 library-17-7
```
1.   > docs/components/button.md
2.   # Button 按钮
3.   ## 基础按钮
4.   :::demo 按钮的 `type` 分别为 `default`、`primary`、`success`、`warning` 和 `error`
5.   ```html
6.   button/basis
7.   ```
8.   :::
9.
10.  > .vitepress/config.mts
11.  md.use(MdContainer, 'demo', {
12.    render(tokens: string, idx: string){
13.      console.log('tokens', tokens)
14.      console.log('自定义容器', idx)   // 6、8
15.    }
16.  })
```

在 components/button.md 文件中定义上述内容，其中，":::"表示一个特殊的标记，用于创建自定义容器或分隔不同的内容部分。第 4 行的:::demo 表示自定义容器的开始，第 8 行的:::则表示自定义容器的结束。然后在 config.mts 文件中打印 tokens 集合和自定义容器索引 idx（第 13、14 行），这里是终端输出，并非浏览器控制台，如图 17-12 和图 17-13 所示。

图 17-12　打印 tokens 集合

图 17-13　打印自定义容器索引 idx

图 17-12 是 markdown-it-container 插件解析整个 Markdown 文件的 tokens 集合，也是一个数组对象，包含"# Button 按钮"、"## 基础按钮"和":::demo"数据。图 17-13 中为自定义容器索引，也就是说，tokens 集合的第 6 个和第 8 个对象是自定义容器，通过索引 idx 便可获取所需的数据，如图 17-14 所示。

```
(6)
Token$3 {
    type: 'container_demo_open',
    tag: 'div',
    attrs: null,
    map: [ 2, 6 ],
    nesting: 1,
    level: 0,
    children: null,
    content: '',
    markup: ':::',
    info: 'demo 按钮的…
',
    meta: null,
    block: true,
    hidden: false
},
```

```
(8)
Token$3 {
    type: 'container_demo_close',
    tag: 'div',
    attrs: null,
    map: null,
    nesting: -1,
    level: 0,
    children: null,
    content: '',
    markup: ':::',
    info: '',
    meta: null,
    block: true,
    hidden: false
}
```

图 17-14 tokens 自定义容器的索引数据

17.3 UI 组件库解析

UI 组件库解析是指对 UI 组件库的组件进行程序自动化解码，包括浏览组件库的文档、阅读组件的说明和 API 文档、研究示例代码等。通过对组件库文件进行解析，开发人员可以更好地掌握组件的功能、用法和配置方法，从而提高开发效率。

17.3.1 定义文档组件

UI 组件库中有很多组件，每种组件有各种不同的属性和方法，如 button 组件有基本类型、禁用类型、图标类型、边框类型等。为了在组件库文档中展示不同类型的组件，需要将每种类型的组件定义成独立的文件，通过路径读取文件，如代码清单 library-17-8 所示。

代码清单 library-17-8
```
1.  > docs/examples/button/basis.vue
2.  <template>
3.      <a-button>默认</a-button>
4.      <a-button type="primary">主要</a-button>
5.      <a-button type="success">成功</a-button>
6.      <a-button type="warning">警告</a-button>
7.      <a-button type="error">错误</a-button>
8.  </template>
9.
10. > docs/examples/button/disabled.vue
11. <template>
12.     <a-button disabled>默认</a-button>
13.     <a-button type="primary" disabled>主要</a-button>
14.     <a-button type="success" disabled>成功</a-button>
15.     <a-button type="warning" disabled>警告</a-button>
16.     <a-button type="error" disabled>错误</a-button>
17. </template>
18.
19. > docs/examples/button/round.vue
```

```
20.  <template>
21.      <a-button round>默认</a-button>
22.      <a-button type="primary" round>主要</a-button>
23.      <a-button type="success" round>成功</a-button>
24.      <a-button type="warning" round>警告</a-button>
25.      <a-button type="error" round>错误</a-button>
26.  </template>
```

在 docs 目录下新建 examples 目录作为 UI 组件库文档的演示包，在 examples 目录中便可以建立对应组件的目录。例如，button 目录对应 button 组件、checkbox 目录对应 checkbox 多选框组件，以此类推。然后在各个组件目录下建立不同类型的组件效果，如 button 组件有基础的、圆角的、禁用的，分别对应 3 个文件 basis.vue、round.vue、disabled.vue。

17.3.2　读取容器信息

容器信息是指由 markdown-it-container 解析 Markdown 文件中:::demo 指定范围的内容信息，并结合 tokens 获取自定义容器的数据，如代码清单 library-17-9 所示。

代码清单 library-17-9
```
1.  > .vitepress\config.mts
2.  md.use(MdContainer, 'demo', {
3.    render(tokens: string, idx: string){
4.      if (tokens[idx].nesting === 1) {
5.        // 使用正则表达式捕获组中的描述内容,即 ::: demo xxx 中的 xxx
6.        const info = tokens[idx].info.trim().match(/^demo\s*(.*)$/);
7.        const description = info && info.length > 1 ? info[1] : "";
8.        // 获取路径
9.        const nextToken = tokens[idx + 1]; // 下一个 tokens
10.       const componentPath = nextToken.type === "fence" ? nextToken.content : "";
11.       return '<Demo>';
12.     } else {
13.       return '</Demo>';
14.     }
15.   }
16. })
```

1. 获取:::demo 后跟随的描述文本

在 markdown-it-container 插件解析的 Markdown 文件中，我们已经了解到 tokens[idx]可以获取自定义容器"开始"和"结束"范围的数据，如图 17-14 所示。在 components/button.md 文件中，":::demo"标记后面的文本"demo 按钮的……"位于属性 info 中，由于属性 info 为字符串类型，因此可以使用正则表达式获取（第 6 行）。tokens[idx].info.trim()去除了属性 info 文本两端的空格，再使用 match(/^demo\s*(.*)$/)匹配以"demo"字符开头之后的所有内容，最后的返回结果是一个数组，如果配置不成功，则返回 null。

tokens 自定义容器索引对象（变量 info）得到的内容如图 17-15 所示。可以看出，数组下标为 1 的是":::demo"字符后跟随的内容，因此只需判断在变量 info 中存在数据并且长度大于 1

时，获取下标为 1 的内容，并赋给变量 description（第 7 行）。

```
[
  'demo 按钮的 `type` 分别为 `default`、`primary`、`success`、`warning` 和 `error` 。',
  '按钮的 `type` 分别为 `default`、`primary`、`success`、`warning` 和 `error` 。',
  index: 0,
  input: 'demo 按钮的 `type` 分别为 `default`、`primary`、`success`、`warning` 和 `error` 。',
  groups: undefined
]
```

图 17-15　tokens 自定义容器索引对象

2. 获取文档组件路径

获取文档组件路径是指获取位于 examples 目录下的组件，例如，button/basis 是指获取 button 目录下的 basis.vue 组件、button/disabled 是指获取 button 目录下的 disabled 组件，其他组件以此类推。

components/button.md 文件中的```html 定义了文档组件的路径，根据图 17-12，在 tokens 集合中，下标为 "6" 的是自定义容器的开始，下标为 "7" 的是```html 定义的文档组件路径。因此，只需根据自定义容器的开始获取下一组对象（第 9 行），然后使用三元运算判断 type 为 "fence" 时，获取 content 属性的值，否则为空字符，最后赋给变量 componentPath（第 10 行），如图 17-16 所示。

button/basis

图 17-16　```html 定义的文档组件路径

17.3.3　读取文档组件

读取文档组件是指解析 Markdown 文件中```html 定义的文档组件路径，结合 node.js 的模块获取本地文件，读取组件源码并传给 Demo 全局组件渲染代码，如代码清单 library-17-10 所示。

代码清单 library-17-10
```
1.  > .vitepress\config.mts
2.  import path from 'path'
3.  import fs from 'fs'
4.  md.use(MdContainer, 'demo', {
5.    render(tokens: string, idx: string){
6.      if (tokens[idx].nesting === 1) {
7.        // 获取路径
8.        const nextToken = tokens[idx + 1]; // 下一个 tokens
9.        const componentPath = nextToken.type === "fence" ? nextToken.content : "";
10.       // 读取文件
11.       let source = ''
12.       if (componentPath) {
13.         let file = path.resolve(__dirname, "../examples", `${componentPath}.vue`);
14.         file = file.replace(/\s+/g, "");
15.         source = fs.readFileSync(file, "utf-8");
```

```
16.         console.log('source', source)
17.     }
18.     console.log('source', source)
19.     return '<Demo>';
20.   } else {
21.     return '</Demo>';
22.   }
23. }
24. })
```

在读取系统本地文件和操作系统文件前，首先需要引入 node.js 的两个核心模块，分别是 path（处理文件和目录的路径）和 fs（文件系统操作）（第 2、3 行）。在上述代码中，优先判断变量 componentPath 是否存在（第 12 行），如果存在，则使用 path.resolve 方法组合参数获取 componentPath 文件的路径，本地文件路径结果如图 17-17 所示。

```
D:\ui-library\docs\examples\button\basis.vue
```

图 17-17　path.resolve 方法获取本地文件路径

接着使用正则表达式清除路径中的所有"空格"（第 14 行），然后调用 fs 核心模块的同步方法 readFileSync 读取文件的内容，将读取的内容赋给变量 source，最终打印出的结果便是 basis.vue 文件的源码，如图 17-18 所示。

```
<template>
    <a-button>默认</a-button>
    <a-button type="primary">主要</a-button>
    <a-button type="success">成功</a-button>
    <a-button type="warning">警告</a-button>
    <a-button type="error">错误</a-button>
</template>
```

图 17-18　basis.vue 文件的源码

17.3.4　渲染组件

从 markdown-it-container 插件解析 Markdown 文件到 node.js 读取文件，我们已经完成了 Markdown 文件数据的处理过程，并满足了渲染 UI 组件及演示代码的需求，下面将获取的数据传入全局<Demo>渲染。

1. 插槽渲染"描述"和"源码"

通过 node.js 的 path 和 fs 两个核心模块获取的数据来渲染"描述"和"源码"，将数据通过组件插槽的方式传入<Demo>组件渲染，如代码清单 library-17-11 所示。

代码清单 library-17-11
```
1. > .vitepress\config.mts
2. md.use(MdContainer, 'demo', {
3.   render(tokens: string, idx: string){
```

```
4.      if (tokens[idx].nesting === 1) {
5.        ...省略代码
6.        if (componentPath) {...省略代码
7.        return `<Demo>
8.          <template #source><pre v-pre><code class="html">${md.utils.escapeHtml(source)}</code></pre></template>
9.          <template #description>${description ? `${md.render(description)}` : ""}</template>
10.         `;
11.     } else {
12.       return '</Demo>';
13.     }
14.   }
15. })
16.
17. > .vitepress/components/demo/index.vue
18. <template>
19.   <div><slot name="description" /></div>
20.   <div><slot name="source" /></div>
21. </template>
```

在<Demo>组件中定义具名插槽<slot name="description" />和<slot name="source" />，用于接收变量 description 和 source 并进行渲染（第 19、20 行）。在 config.mts 文件中需要注意的是，对于变量 source，需使用 md.utils.escapeHtml 将 HTML 转为普通字符，以便在 Markdown 文件中安全地插入 HTML 代码，而不会被解释器解释为 HTML 标签，并且使用<pre>和<code>标签将 HTML 字符格式化，保持原有的格式（第 8 行）。调用 md.render 再次解析变量 description，使之生成具有一定格式的字符（第 9 行）。然后在<Demo>组件的 description 和 source 插槽中分别渲染"描述"和"源码"（第 19、20 行），渲染效果如图 17-19 所示。

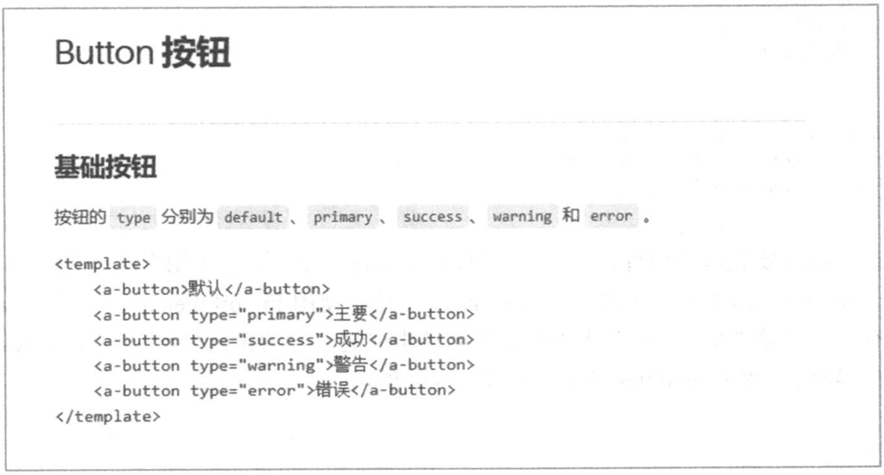

图 17-19　在<Demo>组件中定义 description 和 source 插槽渲染"描述"和"源码"

2. 渲染演示组件

将获取的本地文件路径传入<Demo>组件，在该组件中根据路径获取对应的 vue 文件组件，再使用 Vue.js 3 的动态组件 component 渲染演示组件，如代码清单 library-17-12 所示。

代码清单 library-17-12

```
1.  > .vitepress/components.js
2.  const modulesFiles = import.meta.glob('../examples/*/*.vue', { eager: true })
3.  /** 自动化处理 */
4.  let modules = {}
5.  for (const [key, value] of Object.entries(modulesFiles)) {
6.    var keys = key.split('/')
7.    const name = keys.slice(1).join('/')
8.    modules[name] = value.default
9.  }
10. export default modules
11.
12. > .vitepress/components/demo/index.vue: template
13. <div class="examples-container">
14.     <div class="description"><slot name="description" /></div>
15.     <div class="examples-body">
16.         <div class="examples-inner"><component :is="demo" /></div>
17.         <div class="source-inner"><slot name="source" /></div>
18.     </div>
19. </div>
20.
21. > .vitepress/components/demo/index.vue: script
22. import { computed } from "vue"
23. import modules from "../../components"
24. const props = defineProps({
25.     path: {
26.         type: String,
27.         default: ''
28.     }
29. })
30. const demo = computed(() => {
31.     const key = `examples/${props.path}.vue`
32.     return modules[key];
33. });
```

渲染组件前需要先获取组件，由于 UI 组件的类型比较多，可以采用自动化处理的方式导入所有组件。在.vitepress 目录下新建 components.js 文件，使用 import.meta.glob 导入 examples 目录下的所有组件（第 2 行），然后将所有组件循环生成键值对的模式赋给变量 modules（第 4~9 行），再将其导出。变量 modules 的对象如图 17-20 所示。

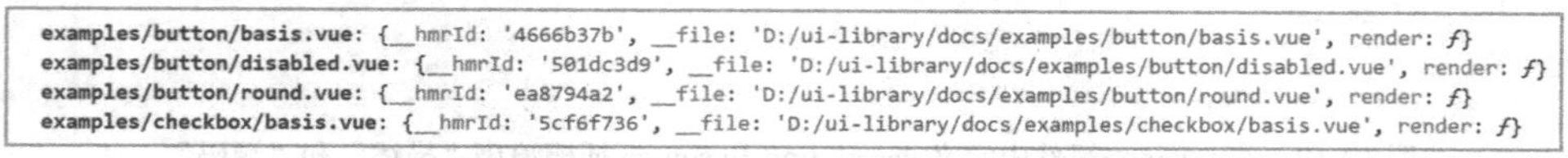

图 17-20　变量 modules 的对象

在图 17-20 中，examples/button/basis.vue、examples/button/disabled.vue、examples/button/round.vue 和 examples/checkbox/basis.vue 是不同 UI 组件的 key，只需匹配上对应的 key，即可渲染指定的组件。

将 components.js 引入全局组件<Demo>（第 23 行），为 defineProps 对象定义 path 属性来接收组件路径（第 25~28 行），使用 computed 计算属性生成与 components.js 对应的 key（第 31 行），将对应 key 的组件返回给变量 demo（第 30 行），最后将变量 demo 与动态组件的 is 属性绑定（第 16 行），UI 组件的渲染效果如图 17-21 所示。

图 17-21　UI 组件的渲染效果

17.3.5　代码高亮

UI 组件库文档的代码高亮包括对关键字、变量、字符串等代码组成部分进行颜色、加粗或其他视觉上的突出显示，可以帮助开发人员快速识别和理解代码结构，从而更有效地使用和定制 UI 组件库中的相关代码。同时，代码高亮也有助于突出代码中的语法和语义，帮助开发人员遵循最佳实践方式，并避免常见的错误。代码高亮的实现方式将采用第三方插件 prism，如代码清单 library-17-13 所示。

代码清单 library-17-13

```
1.  > .vitepress/components/demo/index.vue: script
2.  import { computed, onMounted } from "vue"
3.  import prism from 'prismjs'
4.  import "prismjs/themes/prism-tomorrow.min.css"
5.
6.  onMounted(() => {
7.      prism.highlightAll()
8.  })
9.
10. > .vitepress/config.mts
11. <template #source><pre><code class="language-html">${md.utils.escapeHtml(source)}</code></pre></template>
```

在使用第三方插件 prism 前，需要先安装依赖包至开发环境，指令为 npm install prismjs@1.29.0 -D。为全局组件<Demo>引入 prism 组件包和样式（第 3、4 行），然后在生命周期函数 onMounted 完成组件渲染后执行 prism.highlightAll()（第 7 行）。

最关键的位置在 config.mts 文件中，需要为<code>标签添加类名 language-html，指定 prismjs 对象以什么类型的语言高亮代码块，效果如图 17-22 所示。

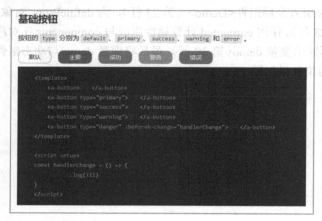

图 17-22　代码高亮

从图 17-22 中可以看到，源码已经自动区分了不同的颜色，并且以黑色为背景。这里呈现的是 prismjs 插件自身的 CSS 效果，如果想修改高亮样式，可以在主题入口 theme/index.ts 文件中引入自定义的样式文件，对指定的类名重写样式，如代码清单 library-17-14 所示，渲染效果如图 17-23 所示。

代码清单 library-17-14

```
1.  > .vitepress/theme/index.ts
2.  ...省略代码
3.  import './highlight.scss'
4.  export default {...省略代码} satisfies Theme
```

图 17-23　自定义代码块高亮样式

提示：代码高亮效果可通过 css 文件 highlight.scss 查看，也可通过控制台查看相应的类名，修改不同的颜色。

17.3.6 展开/收起源码

UI 组件库中的展开、收起功能通常作用于组件源码的信息，可以让用户通过点击按钮或其他交互元素来展开或收起源码，从而提高页面的可读性和用户体验。由于组件库文档的整体结构由全局组件<Demo>结合 markdown-it-container 开发，可以实现完全自由布局的效果，因此根据 UI 组件库设计稿中的文档实现布局即可，如代码 library-17-15 所示，如图 17-24 所示。

代码清单 library-17-15

```
> docs/.vitepress/components/demo/index.vue: template
<div class="examples-container">
    <!-- 描述 -->
    <div class="description"><slot name="description" /></div>
    <!-- 演示主体 -->
    <div class="examples-body">
        <!-- 组件渲染 -->
        <div class="examples-inner"><component :is="demo" /></div>
        <!-- 图标元素 -->
        <ul class="examples-control">
            <li @click="copy" @mouseleave="isCopySuccess = ''">
                <i class="iconfont" :class="iconCopy"></i>
            </li>
            <li @click="toggleSource">
                <i class="iconfont icon-daima"></i>
            </li>
        </ul>
        <!-- 组件源码 -->
        <div v-if="source" class="source-inner"><slot name="source" /></div>
    </div>
</div>

> docs/.vitepress/components/demo/index.vue: script
import { computed, onMounted, ref, useSlots, nextTick } from "vue"
import Clipboard from 'clipboard'
const slots = useSlots()
const source = ref(false)
const isCopySuccess = ref('')
// 复制的图标元素
const iconCopy = computed(() => {
    if(isCopySuccess.value === true) { return 'icon-chenggong1' }
    if(isCopySuccess.value === false) { return 'icon-chahao' }
    return 'icon-fuzhi_line'
})
// 复制组件源码
const copy = async (event) => {
    const clipboard = new Clipboard(event.target, {
```

```
38.         text: () => slots.source()[0]?.children[0]?.children
39.       })
40.       clipboard.on('success', () => {  // 复制成功
41.         isCopySuccess.value = true
42.         clipboard.destroy()
43.       })
44.       clipboard.on('error', () => {    // 复制失败
45.         isCopySuccess.value = false
46.         clipboard.destroy()
47.       })
48.       clipboard.onClick(event)
49.     }
50.     // 展开、收起代码
51.     const toggleSource = () => {
52.       source.value = !source.value
53.       source.value && nextTick(() => prism.highlightAll())
54.     }
```

<Demo>组件的 template 模块分为 4 个块，分别是：描述、组件渲染、图标元素和组件源码。其中的描述和组件渲染已经完成（第 4、8 行），图标元素可以使用阿里图库 iconfont 完成（第 10~17 行），组件源码是单独的一个 DIV 元素（第 19 行）。

图标元素分别是：代码和复制。

◎ 代码图标按钮（第 14~16 行）的点击事件绑定了函数 toggleSource，该函数中控制变量 source 在 true 与 false 之间切换（第 52 行），接着判断变量 source 的值为 true 时，调用 prism.highlightAll()使代码高亮（第 53 行）。最后将变量 source 绑定至"组件源码"的 DIV 元素即可实现展开收起效果（第 19 行）。

◎ 复制功能使用了第三方插件 clipboard，执行指令 npm install clipboard@ 2.0.11 -S 安装依赖包，并将其引入（第 25 行）。复制图标按钮（第 11~13 行）的点击事件绑定了函数 copy，在该函数中调用 Clipboard 的对象复制功能（第 37~39 行），将 slots 插槽的内容传给属性 text 即可复制成功（第 38 行）。

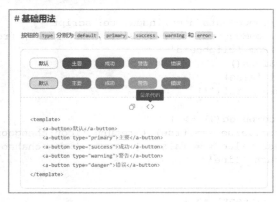

图 17-24　组件展开/收起效果

提示：请在本书配套代码中的分支 docs 中查看<Demo>组件整体结构的 CSS 样式属性效果。

17.4 撰写组件库文档

撰写组件库文档是维护和推广组件库的关键一步。组件库文档有助于为用户提供全面的支持和指导，使用户理解组件库的功能和用法，进而实现标准化使用。由于我们的 UI 组件库文档是由 VitePress 搭建的，并且 VitePress 提供了内置的 Markdown，因此可以直接采用 Markdown 语法撰写组件库文档。

17.4.1 Markdown 语法

Markdown 是一种轻量级标记语言，用于快速、直观地编写格式化文档。它允许使用简单的语法和符号来标记文本，以便将其转换为规整的格式。Markdown 的常用语法如表 17-1 所示。

表 17-1 Markdown 的常用语法

类型	Markdown	HTML
标题	# 一级标题	<h1>一级标题</h1>
	## 二级标题	<h2>二级标题</h2>
	### 三级标题	<h3>三级标题</h3>
	#### 四级标题	<h4>四级标题</h4>
	##### 五级标题	<h5>五级标题</h5>
	###### 六级标题	<h6>六级标题</h6>
文本	**粗体**、__粗体__	粗体
	斜体、_斜体_	斜体
	\`\`\`定义代码文本\`\`\`	<code>定义代码文本</code>
	~~删除线~~	<s>删除线</s>
链接	[链接文本](http://www.example.com)	<a href="http://www.example.com"> 链接文本 </a>
	[链接到同一页面中的锚标记](#anchor)	<a href="#anchor"> 链接到同一页面中的锚标记 </a>
	[链接到另一个页面中的锚标记](page_url#anchor)	<a href="page_url#anchor"> 链接到另一个页面中的锚标记 </a>

17.4.2 Markdown 扩展功能

VitePress 内置了一些 Markdown 扩展功能，通过内置语法可以实现更丰富的文档编写效果。

其中一些内置的 Markdown 扩展功能如图 17-25~图 17-27 所示。

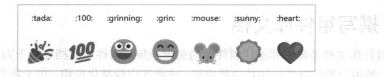

图 17-25　表情符号

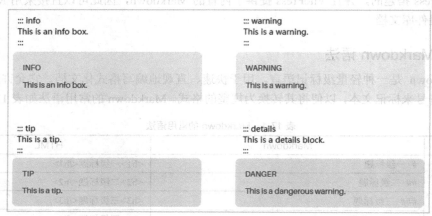

图 17-26　自定义容器

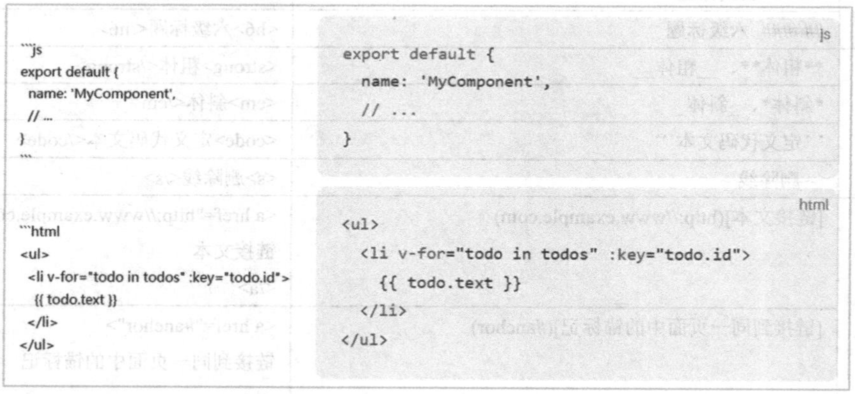

图 17-27　代码高亮格式化

提示：如需了解更多 VitePress 内置的 Markdown 扩展功能，可查看 VitePress 官网，见链接 17-3。

17.4.3　Markdown 表格

在撰写 UI 组件库文档时，Markdown 用得最多的是"表格"功能。使用表格布局可以实现有序的页面结构、清晰的数据呈现、多样的页面布局，从而提高页面的可读性和美观度。Markdown 的表格语法如代码清单 library-17-16 所示。

代码清单 library-17-16
```
1.  > docs/components/button.md: markdown
2.  ## 圆形按钮
3.  ...省略代码
4.
5.  ## Button 属性
6.  |属性|说明|类型|默认值|
7.  |:-|:-|:-:|:-|     // :- 左对齐、-:- 居中对齐、-: 右对齐
8.  |round  |圆角按钮|```Boolean```|false|
9.  |block  |块级按钮|```Boolean```|false|
10. |loading|加载状态|```Boolean```|false|
```

Markdown 的表格语法以左右两侧"|"为一组表示一列，在上述代码清单中共定义了 4 行 4 列（第 6~10 行）。第 1 行定义表头，第 2 行定义文本对齐方式，第 3 行开始定义表格主体内容，渲染效果如图 17-28 所示。

图 17-28　Markdown 表格

VitePress 在生成表格时并不会生成样式效果，如果需要定义表格的样式，需要添加 css 文件，引入 index.ts 文件，如代码清单 library-17-17 所示，渲染效果如图 17-29 所示。

代码清单 library-17-17
```
1.  > docs\.vitepress\theme\index.ts:
2.  import './style.css'
3.  import './highlight.scss'
4.  import './table.scss'
```

图 17-29　定义 Markdown 表格的样式

提示：table.scss 定义了 table 表格样式，详细的样式可通过本书配套代码中的分支 library-17-17 查看。

本章小结

UI 组件库文档是对所有组件的应用描述，包含组件的属性、方法、注意事项、更新事项等。UI 组件库文档通过 VitePress 构建，并使用 markdown-it-container 自定义解析 Markdown 文件，实现不同组件的预览、代码高亮和组件文档的编写等。

第18章
UI组件库的打包和发布

UI 组件库的打包是指"将开发完成的业务代码处理成可在生产环境中运行,并供用户在浏览器上使用"的过程。浏览器可以视为用户使用的生产环境,除浏览器外,常见的生产环境还有 Node.js。Node.js 通常是现代前端开发过程所运行的环境,例如我们常在 Node.js 环境下使用 Vue 和 React,相当于 Node.js 是它们的生产环境。Node.js 在 13.2.0 版本之前仅支持 CommonJS 模块,在 13.2.0 版本之后才开始支持 ESM 模块。为了让开发人员在不同版本的 Node.js 环境下运行 UI 组件库,需要提供 CommonJS 和 ESM 两种模式来打包 UI 组件库。

除上述环境外,打包 UI 组件库后的代码会进行压缩处理,需要通过 source map 文件映射回未压缩的代码,方便开发人员调试及定位代码的出错位置,最后还要将 scss 文件打包成传统的 css 文件,用于提供全量 UI 组件库的 CSS 样式。在 UI 组件库的打包过程中,需要完成的工作如下。

(1)提供浏览器端的代码包,可以是 UMD 或 IIFE。
(2)提供 Node.js 环境的 CommonJS 模块和 ESM 模块代码包。
(3)提供全局引入的 CSS 样式,并按需加载 CSS 样式。
(4)提供 source map 文件。
(5)对 UI 组件库打包的代码进行压缩。

提示:由于 UI 组件库采用 Javascript 技术开发,因此在本书中对 TypeScript 的打包讲解比较少。

18.1 了解 Rollup

Rollup 是一个用于 JavaScript 模块的打包工具,它将小的代码片段编译成更大、更复杂的代码,例如库或应用程序。它使用 ES6 版本的 JavaScript 中包含的新标准化代码模块格式,而不是以往的 CommonJS 和 AMD 等特殊解决方案。

Rollup 的配置简单直观,能够生成轻量、高效的构建文件,尤其适用于构建 JavaScript 库或框架。它还支持代码拆分、按需加载等功能,有助于优化和提升前端应用的性能。

Rollup 具有以下特点。

◎ **高效**:Rollup 通过静态分析的方式,只打包必要的代码和依赖,打包过程更高效,打包后的产物也更小。
◎ **可扩展性强**:Rollup 支持各种插件和加载器扩展其功能,可以轻松地与 Babel、TypeScript、ESLint 等工具集成。

- ◎ **模块化**：Rollup 基于 ES6 模块，支持使用 ES6 模块化的语法和特性，例如 import 和 export。
- ◎ **代码分割**：Rollup 支持将代码分割成多个小块，并支持按需加载或并行加载，以优化性能。
- ◎ **Tree Shaking**：Rollup 可以消除不必要的代码和依赖，以减小最终的产物大小。
- ◎ **多格式输出**：AMD、CommonJS、ES6 modules、UMD、SystemJS loader、iife 等。

使用 Rollup 打包 UI 组件库只需进行全局安装，在终端执行指令 npm install --global rollup@4.3.1。

接下来的 UI 组件库打包过程将围绕 Rollup 的 JavaScript API 特性展开，实现 UMD、ES 及 CJS 这 3 种模块的打包。

提示：Rollup 中文官方网站见链接 18-1。

18.1.1 初始化 Build 打包目录

Rollup 有以下两种配置方式。

- ◎ **配置文件+命令行**：可以提供一个可选的 Rollup 配置文件，以简化命令行的使用，并启用高级 Rollup 功能。
- ◎ **JavaScript API**：一个可以在 Node.js 中使用的 JavaScript API。它很少会被用到，除非正在扩展 Rollup 本身或者需要编程式地打包等特殊用途，否则应该使用命令行 API。

Rollup 官网中是这样描述的："虽然配置文件提供了一种简单的 Rollup 配置的方式，但它们也限制了 Rollup 可以被调用和配置的方式。特别是当你正在将 Rollup 嵌入另一个构建工具中，或者想将其集成到更高级的构建流程中时，直接从脚本中以编程方式调用 Rollup 可能更好。"

我们可以这样来理解上面这句话：使用配置 Rollup 的方式（指"配置文件+命令行"的方式）会有很多有限制，将导致有些需求无法被实现；如果使用"JavaScript API"脚本编程的方式，可以实现更高级的功能。由于在打包 UI 组件库时需要读取指定文件的路径或者一些其他的插件处理逻辑，那么要使用 JavaScript API 的方式扩展更复杂的功能，因此选择"JavaScript API"的方式。

为了能够更好地管理 UI 组件库的打包业务逻辑，可以在 ui-library 目录下新建 build 目录，并在目录中执行 npm init -y，为 build 生成 package.json 文件作为一个打包工具包，如代码清单 library-18-1 所示。

代码清单 library-18-1
```
1.  > ui-library/build/package.json
2.  {
3.      "name": "build",
4.      "version": "1.0.0",
5.      "description": "",
6.      "main": "index.js",
```

```
7.    "scripts": {
8.      "test": "echo \"Error: no test specified\" && exit 1"
9.    },
10.   "keywords": [],
11.   "author": "",
12.   "license": "ISC"
13. }
```

18.1.2　Rollup 的基础配置

UI 组件库是使用 Vue.js 3 开发的，打包 UI 组件库是将 Vue 文件编译成 Javascript 文件的过程，如果项目中使用了 TypeScript，同样需要将以.ts 为后缀的文件编译成 Javascript 文件。在 Rollup 中，推荐使用@vitejs/plugin-vue 插件识别.vue 文件，使用@rollup/plugin-node-resolve 插件解析 npm 包。对于 TypeScript，则可以使用 rollup-plugin-esbuild 插件进行编译。

根据上述说明，优先在 build 目录下安装上述 3 个插件，分别执行指令 npm install @vitejs/plugin-vue@5.0.3、@rollup/plugin-node-resolve、@15.2.3 rollup-plugin-esbuild@6.1.1 -D，将其安装至开发环境中。

Rollup 打包主要有 4 个步骤。

（1）配置打包入口文件。

（2）配置所需插件。

（3）配置输出文件格式。

（4）打包输出目录。

在 build 目录下新建 src 目录及 umdBuild.js 文件，用于实现 UMD 格式的打包，如代码清单 library-18-2 所示。

代码清单 library-18-2

```
1.  > ui-library/build/src/umdBuild.js
2.  import { rollup } from "rollup";
3.  import { nodeResolve } from "@rollup/plugin-node-resolve";
4.  import vue from "@vitejs/plugin-vue";
5.  import esbuild from "rollup-plugin-esbuild";
6.  // umd 打包
7.  const umdBuildEntry = async () => {
8.    const writeBundles = await rollup({
9.      input: "",       // 配置打包入口文件
10.     plugins: [       // 配置插件
11.       vue(),
12.       nodeResolve(),
13.       esbuild(),
14.     ],
15.     external: ["vue"], // 排除不进行打包的 npm 包
16.   });
17.   writeBundles.write({});  // 配置输出文件格式，属性为 output
18. };
```

在 umdBuild 文件中引入 rollup 及上述 3 个插件 @rollup/plugin-node-resolve、@vitejs/plugin-vue 和 rollup-plugin-esbuild（第 2~5 行）。声明一个异步函数 umdBuildEntry（第 7 行），在该函数中调用 rollup 对象，并配置 rollup 对象的基础属性 input、plugins 和 external（第 9、10、15 行），然后赋给变量 writeBundles（第 8 行），最后调用 rollup 的 write 方法生成 output 输出（第 17 行）。

18.1.3 配置打包路径

在打包 UI 组件库时要配置打包文件的入口、插件、格式和输出目录，如代码清单 library-18-3 所示。

代码清单 library-18-3
```
1.  > ui-library/build/src/common.js
2.  // 输出包的目录
3.  const outputPkgDir = "azong"
4.  // node.js 核心方法 url、path
5.  import { fileURLToPath } from "url";
6.  import { resolve, dirname } from "path";
7.  export const filePath = fileURLToPath(import.meta.url);
8.  export const dirName = dirname(filePath);
9.  export const rootDir = resolve(dirName, "..", "..");      // 获取UI组件库的"根目录"
10. export const pkgRoot = resolve(rootDir, "packages");        // 获取UI组件包的目录
11. export const outputDir = resolve(rootDir, outputPkgDir)    // azong
12. export const outputEsm = resolve(rootDir, outputPkgDir, "es")  // es
13. export const outputCjs = resolve(rootDir, outputPkgDir, "lib") // lib
14. export const outputUmd = resolve(rootDir, outputPkgDir, "dist") // umd 全量打包
```

1. 定义 UI 组件库打包文件的入口

定义 UI 组件库打包文件入口，也就是入口文件路径。打包 UI 组件库包括 UMD、ESM、CJS 格式以及 CSS 样式，这些打包动作都需要指定文件路径。因此，可以在 src 目录下新建 common.js 公共文件获取文件路径，并使用 node.js 核心对象 url 和 path 中的 fileURLToPath、resolve 和 dirname 方法（第 5、6 行）。

◎ fileURLToPath

import.meta.url 获取当前文件的完整路径，使用 fileURLToPath 对象将获取的文件路径解码为字符串，然后赋给变量 filePath（第 7 行），如图 18-1 中的序号①所示。

◎ dirname

dirname 对象是返回路径中的目录部分，也就是去除变量 filePath 的文件名称，保留 src 目录的路径（第 8 行），如图 18-1 中的序号②所示。

◎ resolve

resolve 对象是拼接路径，可以将多个参数拼接成指定的路径。如果需要向上一层，则可以使用 ".."（第 9 行）。通过变量 dirName 拼接两次 ".."，便可使路径 E:\ui-library\src\common.js 向上两层，变为 E:\ui-library，然后将获取的 ui-blirary 根目录赋给变量 rootDir，如图 18-1 中的

序号③所示。

◎ packages

UI 组件库已经将所有的组件汇集到了 packages 目录的 index.js 文件中，这就是打包 UI 组件库的入口。因此，可以再使用 resolve 方法拼接参数 "packages"，使目录指向 packages 目录，并将打包入口路径赋给变量 pkgRoot（第 10 行），如图 18-1 中的序号④所示。

```
① filePath D:\ui-library\build\src\common.js
② dirName  D:\ui-library\build\src
③ rootDir  D:\ui-library
④ pkgRoot  D:\ui-library\packages
```

图 18-1　文件路径

2. 定义 UI 组件库的输出目录

在打包 UI 组件库时，一个清晰且逻辑性强的输出目录结构是非常重要的。这样不仅可以帮助用户更容易地了解和使用 UI 组件库，还能提升库的可维护性。在输出打包的 UI 组件库时，需要确定最终的"包"的名称，也就是确定输出目录。在代码清单 library-18-3 中，定义了以"azong"为包的目录并赋给变量 outputPkgDir（第 3 行）。

本章开头提到，UI 组件库的包需要提供 UMD、CommonJS 和 ESM 这 3 种模块，还需要"全量"和"按需加载"的 CSS 样式文件，因此同样要预先定义好这几个目录（第 11~14 行），如图 18-2 所示。

```
① outputDir D:\ui-library\azong
② outputEsm D:\ui-library\azong\es
③ outputCjs D:\ui-library\azong\lib
④ outputUmd D:\ui-library\azong\dist
```

图 18-2　UI 组件库包的目录结构

18.2　UMD 打包

UMD（Universal Module Definition）打包是一种将 JavaScript 库或模块打包成可以在不同环境中使用的通用格式的方法。UMD 打包同时兼容 CommonJS、AMD 和全局变量的使用方式，因此可以在项目的<script>中引入通过 UMD 打包后的产物，直接在浏览器中以访问全局变量的方式使用。

18.2.1　输出 UMD 组件包

完成了 UI 组件库打包文件入口的路径后，在打包时，只需要配置打包文件格式的模式以及打包后产物输出的目录即可,如代码清单 library-18-4 所示。

代码清单 library-18-4
```
1.  > ui-library/build/src/umdBuild.js
2.  import { resolve } from "path";
3.  import { pkgRoot, outputUmd } from "./common.js";
4.  // umd 打包
5.  export const umdBuildEntry = async () => {
6.    const writeBundles = await rollup({
7.      input: resolve(pkgRoot, "index.js"),   // 配置打包入口文件
8.      plugins: [   // 配置插件
9.        vue(),
10.       nodeResolve({ extensions: ['.ts'] }),
11.       esbuild(),
12.     ],
13.     external: ["vue"],   // 排除不进行打包的 npm 包
14.   });
15.   writeBundles.write({
16.     format: "umd",
17.     file: resolve(outputUmd, "index.full.js"),
18.     name: "AzongUI",
19.     globals: {
20.       vue: "Vue",
21.     },
22.   })
23. };
24. // 执行打包
25. umdBuildEntry()
```

变量 pkgRoot 的路径是"E:\ui-library\packages",也就是 UI 组件库的入口。为了配置入口文件的属性 input,可以再次使用 resolve 方法拼接"index.js",形成完整的路径"E:\ui-library\packages\index.js"(第 7 行),同时需要配置插件@rollup/plugin-node-resolve 的属性 extensions 识别以.ts 为后缀的文件(第 10 行),接着调用 writeBundles 的 write 方法并配置以下参数。

◎ format: 指定生成的包的格式。这里是"全量打包",也就是通用模块,定义格式为"umd"(第 16 行)。

◎ file: 生成的文件。同样使用 reslove 拼接 common.js 的变量 outputUmd,生成指定文件(第 17 行)。file 属性配置的生成的文件路径为 E:\ui-library\azong\dist\index.full.js。

◎ name: 自定义包的全局变量名称,也是打包后的产物可访问的变量名称。将属性 format 指定为 iife 或 umd 打包格式,属性 name 是必填项(第 18 行)。

◎ globals: 定义 UI 组件库打包后所要依赖的变量。目前,我们开发的 UI 组件库使用的是 Vue.js 3,因此需要告诉 Rollup,Vue 是外部依赖的,vue 模块的全局变量是"Vue"(第 19~21 行)。

参数配置完成后,需要调用异步函数 umdBuildEntry 执行打包逻辑(第 25 行)。然后进入

build/src 目录，并在终端执行命令行 node ./umdBuild.js，打包生成 UMD 格式的组件包，如图 18-3 所示。

图 18-3　打包 UI 组件库，生成 UMD 组件包

18.2.2　测试 UMD 组件包

UMD 组件包是一种在不同环境中运行的包模式，它可以被用在浏览器端和 Node.js 等环境中，下面来测试使用 Rollup 打包输出的 UMD 组件包在浏览器端运行的效果，如代码清单 library-18-5 所示。

```
代码清单 library-18-5
1.  > examples/test.html
2.  <!DOCTYPE html>
3.  <html lang="en">
4.  <head>
5.    <meta charset="UTF-8" />
6.    <link rel="icon" type="image/svg+xml" href="/vite.svg" />
7.    <meta name="viewport" content="width=device-width, initial-scale=1.0" />
8.    <script src="https://unpkg.com/vue@3.4.15/dist/vue.global.js"></script>
9.    <script src="../azong/dist/index.full.js"></script>
10.   <title>浏览器引用组件库包</title>
11. </head>
12. 
13. <body>
14.   <div id="app">
15.     <a-button type="primary">{{ message }}</a-button>
16.   </div>
17.   <script>
18.     const { createApp, ref } = Vue
19.     const App = {
20.       setup() {
21.         const message = ref('浏览器引用组件库包')
22.         return {
23.           message,
24.         }
25.       },
26.     }
27.     const app = Vue.createApp(App)
28.     app.use(AzongUI)
```

```
29.         app.mount('#app')
30.     </script>
31. </body>
32. </html>
```

在 examples 演示库目录下新建 test.html 文件，使用<script>引入外部 Vue.js 3 资源及打包后的 UMD 包（第 8、9 行）。接着挂载打包后的 UI 组件包，也就是属性 name 指定的名称 AzongUI（第 28 行）。然后测试 button 组件是否生效（第 15 行），渲染效果如图 18-4 所示。

图 18-4 button 组件

button 组件已经被成功渲染，并且传入的插槽内容也可以正常渲染，生成了类名，这表示 UMD 组件包可以正常使用。

18.3 ESM、CJS 模块化打包

ESM（ECMAScript Modules）和 CJS（CommonJS）是 JavaScript 中使用的不同模块。ESM 是现代浏览器和 Node.js 支持的标准模块，CJS 是传统意义上在 Node.js 中使用的模块系统。

ES Modules 是 ECMAScript 标准中定义的模块化规范，它是现代 JavaScript 开发中推荐的模块化方案。ES Modules 分别使用 import 和 export 关键字来导入和导出模块。

CommonJS 是 Node.js 中使用的模块化规范，也可以在其他环境中使用，如使用 Rollup、Browserify、Webpack 等工具进行打包。CommonJS 分别使用 require 和 module.exports 来导入和导出模块。

18.3.1 ESM、CJS 打包输出

UMD 包属于全量模式打包，也就是将所有的组件打包为一份 JS 文件，通过在浏览器中使用<script>标签引入组件。经过 UMD 打包的文件大，并且无法支持按需加载。为了使打包的组件库支持按需加载模式，需要使用 ESM 和 CJS 打包模式实现按需加载，也可以在打包过程中实现 Tree shaking（去除 JS 中无用的代码）优化。如代码清单 library-18-6 所示。

代码清单 library-18-6

```
1.  > build\src\moduleBuild.js
```

```
2.  import glob from "fast-glob";
3.  import { pkgRoot, outputEsm, outputCjs } from "./common.js"
4.  // 重写@import 关键字的路径
5.  const compileStyleEntry = () => {
6.    const themeEntryPrefix = `@ui-library/theme/src/`
7.    return {
8.      name: 'compile-style-entry',
9.      resolveId (id) {
10.       // if(id.startsWith(themeEntryPrefix)) {
11.       //   console.log('id', id.replaceAll('@ui-library', 'azong'))
12.       // }
13.       // 匹配是否满足以@ui-library/theme/src/开头的字符
14.       if (!id.startsWith(themeEntryPrefix)) return
15.       return {
16.         // 将@ui-library/theme/src/字符替换成 azong/theme/src/
17.         id: id.replaceAll(themeEntryPrefix, `${outputPkgDir}/theme/src/`),
18.         external: 'absolute',
19.       }
20.     }
21.   }
22. }
23.
24. // UMD 打包
25. export const moduleBuildEntry = async () => {
26.   const input = await glob("**/*.{js,ts,vue}", {
27.     cwd: pkgRoot,
28.     absolute: true, // 返回绝对路径
29.     onlyFiles: true, // 只返回文件的路径，不需要目录
30.   })
31.   const writeBundles = await rollup({
32.     input,     // 配置打包入口文件
33.     plugins: [    // 配置插件
34.       compileStyleEntry(),
35.       vue(),
36.       nodeResolve({ extensions: ['.ts'] }),
37.       esbuild(),
38.       postcss({
39.         pextract: true,  // css 通过链接引入
40.       }),
41.     ],
42.     external: [ // 去除不进行打包的 npm 包
43.       'vue',
44.       '@vue/shared',
45.       'async-validator',
46.     ],
47.   });
48.   writeBundles.write({
49.     format: "esm",
50.     dir: outputEsm,
51.     preserveModules: true,
```

```
52.      entryFileNames: `[name].mjs`,
53.      sourcemap: true,
54.    })
55.    writeBundles.write({
56.      format: "cjs",
57.      dir: outputCjs,
58.      preserveModules: true,
59.      entryFileNames: `[name].cjs`,
60.      sourcemap: true,
61.    });
62. };
```

由于 ESM 和 CJS 打包组件库的配置、入口等参数与 UMD 包一致，因此可以将 umdBuild.js 文件复制一份，将文件名改为 moduleBuild.js，然后将异步函数名称改为 moduleBuildEntry（第 25 行）。

1. 打包入口文件

在开发 UI 组件库的过程中用到了 3 种后缀类型的文件，分别是 .js、.ts、.vue。因此，在打包组件库时将使用 fast-glob 插件遍历 .js、.ts、.vue 这 3 种类型的系统文件，为获取的类型文件赋值变量 input（第 26~30、32 行），获取的文件是 Array 数据的集合，如图 18-5 所示。

使用 fast-glob 插件需安装依赖包：npm install fast-glob@3.3.2 -D，然后将其引入（第 2 行）。

```
'D:/ui-library/packages/components.js',
'D:/ui-library/packages/index.js',
'D:/ui-library/packages/make-installer.js',
'D:/ui-library/packages/components/index.js',
'D:/ui-library/packages/hook/config.js',
'D:/ui-library/packages/hook/index.js',
'D:/ui-library/packages/icons/index.js',
'D:/ui-library/packages/utils/componentsType.js',
'D:/ui-library/packages/utils/index.js',
'D:/ui-library/packages/utils/install.js',
'D:/ui-library/packages/utils/themeType.js',
```

图 18-5　使用 fast-glob 插件遍历 .js、.ts、.vue 文件

2. 重写 @import 路径

重写 @import 路径是指重写每个组件 style/index.js 引用样式的路径。以 button 组件为例，button 组件引用样式的路径是 import '@ui-library/theme/src/button.css'，其中，@ui-library 是我们在本地开发组件库时自定义的"包"名称。如果将其发布至 npm，就无法正确找到指定路径的 css 文件。

在 18.1.3 节中已经定义了变量 outputPkgDir 的值为 azong，即打包组件库后的目录名称（组件库的根目录），并且包含组件库所需要的其他文件目录，如 es、lib、dist、theme 等。其中，theme 目录是每个组件的独立样式文件，需要将每个组件引用样式文件的路径指向 theme 目录。因此，自定义方法 compileStyleEntry（第 5 行），定义要匹配的字符 "@ui-library/theme/src"，并赋给变量 themeEntryPrefix（第 6 行）。然后使用 Rollup 的钩子函数 resolveId 获取路径（第 9 行），

resolveId 的参数 id 是获取到的路径，如图 18-6 所示。

```
id @ui-library/theme/src/initRoot.scss
id @ui-library/theme/src/button.scss
id @ui-library/theme/src/initRoot.scss
id @ui-library/theme/src/col.scss
id @ui-library/theme/src/initRoot.scss
```

图 18-6　resolveId 的参数 id

接下来使用 startsWith 方法判断路径中是否包含 "@ui-library/theme/src/"。如果不是，则阻止执行（第 14 行）；如果是，则将 "@ui-library/theme/src/" 替换为 "${outputPkgDir}/theme/src/"（第 17 行）。最终，import '@ui-library/theme/src/button.scss' 变成了 import 'azong/theme/src/button.scss'。最后将 compileStyleEntry 方法作为插件添加到 plugins 对象中（第 34 行）。

3. writeBundles.write 方法的参数配置

writeBundles.write 方法的参数配置与 UMD 打包配置类似。

◎ format：打包格式，分别设置为 esm、cjs（第 49、56 行）。
◎ preserveModules：esm 和 cjs 是模块化打包，如果要使打包后的组件库模块结构和源码的模块结构保持一致，可以设置参数的值为 true（第 51、58 行）。
◎ dir：打包后组件输出的目录。也就是打包的 esm 和 cjs，分别使用 common.js 文件定义的变量 outputEsm 和 outputCjs（第 50、57 行）。
◎ entryFileNames：入口文件名称。[name]用于打包生产以.mjs 和.cjs 结尾的文件。
◎ input：打包入口文件。

完成参数配置后，进入 build/src 目录，在终端执行指令 node ./moduleBuild.js，生成 azong 目录以及 es 和 lib 模块包的组件，引入重写的组件样式文件路径，如图 18-7 和图 18-8 所示。

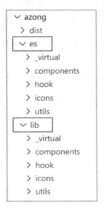

图 18-7　es、lib 模块包组件

```
import 'azong/theme/src/initRoot.scss';
import 'azong/theme/src/button.scss';
```

图 18-8　button 组件引入重写的样式文件路径

18.3.2　测试模块化组件包

完成组件库打包后，可以在本地模拟 npm 包测试，无须发布至 npm 官网。在本地模拟 npm 包中，通过 npm link 命令与全局 node_modules 包建立全局链接，如代码清单 library-18-7 所示。

代码清单 library-18-7

```
1.  > azong/package.json
2.  {
3.    "name": "azong",
4.    "version": "1.0.0",
5.    "description": "",
6.    "main": "./lib/index.cjs",
7.    "module": "./es/index.mjs",
8.    "directories": {
9.      "lib": "lib"
10.   },
11.   "scripts": {
12.     "test": "echo \"Error: no test specified\" && exit 1"
13.   },
14.   "keywords": [],
15.   "author": "",
16.   "license": "ISC"
17. }
```

1. 建立全局链接

每个 npm 包中都有一个 package.json 文件，本地模拟 npm 包也不例外。进入 azong 目录，执行 npm init -y 命令，自动生成 package.json 文件（第 2~17 行）。在 package.json 文件中定义 main 和 module 两个关键字，分别指向打包后的 lib/index.cjs 和 es/index.mjs（第 6、7 行），然后执行命令 npm link。

npm link 命令可以为一个任意位置的 npm 包与全局的 node_modules 建立链接，在系统中做快捷映射，建立链接之后即可在本地进行模块测试，如图 18-9 所示。

```
PS D:\ui-library\azong> npm link
up to date, audited 4 packages in 3s
found 0 vulnerabilities
```

图 18-9　建立链接

2. 本地测试

安装打包后的组件包，执行命令 npm link azong。其中，azong 是 package.json 文件属性 name 的值，也就是包的名称，如代码清单 library-18-8 所示。

代码清单 library-18-8

```
1.  > examples/src/main.js
2.  import { createApp } from 'vue'
3.  import App from './App.vue'
4.  // UI 组件库
5.  import UILibrary from "azong";
6.  const app = createApp(App)
7.  app.use(UILibrary)
8.  app.mount('#app')
9.
10. > examples/src/App.vue
11. <template>
12.   <a-button type="primary">主要</a-button>
13.   <a-button type="success">成功</a-button>
14. </template>
```

3. 全局引入

在 examples 演示包中完成 azong 组件包的安装后，在 main.js 文件中引入并调用 app.use 方法完成全局注入（第 5、7 行）。然后在 App.vue 文件中测试是否生成了 npm 包（第 12、13 行），渲染效果如图 18-10 所示。

图 18-10　本地测试 npm 包

4. 按需引入

除了测试全局注入的效果，还要测试按需引入的效果。按需加载不需要在 main.js 中全局注入，而是在需要使用的组件中使用 import 关键字导入，如代码清单 library-18-9 所示，效果与图 18-10 一致。

代码清单 library-18-9

```
1.  > examples/src/App.vue
2.  <template>
3.    <a-button type="primary">主要</a-button>
4.    <a-button type="success">成功</a-button>
5.  </template>
6.  <script setup>
7.    import { AButton } from 'azong';
8.  </script>
```

18.4　Gulp 打包 scss 文件

scss 文件基本上是现代前端开发的标配，本书的 UI 组件库同样使用 scss 文件编写组件样式。但浏览器端并不支持 scss 文件，因此需要将 scss 文件编译成 css 文件。

gulp-sass 是一个 Gulp 插件，用于将 scss 文件编译成 css 文件。

gulp-sass 的优势如下。

- 高效性：gulp-sass 可以快速地将 scss 文件编译成 css 文件，提高开发效率。
- 实时预览：结合浏览器自动刷新等工具，实时预览更改后的样式。
- 易于配置：灵活的选项和参数允许开发人员根据需求定制编译过程。

使用 gulp-sass 需要先安装 gulp 和 gulp-sass 依赖包，执行以下指令。

- gulp：npm install gulp@4.0.2 -D
- gulp-sass：npm install gulp-sass@5.1.0 -D

18.4.1　全量打包 CSS

全量打包 CSS 是指将所有组件的 css 文件合并为一个单独的文件。全量打包的优势在于减少 HTTP 请求次数，提高页面加载速度，并简化管理和部署过程。然而，需要注意的是，全量打包 CSS 可能导致文件体积过大，反而影响网页性能。因此，在全量打包 CSS 时，应考虑对 css 文件进行压缩和优化，以确保在减少请求次数的同时，保持较小的文件大小和高效的加载速度。

在 build 目录下新建 styleBuild.js 文件，用于打包 CSS 样式，并引入 gulp 和 gulp-sass，如代码清单 library-18-10 所示。

代码清单 library-18-10

```
> build/src/styleBuild.js
import gulp from "gulp";
import dartSass from "sass";
import gulpSass from "gulp-sass";
import autoprefixer from "gulp-autoprefixer";
import cleanCSS from "gulp-clean-css";
import gulpConcat from "gulp-concat";
import { rootDir, pkgRoot, outputDir, outputUmd } from "./common.js";
/**
 * 全量打包 CSS
 */
const buildScssFull = async () => {
  const sass = gulpSass(dartSass);
  await new Promise((resolve) => {
    gulp.src(`${pkgRoot}/theme/src/index.scss`)        // 指定打包入口
      .pipe(sass.sync())                                // 编译
      .pipe(autoprefixer({ cascade: false }))           // 浏览器兼容
      .pipe(cleanCSS())                                 // 压缩
      .pipe(gulpConcat('index.min.css'))                // 合并到指定文件
```

```
20.        .pipe(gulp.dest(outputUmd))                // 输出到指定目录 dist
21.        .on("end", resolve);                       // 监听流完成
22.    });
23. };
24.
25. export const buildStyle = async () => {
26.    await Promise.all([buildScssFull()]);
27. };
28. // 执行打包
29. buildStyle()
```

由于我们安装的 gulp sass 为 4.0.2 版本，该版本不再自带默认的 sass 编译器，因此需要安装 sass 编译器的包。进入 build 目录，在终端执行指令 npm install sass@1.71.1 -D，安装完成后引入 styleBuild.js 文件（第 3 行）。接着定义方法 buildScssFull，作为全量打包 CSS 的函数（第 12 行），首先要让 gulp sass 支持编译 sass，只需要将 dartSass 传入 gulpSass（第 13 行）。

◎ gulp.src：指定打包入口。可直接指定 packages/theme/src/index.scss（第 15 行）。
◎ sass.sync：执行 scss 文件编译（第 16 行）。
◎ autoprefixer：浏览器兼容器。自动根据所使用的 CSS 属性添加浏览器前缀，如-webkit-、-ms-等（第 17 行）。
◎ cleanCSS：压缩 CSS（第 18 行）。
◎ gulpConcat：将所有编译的文件合并到一个指定的文件中。需要安装并引入 gulpConcat 包，执行指令 npm install gulp-concat@2.6.1 -D（第 7、19 行）。
◎ gulp.dest：输出到指定目录。由于是全量打包 CSS，因此可以将文件输出到 azong/dist 目录下（第 20 行）。

除了全量打包 CSS，UI 组件库还要打包按需加载的 css 文件，因此可以将所有需要编译的函数放在 Promise.all 对象中（第 25-27 行），然后调用 buildStyle()方法（第 29 行）。进入 build/src 目录，执行指令 node ./styleBuild.js，全量打包 CSS 的结果如图 18-11 所示。

图 18-11　全量打包 CSS

提示：第 29 行中调用了 buildStyle()方法，只是为了临时测试，在测试完成后需删除。

18.4.2　按需加载打包 CSS

按需加载打包 CSS 的方式和全量打包 CSS 类似，只是指定的入口和输出的目录不同，如代码清单 library-18-11 所示。

代码清单 library-18-11

```
1.  > build/src/styleBuild.js
2.  import gulp from "gulp";
3.  import dartSass from "sass";
4.  import gulpSass from "gulp-sass";
5.  import autoprefixer from "gulp-autoprefixer";
6.  import cleanCSS from "gulp-clean-css";
7.  import gulpConcat from "gulp-concat";
8.  import { rootDir, pkgRoot, outputDir, outputUmd } from "./common.js";
9.  /**
10.  * 按需加载打包 CSS
11.  */
12. const buildScssModules = async () => {
13.   const sass = gulpSass(dartSass);
14.   await new Promise((resolve) => {
15.     gulp.src(`${rootDir}/packages/theme/src/**/*.scss`)
16.       .pipe(sass.sync()) // 编译
17.       .pipe(autoprefixer({ cascade: false })) // 兼容
18.       .pipe(cleanCSS()) // 压缩
19.       .pipe(gulp.dest(`${outputDir}/theme`))
20.       .on("end", resolve); // 监听流完成
21.   });
22. };
23.
24. export const buildStyle = async () => {
25.   await Promise.all([buildScssFull(), buildScssModules()]);
26. };
27. // 执行打包
28. buildStyle()
```

在 gulp.src 对象中指向 packages/theme 目录，获取所有后缀为.scss 的文件。然后 gulp.dest 对象指定输出目录为 azong/theme，打包结果如图 18-12 所示。

图 18-12　按需加载打包 CSS

可以看到，所有文件均是每个组件的独立 css 文件，此时再回看 18.3.1 节中的"重写@import

路径"，可知重写组件引入 CSS 样式文件路径为 azong/theme 的原因。从图 18-12 中还可以看到 index.css 文件和 common 目录，其中 index.css 是全量打包 CSS，common 目录是在开发 UI 组件库时定义的 Scss 变量和混合指令。这两个文件基本上没什么作用，可以将其删除，如代码清单 library-18-12 所示。

代码清单 library-18-12

```
1.  > build/src/styleBuild.js
2.  //删除文件或者文件夹
3.  import { deleteAsync } from "del"
4.  const buildScssModules = async () => {
5.    const sass = gulpSass(dartSass);
6.    await new Promise((resolve) => {
7.      gulp.src(`${rootDir}/packages/theme/src/**/*.scss`)
8.      ...省略代码
9.    });
10.   // 删除指定文件
11.   deleteFiles()
12. };
13.
14. /**
15.  * 删除指定文件或文件夹
16.  */
17. const deleteFiles = async () => {
18.   await deleteAsync(
19.     [`${outputDir}/theme/index.css`, `${outputDir}/theme/common`],
20.     { force: true }
21.   )
22. }
```

借助 del 包删除文件，需要先安装依赖，执行指令 npm install del@7.1.0 -D，并引入 styleBuild.js 文件（第 3 行）。定义 deleteFiles 方法，在该方法内调用 del 包的 deleteAsync，删除指定路径的文件（第 18~21 行）。必须将 force 属性设置为 true，表示可以跨当前目录删除文件，最后在 buildScssModules 方法完成 CSS 打包后执行 deleteFiles()（第 11 行）。

18.5 Gulp 多任务

Gulp 多任务（multi-task）是指一种功能强大的机制，允许用户定义一组相关的任务进行批量处理，无须单独定义和调用每个任务。这种机制允许用户以模块化的方式组织构建过程，并避免编写重复的代码。

18.5.1 series()和 parallel()

Gulp 执行多任务的常用方式分为串行和并行，分别对应 series()和 parallel()。可以简单理解为：串行是一个一个任务执行，只有上一个任务执行完成，才会进入下一个任务；并行是可以同时执行多个任务。

在打包 UI 组件库时，包含 UMT、ESM、CJS 和 CSS 样式及删除等动作，属于多任务。由于打包时执行的任务并不多，并且各个函数之间不存在相互依赖的关系，因此可以使用串行或并行方式，如代码清单 library-18-13 所示。

代码清单 library-18-13
```
> build/src/index.js
export * from "./umdBuild.js"
export * from "./moduleBuild.js"
export * from "./styleBuild.js"

> build/gulpfile.js
import gulp from "gulp";
import { umdBuildEntry, moduleBuildEntry, buildStyle } from "./src/index.js"
// 执行串行任务
export default gulp.series(umdBuildEntry, moduleBuildEntry, buildStyle)

> build/package.json
"scripts": {
  "start": "gulp --require @esbuild-kit/cjs-loader -f gulpfile.js",
  "test": "echo \"Error: no test specified\" && exit 1"
}
```

在 src 目录下新建 index.js 文件，使用 export 关键字将 umdBuild.js、moduleBuild.js 和 styleBuild.js 文件暴露的方法汇集到 index.js 文件中（第 2~4 行）。在 build 目录下新建 gulpfile.js 文件，并引入 umdBuildEntry、moduleBuildEntry、buildStyle 方法（第 9 行），然后调用 gulp.series 串行执行 umdBuildEntry、moduleBuildEntry、buildStyle（第 10 行）。

最后在 package.json 文件的 scripts 属性中自定义名称 "start"，并使用 @esbuild-kit/cjs-loader 工具运行 gulpfile.js 文件（第 14 行），后续在打包 UI 组件库时便可在 build 目录下执行指令 npm run start。

当在 UI 组件库的根目录 "ui-library" 下直接打包组件库时，无须进入 build 目录。可以在 ui-library/package.json 文件的 scripts 属性中自定义名称 "build"，并执行指令 pnpm -C build start，即可在根目录下执行指令 pnpm run build。

18.5.2 删除组件包

删除组件包是为了确保打包后的 UI 组件库是最新的文件。如果没有删除组件包，那么会因为存在之前打包的组件库而遗留历史文件。为了确保打包后的文件是最新的，需要在 Gulp 多任务中添加删除组件包的逻辑，如代码清单 library-18-14 所示。

代码清单 library-18-14
```
> build/src/files.js
import { deleteAsync } from "del"
import { outputDir } from "./common.js"
// 存在包，则先删除
```

```
5.  export const deletePkg = async () => {
6.    await deleteAsync([outputDir], { force: true })
7.  }
8.
9.  > build/src/index.js
10. export * from "./files.js"
11. ...省略代码
12.
13. > build/gulpfile.js
14. import gulp from "gulp";
15. import { deletePkg, umdBuildEntry, moduleBuildEntry, buildStyle } from "./src/index.js"
16. // 执行串行任务
17. export default gulp.series(deletePkg, umdBuildEntry, moduleBuildEntry, buildStyle)
```

在 src 目录下新建 files.js 文件，并引入 del 包的 deleteAsync 方法（第 2 行）。定义 deletePkg 函数，并在函数中调用 deleteAsync 删除指定的目录 azong，也就是变量 outputDir（第 6 行）。将 files.js 文件引入 src/index.js 文件，然后在 gulpfile.js 文件的 gulp.series 方法中优先执行 deletePkg 任务，删除历史组件包（第 17 行）。

18.5.3 生成 package.json 文件

npm 包和 package.json 文件之间有着密切的关系，package.json 文件是 npm 包的清单和配置文件，它给出了一个 Node.js 项目的基本信息、依赖项和运行脚本，并标识了项目的名称和版本号。npm 包的发布需要 package.json 文件存在，并且必须有 package.json 文件的属性 name 和 version，如果没有，则无法正常执行 npm install 命令。

在打包 UI 组件库时，我们使用 deletePkg 函数删除了组件包的整个目录，重新打包生成的组件库中自然不会存在 package.json 文件。在 18.4.2 节中测试本地 npm 包时，我们手动生成了 package.json 文件，手动生成的方式对开发人员来说并不友好。为了简化手动生成的动作，可以在 packages 目录下新建 package.json 文件，在完成组件库打包后，再将 packages 目录下的 package.json 文件复制到组件库目录中，如代码清单 library-18-15 所示。

代码清单 library-18-15

```
1.  > packages/package.json
2.  {
3.    "name": "azong",
4.    "version": "1.0.0",
5.    "description": "Vue.js 3 高级编程：UI 组件库开发实战，UI 组件库开发，开发自己的组件库",
6.    "main": "./lib/index.cjs",
7.    "module": "./es/index.mjs",
8.    "scripts": {
9.      "test": "echo \"Error: no test specified\" && exit 1"
10.   },
11.   "keywords": [
12.     "UI 组件库",
```

```
13.        "Vue",
14.        "React"
15.    ],
16.    "peerDependencies": {
17.        "vue": "^3.4.0"
18.    },
19.    "author": {
20.        "name" : "A总",
21.        "email" : "409019683@qq.com",
22.        "url" : "http://ui.web-jshtml.cn"
23.    },
24.    "license": "ISC"
25. }
26.
27. > build/src/files.js
28. import gulp from "gulp";
29. import { outputDir, pkgRoot } from "./common.js";
30. // 复制 package.json
31. export const copyPackage = async () => {
32.    await new Promise((resolve) => {
33.      gulp.src(`${pkgRoot}/package.json`)
34.        .pipe(gulp.dest(`${outputDir}`))
35.        .on("end", resolve); // 监听流完成
36.    });
37. }
38.
39. > build/gulpfile.js
40. import gulp from "gulp";
41. import { deletePkg, umdBuildEntry, moduleBuildEntry, buildStyle, copyPackage } from "./src/index.js";
42.
43. export default gulp.series(
44.    gulp.series(deletePkg, umdBuildEntry, moduleBuildEntry, buildStyle, copyPackage)
45. )
```

在 packages 目录下新建 package.json 文件，并写入上述描述组件包的基本信息（第 2~22 行）。在 src/files.js 文件中定义 copyPackage 函数，使用 gupl.src 方法获取指定路径的文件（第 33 行），再使用 gulp.dest 方法将其输出到指定目录（第 34 行）。最后在 gulpfile.js 文件的 gulp.series 方法中执行 copyPackage 方法（第 41、44 行），打包组件库后生成的 package.josn 文件如图 18-13 所示。

图 18-13　打包组件库后生成的 package.josn 文件

18.6　npm 发布

npm 是 Node Package Manager 的缩写，即 Node 包管理器，是 Node.js 的一个包管理和分发工具。

npm 是 Node.js 平台的默认包管理工具，也是世界上最大的软件注册表，包含超过 60 万个包的结构，可以使用户轻松跟踪依赖项和版本。npm 是一个开放源代码的命令行工具，用于安装、更新和管理 Node.js 模块。它允许用户从一个集中的仓库中下载和安装公共的 Node.js 模块。npm 随 Node.js 一起安装，解决了 Node.js 代码部署中的许多问题，如包的管理（包括安装、卸载、更新、查看、搜索、发布等），并详细记录了每个包的信息（包括作者、版本、依赖、授权信息等），从而将开发人员从烦琐的包管理工作中解放出来，专注于功能的开发工作。

18.6.1　package.json 文件

package.json 是一个基于 JSON 格式的文件，它在 Node.js、前端项目、包管理器等项目中扮演着非常重要的角色。package.json 出现在 Node.js 应用程序或者模块的根目录中，充当项目的清单文件。package.json 文件中包含项目的各种必要信息，包括但不限于项目名称、版本、描述、作者、许可证信息，以及更重要的依赖关系管理和脚本定义。

在 18.5.3 节中，我们在 packages 目录下初始化 package.json 文件，并使用 copyPackage 函数将文件复制到打包后的 UI 组件库目录下。初始化的 package.json 文件生成了基本的字段，但对于库开发人员来说，有必要了解更多有关 package.json 文件的字段。

◎ name：包的名称，显示在 npm 平台中的名称，用户使用安装、引用包的名称。
◎ version：包版本号。格式为"主版本号.次版本号.修订号"，如 1.0.0。如果存在相同的版本号，则无法发布至 npm 平台。
◎ private：是否私有。发布 npm 时将其设置为 false。
◎ author：项目的作者信息。
◎ contributors：项目贡献者，由多个人组成。如[{ name:'laohan', email:" }]。
◎ keywords：包的关键词。字符串数组，用于增加包的曝光率。在 npm 平台中搜索，在结果列表中可以看到搜索的包和对应的描述。
◎ description：项目的简要描述，用于增加包的曝光率。作用与 keywords 类似。
◎ main：指定 CommonJS 模块或 ES 模块入口文件（从 Node.js 12.20.0 版本开始支持 ES 模块）。
◎ module：指定 ES 模块入口文件。
◎ scripts：用于定义运行命令的脚本。
◎ dependencies：项目依赖的生产环境包。
◎ devDependencies：项目依赖的开发环境包。
◎ peerDependencies：对等依赖包。

- sideEffects：用于标识某些模块是无副作用的，可以被 Tree-shaking 移除。
- license：项目的许可证信息。开源许可证是一种法律协议，用于明确开源软件的使用规则，可以选择 ISC、MIT。开源协议查询地址见链接 18-2。
- homepage：主页信息，项目首页的 url 地址。
- bugs：问题反馈的 url 地址，可以是 email 地址，一般是项目下的 issues，如 bugs:{ url: http://path/to/bug, email: bug@example.com }。
- respository：项目源代码仓库所在位置，如 respository: { type:'git', url:'git+https://github.com/xxx.git' }。

提示：上述字段是 package.json 文件的常用字段，如需了解 package.json 文件的详细字段，请访问本书的 UI 组件库演示地址，见链接 18-3。

18.6.2　version

verstion 用于描述项目、包、工具的版本号，格式为"主版本号.次版本号.修订号"，如 1.0.0。需要注意的是，npm 平台存在包名称与版本号相同的情况，这时无法发布，需要重新定义版本号。

1. 正式版

正式版也称稳定版，表示可以发布至生产环境，让用户下载安装的版本。

- 主版本号

重大版本变化的情况，并且不兼容低版本。如 ElementUI -> Element Plus，由 2 版本到 3 版本，并且 3 版本不兼容 2 版本，这种情况是重大版本的变化。修订规则：主版本号加 1，其余版本号重置 0。

- 次版本号

以向下兼容的方式添加功能。如 API 的兼容更改、语法变更、使用方式变更等。修订规则：主版本号不变，次版本号加 1，修订号重置 0。

- 修订号

通常用于修复 bug。修订规则：修订号加 1，其他版本号不变。

2. 先行版

先行版表示当前版本并非稳定，而且可能无法满足预期的兼容性需求。先行版本号可以被标注在修订号之后，使用"-"连接符加上版本号类型，如 alpha、beta、rc 等。

- alpha

内测版本。表示开发团队或有限用户体验测试版本。版本号规则：1.0.0-alpha.x。

- beta

公测版本。表示可以对外部的所有用户公开的测试版本。版本号规则：1.0.0-beta.x。

◎ rc

公测修正版。表示该版本相对稳定，基本上不存致致命的 Bug，与即将发布的正式版无较大差别。版本号规则：1.0.0-rc.x。

提示：先行版中仅列举了 alpha、beta、rc，如需其他先行版，可根据所在团队的需要进行修订。先行版是非必需的。语义化版本修订可参考链接 18-4。

18.6.3　peerDependencies

现代前端开发人员应该对 package.json 文件的 dependencies 和 devDependencies 两个属性比较熟悉，这两个属性分别对应"生产环境"和"开发环境"。但属性 peerDependencies 令人比较陌生，特别是对没有开发过组件库、包、工具、插件的前端开发人员来说。peerDependencies 的作用是提示用户需要安装 peerDependencies 所指定依赖的包。可以简单理解为：对于我们开发的组件库、包、工具、插件等，需要指定某个范围版本的依赖包才能正常运行，如果它们不是在指定范围内的版本，则会报错或发出警告。

在 18.5.3 节中，我们手动配置了属性 peerDependencies 需要依赖 vue@3.4.0 版本，此时将所依赖的版本号改为 2.6.0，再发布至 npm（详见 18.6.4 节），再回到 examples 演示包中使用 npm install azong 安装发布的 UI 组件库，便会提示错误信息，如图 18-14 所示。

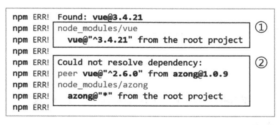

图 18-14　安装依赖包，提示错误信息

其中，标记①说明主项目所需依赖的 vue 版本号是 3.4.21，标记②说明 azong@1.0.9 的包所需依赖的 vue 版本号是 2.6.0。由于主项目依赖的 vue 版本号超出了 azong 所需的 vue 版本号范围，导致无法解析而引起报错。

需要注意的是，这里说到了"范围"，我们必须了解一下它指的是什么，如表 18-1 和表 18-2 所示。

表 18-1　"^"兼容版本（不超过最左边非零数字）

版 本 号	范　　围
^1.2.3	>=1.2.3 ~ <2.0.0-0
^0.2.3	>=0.2.3 ~ <0.3.0-0
^0.0.3	>=0.0.3 ~ <0.0.4-0
^1	>=1.0.0 ~ <2.0.0-0

表 18-2 "~" 最接近匹配版本

版 本 号	范　　围
~1.2.3	>=1.2.3 ~ <1.3.0
~1.2	>=1.2.0 ~ <1.3.0
~1	>=1.0.0 ~ <2.0.0

通过前缀符号"^"和"~"的范围可以看出，在图 18-14 的标记②处，版本号^2.6.0 对应的范围是>=2.6.0 ~ 3.0.0，但主项目依赖的 vue 版本是 3.4.21，超出了 azong 包所需要依赖 vue 的版本号环境，结果就会报错。当然，图中的报错和 npm 版本有关。

◎ npm 1、npm 2 版本

如果出现主项目和工具包共同所需的依赖包版本不兼容的情况，那么会使主项目依赖一个 vue 包，工具包依赖一个 vue 包，如图 18-15 所示。

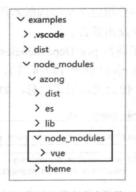

图 18-15　主项目和工具包各自依赖一个 vue 包

◎ npm 3 ~ npm 6 版本

如果出现主项目和工具包共同所需的依赖包版本不兼容的情况，会出现如图 18-16 所示的提示性警告，提示用户 azong@1.0.11 需要对等的 vue@^2.6.0 的依赖包，并需自己手动安装依赖。

```
npm WARN sass-loader@13.3.3 requires a peer of webpack@^5.0.0 but none is i
nstalled. You must install peer dependencies yoursenpm WARN anpm WAnpm WARN
npm WARN azong@1.0.11 requires a peer of vue@^2.6.0 but none is installed.
You must install peer dependencies yourself.

+ azong@1.0.11
```

图 18-16　提示性警告

◎ npm 7 及以上版本

如果出现主项目和工具包共同所需的依赖包版本不兼容的情况，会提示错误信息，如图 18-14 所示，无法完成对包的依赖。

18.6.4 发布组件库

npm 是 Node.js 包的标准发布平台，用于发布、管理各种类型的包。npm 可以通过命令行或工具，让用户在项目中安装、升级、删除包等，以及发布并维护各类工具包。

在 npm 官方网站（见链接 18-5）中注册自己的账号，并在 npm 平台中确认是否存在和 packages.json 文件的 name 属性相同名称的组件包，如果存在，则需要修改 name 的值，确保其是唯一的，如图 18-17 所示。

图 18-17　在 npm 平台中搜索组件包

1. 命令行登录

在根目录终端中执行 pnpm run build 命令打包成功后，会自动生成 azong 组件包。然后进入 azong 目录，在终端中执行命令 npm login，登录 npm 平台，在登录过程中要输入账号、密码、邮箱、验证码等，如图 18-18 所示。

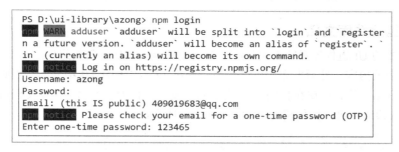

图 18-18　登录 npm 平台

2. 浏览器登录

如果你安装的是 9 版本及以上的 npm，那么在终端执行 npm login 命令时，会提示你按下 "ENTER" 键，使用浏览器登录，如图 18-19 所示。

图 18-19　浏览器登录

3. 发布确定镜像源

发布确定镜像源是为了确定发布的平台，由于在开发项目过程中可能使用了非 npm 官方的镜像源，如淘宝 npm 镜像、阿里云 npm 镜像、华为云 npm 镜像等。因此，为了将项目顺利发布至 npm 官方平台，可以先执行命令行 npm config set registry= http://registry.npmjs.org，将镜像源改到 npm 官方平台。然后执行命令行 npm login，登录成功后，再执行 npm publish 发布命令行，发布成功的界面如图 18-20 所示。

```
=== Tarball Details ===
name:          azongtest
version:       1.0.0
filename:      azongtest-1.0.0.tgz
package size:  171.9 kB
unpacked size: 869.2 kB
shasum:        a9689bd60de2849cbb79bac2d59c3c2558e2a651
integrity:     sha512-J8GSGn66UTG0b[...]GNKUwJfAebpFw==
total files:   496

Publishing to https://registry.npmjs.org/
+ azong@1.0.0
```

图 18-20　发布成功

提示：npm 发布成功后，可登录 npm 官网查看已成功发布的组件包，见链接 18-5。

18.6.5　打包组件库文档

发布新版本的 UI 组件库意味着组件库中有功能调整或增加了新的组件，需要在 UI 组件库文档中更新相应的内容，然后发布到生产环境。在第 17 章中构建 UI 组件库文档时，通过相对路径的方式引入组件库，但现在我们已经通过 npm 发布组件库，因此组件库文档可以通过 npm install 的方式依赖组件包，如代码清单 library-18-16 所示。

代码清单 library-18-16
```
1.  > docs\.vitepress\theme\index.ts
2.  // UI 组件库
3.  import UILibrary from "azong"
4.  import "azong/dist/index.min.css"
5.  import './style.css'
```

进入 docs 目录，在终端执行指令 npm install azong，安装发布的依赖包，安装成功后引入组件包及全量样式（第 3、4 行）。接着在终端执行 package.json 文件属性 scripts 的打包指令 npm install docs:build。打包完成后，会在.vitepress 目录下生成 dist 文件，该目录便是打包后的组件库文档的所有文件。然后可以继续执行预览指令 npm run docs:preview，预览打包成功的效果，如图 18-21 所示。

提示：关于 UI 组件库部署生产环境，本书不做介绍。如需了解部署生产环境，可前往链接 18-3。

图 18-21　UI 组件库文档打包成功预览

18.6.6　按需引入组件样式

按需引入组件样式用于在按需加载组件时，自动载入对应组件 src/style 目录下的 index.js 文件，也就是 18.3.1 节中"重写 @import 路径"部分的 scss 文件，如代码清单 library-18-17 所示。

```
代码清单 library-18-17
1.  > examples/vite.config.js
2.  import { createStyleImportPlugin } from 'vite-plugin-style-import'
3.  export default defineConfig({
4.    plugins: [
5.      vue(),
6.      createStyleImportPlugin({
7.        libs: [
8.          {
9.            libraryName: 'azong',
10.           esModule: true,
11.           resolveStyle: (name) => {
12.             const path = name.split('-')  // a-button
13.             return `azong/es/components/${path[1]}/src/style/`
14.           },
15.         },
16.       ],
17.     }),
18.   ],
19. })
```

借助第三方插件 vite-plugin-style-import 实现按需载入组件的 CSS 样式，并用到 npm i vite-plugin-style-import@2.0.0 -D 和 npm i consola@3.2.3 -D 两个依赖包。

在 vite.config.js 文件中引入插件 vite-plugin-style-import，并解构 createStyleImportPlugin 对象（第 2 行），在 plugins 插件对象中使用 createStyleImportPlugin（第 6 行）。其中，要注意

libraryName 属性，该属性用于指定某一个 UI 组件库的名称（第 9 行）。属性 resolveStyle 是返回指定目录的路径，其中，参数 name 用于按需加载组件名称（第 11 行），例如，图 18-22 所示为按需加载 button 组件。

```
import { AButton } from "azong";
```

图 18-22 按需加载 button 组件

图 18-22 中引入的是 button 组件，那么 resolveStyle 方法的参数 name 就是 a-button，然后使用 split 方法指定 "-" 进行分割（第 12 行），并获取下标为[1]的字符，也就是 button。最后返回指定组件的路径样式（第 13 行），如图 18-23 所示。组件库中 button 组件的目录结构如图 18-24 所示。

```
azong/es/components/button/src/style/
```

图 18-23 组件的路径样式

图 18-24 组件库中 button 组件的目录结构

本章小结

本章使用 Rollup 方式打包 UI 组件库，实现"全量"和"按需加载"两种模式，并通过 Gulp 实现多任务自动打包。最后将 UI 组件库发布至 npm 平台，并在组件库文档中依赖发布的组件库，打包组件库文档，并实现预览。